PSAM 12

Probabilistic Safety Assessment and Management
22–27 June 2014 • Sheraton Waikiki, Honolulu, Hawaii, USA

Proceedings of the Probabilistic Safety Assessment and Management (PSAM) 12 Conference

Volume 12 - Thursday PM II

PSAM 12

Probabilistic Safety Assessment and Management

22 - 27 June, 2014

Sheraton Waikiki, Honolulu, Hawaii USA

CONFERENCE PROCEEDINGS

Volume 12

Thursday PM II

Foreword

It is was our honor to welcome you to Honolulu, Hawaii, for the twelfth rendition of the Probabilistic Safety Assessment and Management (PSAM) Conference. The planning for PSAM Honolulu began back in 2007 (before PSAM 9 in Hong Kong), when we looked at several locations around the United States, included Arizona, California, Boston, and even considered locations in Oceania. Based upon the feedback both during and after the conference, PSAM 12 proved to be a great success.

We would like to thank all of the volunteers, those that served before, during, and after the Conference. Members of the Technical Program Committee, the Organizing Committee, the session chairs, and the presenters have our gratitude for making PSAM 12 the most memorable PSAM yet.

This publication represents the technical proceedings for the Conference. Due to the large number of published papers (a total of 391), we have subdivided the technical content (papers) into five volumes, one for each day of the conference.

On behalf of the International Association for Probabilistic Safety Assessment and Management Board of Directors, we hope that this publication will provide a valuable technical resource in addition to a reminder of the memorable stay in the Hawaiian Islands.

Dr. Curtis Smith
Technical Program Chairs

Dr. Todd Paulos
General Chair

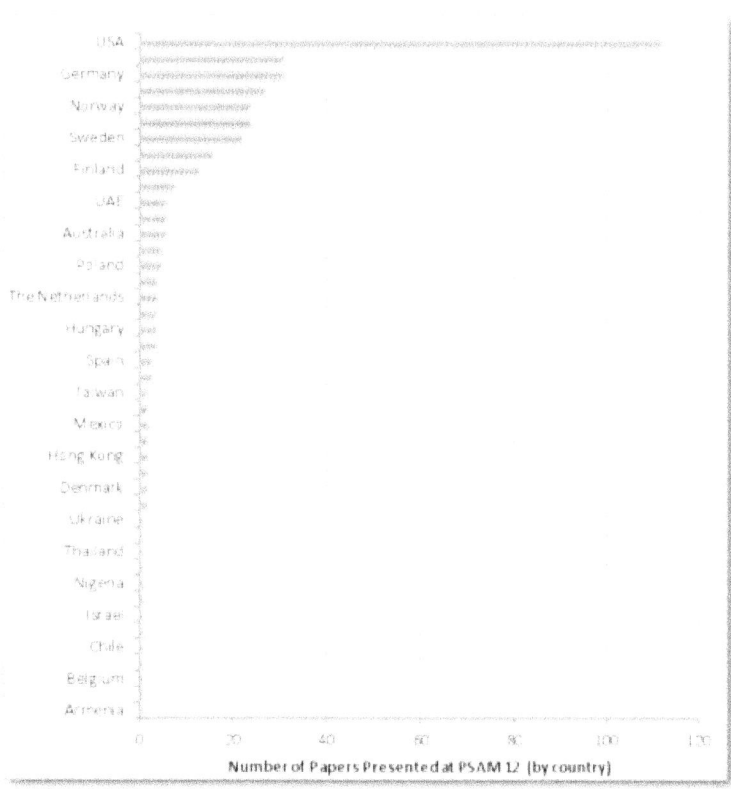

Sponsors

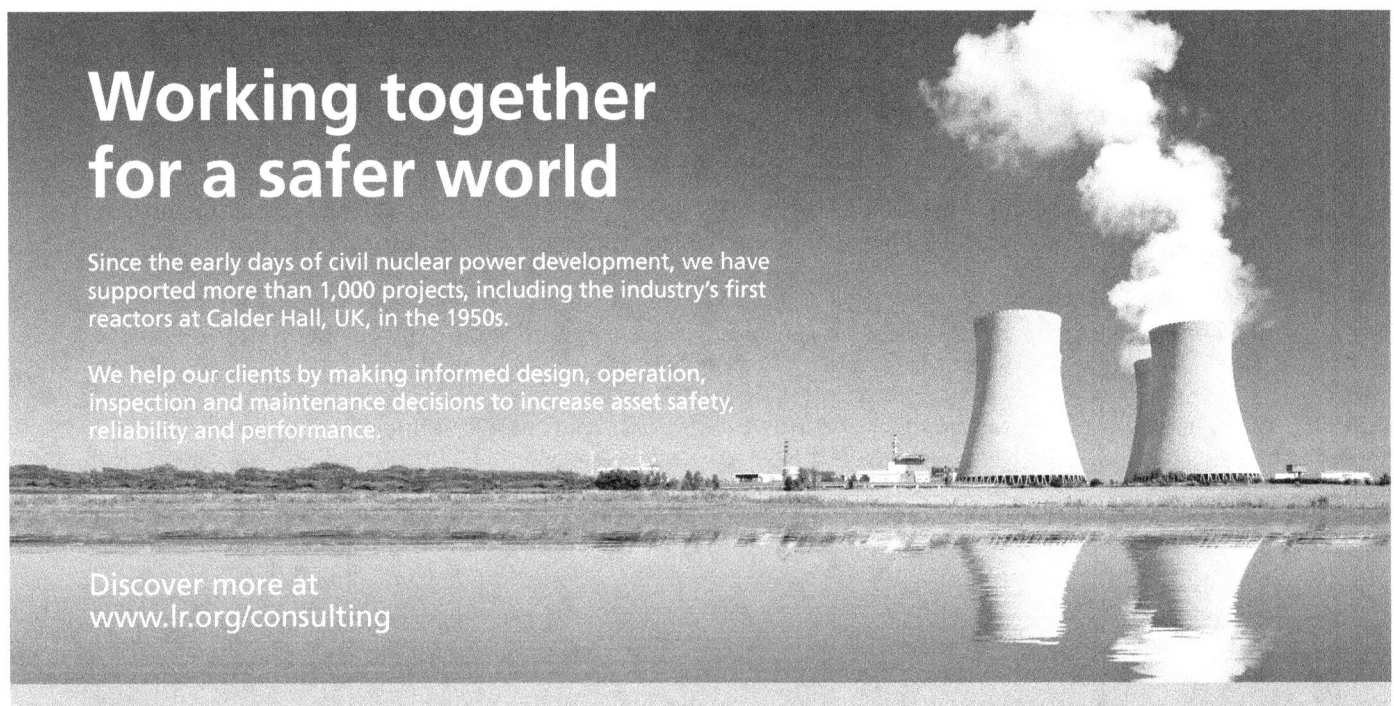

Sponsors

EPRI Assesses Seismic Resistance of Electronic Components

Together... Shaping the Future of Electricity

Technical Program Committee

Technical Program Chair: Curtis Smith, INL USA

Assistant Technical Program Chairs: Steve Epstein, Lloyd's Register Japan
Vinh Dang, PSI Switzerland
Ted Steinberg, QUT Australia

We would like to thank the members of the PSAM 12 Technical Program Committee. These individuals helped to make PSAM 12 a success by reviewing abstracts, technical papers, organizing sessions, and providing technical leadership for the conference.

Technical Committee Members:

Roland Akselsson

S. Massoud (Mike) Azizi

Tito Bonano

Ronald Boring

Roger Boyer

Mario Brito

Kaushik Chatterjee

Vinh Dang

Claver Diallo

Nsimah Ekanem

Steve Epstein

Fernando Ferrante

Federico Gabriele

Ray Gallucci

S. Tina Ghosh

David Grabaskas

Katrina Groth

Seth Guikema

Steve Hess

Christopher J. Jablonowski

Moosung Jae

Jeffrey Joe

Vyacheslav S. Kharchenko

James Knudsen

Zoltan Kovacs

Ping Li

Harry Liao

Francois van Loggerenberg

Jerome Lonchampt

Soliman A. Mahmoud

Diego Mandelli

Donoval Mathias

Zahra Mohaghegh

Thor Myklebust

Cen Nan

Mohammad Pourgolmohammad

Marina Roewekamp

Clayton Smith

Shawn St. Germain

Ted Steinberg

Kurt Vedros

Smain Yalaoui

Robert Youngblood

Enrico Zio

Organizing Committee

General Chair: Dr. Todd Paulos

General Vice Chair: Prof. Stephen Hora, USC

Technical Program Chair: Curtis Smith, INL USA

Webmaster, Registration,
Support for Papers/Abstracts
Submission and Review: Hanna Shapira, TICS

Table of Content

Table of Content

Study on Next Generation Seismic PRA Methodology
Part II: Quantifying Effects of Epistemic Uncertainty on Fragility Assessment

Akemi Nishida[a*], Tsuyoshi Takada[b], Itoi Tatsuya[b], Osamu Furuya[c] and Ken Muramatsu[c]
[a] Japan Atomic Energy Agency, Tokyo, Japan
[b] University of Tokyo, Tokyo, Japan
[c] Tokyo City University, Tokyo, Japan

Abstract: This study focused on uncertainty-assessment frameworks, utilization of expertise, and on developing relevant software to improve reliability of Seismic Probabilistic Risk Assessment (SPRA) and to promote further use of SPRA, develops methodology for quantification of uncertainty associated with final results from SPRA in the framework of risk management of Nuclear Power Plant (NPP) facilities. This research aimed to contribute to the development of probabilistic models for uncertainty quantification- and software (1); to the aggregation of expert opinions on structure/equipment fragility estimation and development of implementation guidance on epistemic uncertainty (2); and to the study of applicability of newly proposed SPRA models to plant models (3). In particular, we focused on the second goal. There were two different groups of experts used: those in the field of civil engineering, and those in the fields of mechanical engineering. With these groups, we conducted a pilot study on the use of expert-opinion elicitation for identification and quantification of parameters of fragility assessment. Sensitivity analysis was performed by using a reactor-building model, and results were provided to experts for expert-opinion elicitation.

Keywords: PRA, Reactor-Building Model, Equipment & Piping, Seismic Ana[*]lysis, Sensitivity Analysis.

1. INTRODUCTION

The evaluation of seismic safety is performed by quantifying and identifying the various uncertainties in probabilistic seismic risk evaluation of nuclear power plant (NPP) facilities. The level 1 Probabilistic Risk Assessment (PRA) is estimated by distinguishing three steps of onsite seismic hazard evaluation, fragility evaluation of buildings and equipment, and systematic analysis in paying attention to the interface between each step [1]. For the evaluation process, uncertainty is classified into aleatory uncertainty (i.e., randomness) and epistemic uncertainty (i.e., lack of knowledge). Upon evaluation, these uncertainties are generally quantified based on engineering judgment- and experience, due to lack of data. Especially the importance of epistemic uncertainty must be emphasized here.

The Senior Seismic Hazard Analysis Committee's (SSHAC) method [2, 3] for earthquake hazard assessment was proposed and implemented in the United States. This method employs a graded approach, which considers the difficulty, complexity, and significance of results from hazard evaluation of target sites. For the most in-depth (level 4) evaluation, a logic tree is adopted to summarize different opinions on serious subjects obtained from multiple experts, in addition to relevant literature review.

Fragility evaluation of buildings and equipment is divided in two parts: response evaluation of the soil, buildings, and equipment in the upper region of the engineering-base surface (1), and strength evaluation associated with damage modes of buildings and equipment components (2). In this study, we performed an assessment of the uncertainty in the seismic response evaluation of soil in the upper region of the engineering-base surface, of the uncertainty in inputs of the reactor building, and of the uncertainty in response evaluation of buildings and equipment. Because these methods were developed for earthquake hazard assessment, this is the first time they were applied to fragility

---

[*] *nishida.akemi@jaea.go.jp*

evaluation in Japan. Specialized knowledge extracted during the process is adjusted and integrated by using a logic tree to categorize important subjects.

2. PREPARATION FOR EXPERT OPINION EXTRACTION

2.1. Target Problem

Epistemic uncertainty with regard to fragility analysis is assessed by extracting the knowledge of experts, in order to identify the sources and ranges of uncertainty. Through this assessment process, credibility of fragility analysis will be enhanced. Finally, a guideline for treatment of epistemic uncertainty will be proposed using the knowledge obtained in this study.

Our main target is earthquake response analysis using the Sway-Rocking (SR) model shown in Figure 1, which is a standard model used in the fragility analysis of buildings and equipment. In particular, we focused on the uncertainty of one- dimensional wave propagation theory (i.e., equivalent to linear analysis) indispensable for soil analysis. Furthermore, to extract unbiased opinions from experts, we selected a concrete case-study site and model plant, and then presented case-study site information, uncertainty-research results, and an example of response analysis of the selected model plant to experts beforehand. The Peak Ground Acceleration (PGA) of input ground motions is assumed to be twice that of "Ss," which is the basic earthquake ground motion for seismic assessment of NPP in Japan. We anticipate that the building will respond in a near-linear range and that part of the soil will respond nonlinearly under these conditions.

2.2. Opinion Extraction Method

Characteristics of various opinion extraction methods are shown in Figure 2. Having considered this figure, we adopted the appropriate method using both a questionnaire and group discussion. Specifically, civil engineering experts were consulted for reactor building and soil behavior under EQ, and mechanical engineering experts were consulted for equipment and piping behavior under EQ. Because the target range was limited, a questionnaire was used to maintain independence of each expert's opinion and to gather opinions from broad set of perspectives. Extracted opinions were then disclosed to consulted experts, and expert opinions were exchanged led by the Technical Integrator (TI). The TI acts as facilitator in the exchange of opinions, allowing experts to give opinions in a less constrained way so that subjects may be explored by in-depth discussion.

In this research, both the used questionnaire and gathered expert opinions are presented, and workshops during which to exchange expert opinions are arranged two to three times a year. Selection of experts was based on the following criteria.

1) Experts should be unbiased toward target subjects and projects.
2) Experts should have sufficient field-specific knowledge- and ability to form expert opinions on issues relating to that field,
3) For balanced consensus formed by a variety of expert opinions, the panel of experts should represent all areas of research directly related to such issues.

The selected civil engineering experts were highly experienced in the fields of nuclear seismic research- and design, and had a wealth of knowledge on soil-structure interaction, soil analysis, and earthquake-resistant building research. The mechanical engineering experts were highly experienced in the field of seismic design and assessment of nuclear facilities, and were selected from electric power companies, professional manufacturers, universities and research institutes. The questionnaire sent to the experts mainly contained questions about the embedded SR model and the one-dimensional wave theory used for fragility assessment of buildings and equipment.

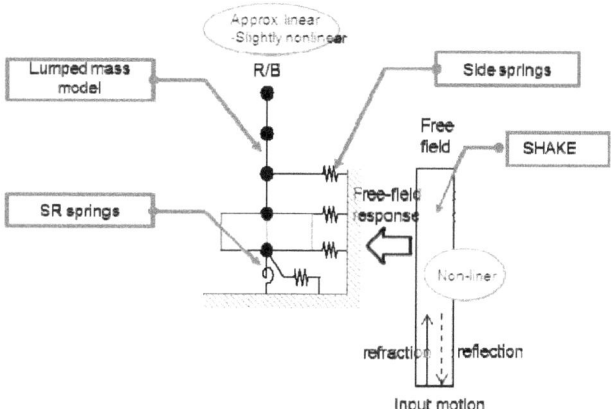

Figure 1: Profile of Embedded SR (Sway-Rocking) model.

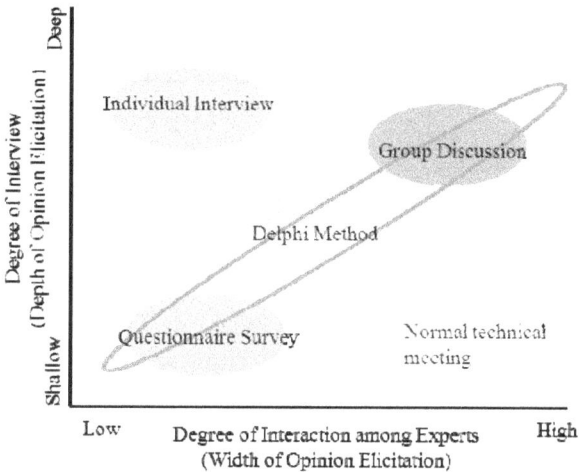

Figure 2: Characteristics of Various Opinion Extraction Methods.

2.3. Questionnaire

<u>(1) Civil Engineering Experts</u>
When asked about the factor of uncertainty in the estimation of the value given by the response evaluation model using the embedded SR model at the beginning of questionnaire, experts responded as shown in Table 1. These responses were related to the treatment of Soil Structure Interaction (SSI) springs during interaction of the building and soil, the evaluation of phenomena such as lift foundation and peeling of the building against a large input, and the treatment of the modeling of the building. Next, we asked experts about the parameters that should be considered to perform sensitivity analysis of the target model plant. As a result, respondents raised the following issues.
1) Input ground motion evaluation
2) Nonlinear interaction (effects of improved and backfilling soil)
3) Precision of soil analysis of large-strain region, the effects of soil stratification
4) Precision of embedded SR model from the viewpoint of equipment response, the effects of treatise of side spring

Table 1: Classification of Epistemic Uncertainties Addressed by Civil-Engineering Experts.

| | Input data collection (1) | Estimation of sensitivity (2) | Analytical method (3) | | Overall |
			Modeling	Discretization	
Building (B)		Local Vibration		A stick model	
SSI (I)		Geometrical nonlinearity (sliding, uplift, detachment) Building-building interaction	Geometrical nonlinearity (sliding, uplift, detachment) Side springs		
Ground (S)	Soil properties Mechanical (dynamic) properties	Nonhomogeneity Non-layered, Backfilling ground, Topographical irregular ground	Damping model of ground Nonlinearity of ground Damping model of ground Equivalent linearization method	Division into sub-layers	
Foundation (F)			Location of bedrock		
Others (O)	Info. Of input motion			Freq. range of analysis	Correlation btw uncertainties

(2) Mechanical Engineering Experts

The factor of uncertainty obtained from the group of mechanical engineering experts was divided into three elements: the analysis model, the mechanical structures, and the interactions between soil, buildings, and equipment structures. These results are shown in Table 2. The uncertainties included in these elements will be extracted from mechanical engineering expert opinions as important factors in equipment-fragility assessment.

Table 2: Example of Opinion Extraction from Mechanical Engineering Experts.

Uncertainty in the mechanical structure	·Non-linearity caused by the support section of the equipment and piping ·The dynamic behavior of the different aspects in the increase of input level ·Material tolerance ·Actual test results ·Preparation method of fragility curve
Uncertainty in the analytical model	·Combination of the response of the coupling of the analysis model ·The difference of floor responses on the same floor by considering building-floor flexibility ·Correlation of equipment on the same floor ·Power (energy) pathway ·Calculation results and actual response ·How the multi-input evaluation is affected by displacement input ·Attenuation evaluation that depends on response acceleration ·Response error in how to tighten bolts
Uncertainty in the interactions between soil, buildings, and mechanical structures,	·Input ground motion ·Physical constant of soil ·Phase characteristics ·Correlation coefficient at multi-axis input ·Three-dimensional coupling

For the strength of the equipment, information from seismic-resistant verification tests of the actual machine size and level conducted in Tadotsu Engineering Experiment Station would be collected and considered. In particular, the piping was selected as static equipment and the large vertical pump was selected as dynamic equipment. Storage condition and treatment method of the recorded data of the experiment were confirmed. As for the functional limits of the load-bearing structures, i.e., the active component and the foundation bolts for joining the mechanical structures and the building structure, we are planning to proceed with the analysis and evaluation, in reference to the analysis methods in the current technical provisions [4].

3. SENSITIVITY- AND RESPONSE ANALYSIS FOR EXPERT-OPINION EXTRACTION

Sensitivity-and response analysis by using constructed embedded SR model and a three-dimensional (3D) model are performed to provide the results to experts for expert-opinion elicitation. Deliverables of our research are a selection guide of uncertainty factors and a comprehensive report of the range of uncertainties investigated in the response analysis. Response- and sensitivity analysis of target buildings and soil of the model plant are carried out with the aim to provide information to experts, facilitating assessment of epistemic uncertainty in the response evaluation of fragility analysis of equipment by expert-opinion elicitation. In this section, we summarize the reactor-building modeling, results of the eigenvalue analysis, and comparisons of analytical results and observed seismic-response records, carried out to validate the model.

3.1. Selection of Model Plant

To survey available plant information and observed earthquake data to construct a reactor-building model, public information on light-water reactor buildings was collected from the Japan Nuclear Energy Safety (JNES) library and the Nuclear and Industrial Safety Agency (NISA)[5, 6]. On the other hand, we also obtained information from the KARISMA benchmark[7], which was carried out by the International Atomic Energy Agency (IAEA) over a period of five years from 2008 until 2013 for the Kashiwazaki-Kariwa Nuclear Power Plant (Unit 7, TEPCO KK7). We decided to become a member of the KARISMA benchmark and received access to the plant's information about the KARISMA benchmark. Then we constructed analytical model of the model plant by using the information of KARISMA benchmark. Because the obtained data only provided us with part of the required information, and were insufficient to create a full model of the target-reactor building, we were forced to supplement used publicly available information[8, 9, 10, 11]. Figure 3 shows a cross-sectional view of the Kashiwazaki-Kariwa NPP Station (Unit 7, TEPCO KK7). The red points show where seismic observation data were recorded.

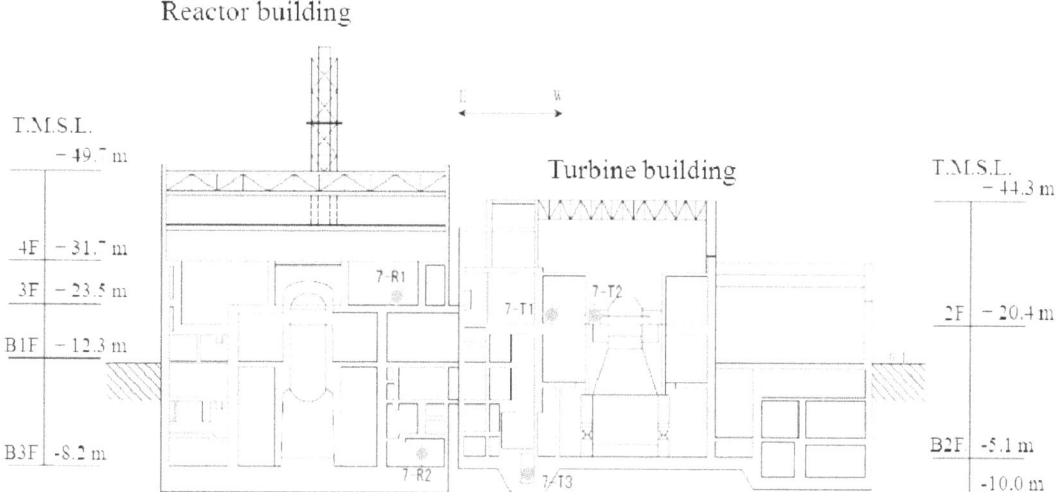

T.M.S.L.: Altitude difference with respect to average sea level in Tokyo Bay
Figure 3: Overview of the Kashiwazaki-Kariwa Nuclear Power Plant Unit 7 (Cross Section)[12].

3.2. Summary of Reactor-Building Modeling

The SR model and the three-dimensional FEM model were created in order to perform the sensitivity analysis. Sensitivity analysis would be performed by using the SR model mainly though the analysis of the three-dimensional FEM model also would be performed for analysis of the results of the SR model. The SR model was created based on the public information and literatures, and confirmed to be consistent with the results of TEPCO. In this section, the summary of construction of the three-dimensional FEM model is mainly described.

(1) Construction of Finite Element Method (FEM) Model

As a first modeling step, the 3D shape-model was created by 3D Computer Aided Design (CAD) software. We created the FEM model based on 3D shape-model. Examples of the 3D shape-model are shown in Figure 4. The FEM model consisted of shell elements, solid elements, and truss or beam elements. We used shell elements for walls and slabs, and solid elements for foundation slab. The total number of elements is 14,334, that of nodes is 11,796, while the total number of degrees of freedom is about 65,000. Furthermore, to determine the weight inputs of the FEM model, the volume of columns, beams, walls, and floors were calculated. We determined the mass density of each floor of the FEM model to correspond to the total mass of each floor of the SR model. Figure 5 shows the FEM model of the reactor building.

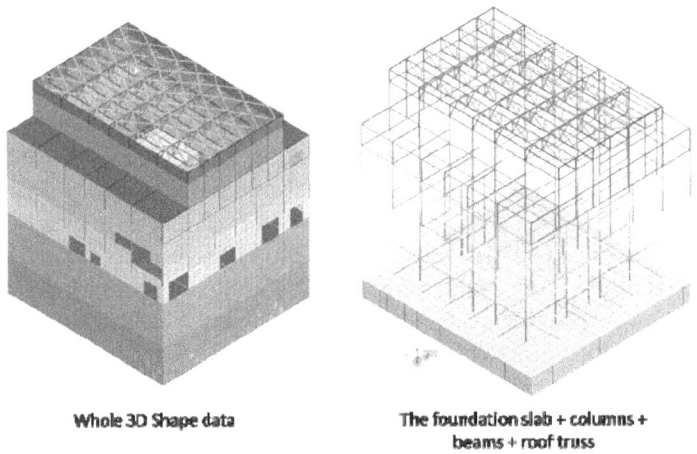

Whole 3D Shape data   The foundation slab + columns + beams + roof truss

Figure 4: Example of 3D Shape Model.

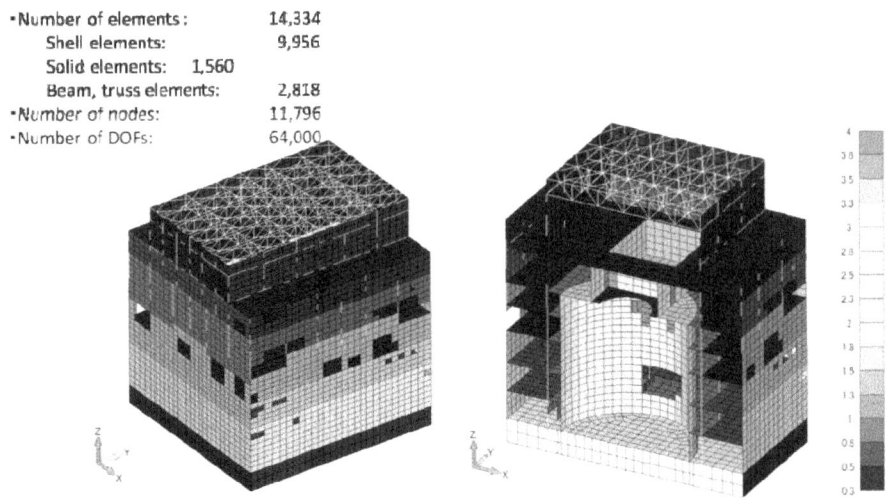

- Number of elements :   14,334
  - Shell elements: 9,956
  - Solid elements: 1,560
  - Beam, truss elements: 2,818
- Number of nodes: 11,796
- Number of DOFs: 64,000

(a)   (b)

Figure 5: Bird's-Eye View (a) and Cross-Sectional View (b) of the FEM Model of Reactor-Building Model
(color ramp indicates thickness of each element in meters).

Figure 6 shows an example of how quake-resistant and auxiliary walls can be arranged. The lower half of Figure 6 shows the walls considered in the SR model by TEPCO. That is, blue indicates modeled area, for which stiffness is estimated as quake-resistant wall, and red indicates modeled area, for which stiffness is estimated as auxiliary wall. We selected walls for 3D modeling of the plant building according to these documents. Constructed 3D models are shown in the upper half of Figure 6. Material properties of roof truss and building are shown in Table 3. The properties of the SSI springs of the FEM model were set in reference to the SR model of the KARISMA benchmark. The springs were arranged by distributing the spring of the SR model along the interacting surfaces of the wall or the base mat.

Table 3: Material Properties of Roof Truss and Building.

	Material	Young's modulus (kN/m²)	Poisson's ratio	Mass density (kg/m³)	Dumping constant :h
Roof truss	Steel	2.05 E+08	0.3	77.0	0.02
Building	Concrete	3.13 E+07	0.2	25.0	0.05

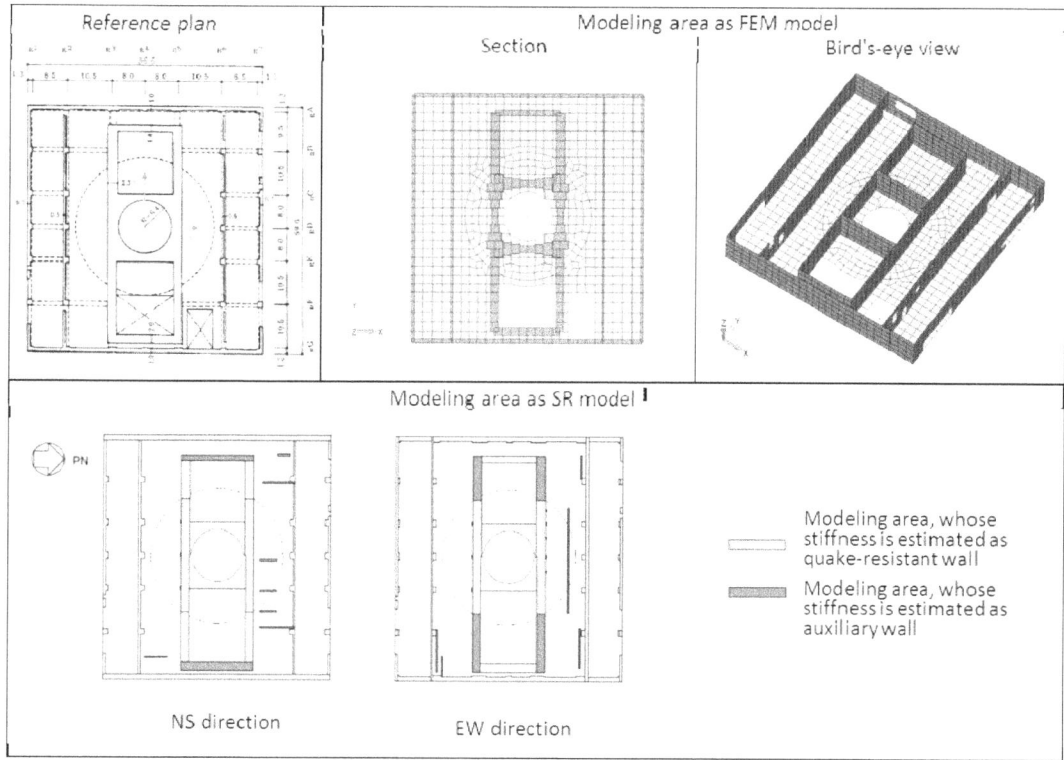

Figure 6: Example of Arrangement of Quake-Resistant and Auxiliary Walls (3rd Floor) [9].

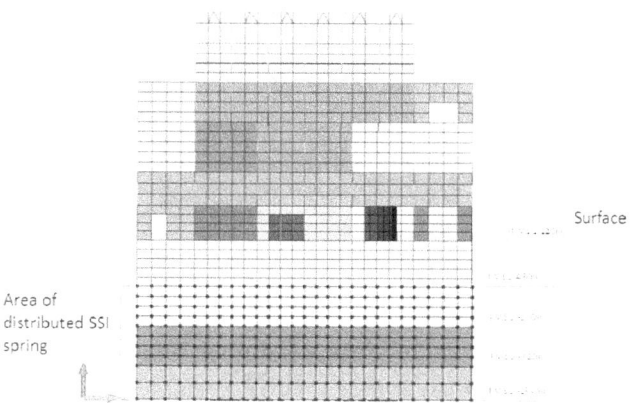

Figure 7: Setting of SSI Springs.

3.3. Validation of the Analytical Model

(1) Eigenvalue Analysis

To validate the constructed analytical models, eigenvalue analysis was performed, and the results of the FEM- and SR model were compared. The obtained natural modes for the FEM- and SR model are shown in Figures 8 and 9, respectively. These figures show the first and second order natural modes of NS direction of the building. Furthermore, the obtained natural frequencies are shown in Table 4. Table 4a shows the analytical results of the FEM model, and Table 4b shows the analytical results of the SR model. It is found that the first and second modes are the primary modes because the sums of the effective mass ratio of these two modes for the SR model and FEM model are 93.9% and 99.6%, respectively.

Table 4: Natural Frequencies of (a) the FEM and (b) the Embedded SR Model (NS direction).

(a)

mode	Natural Frequency (Hz)	Natural Period (s)	Effective Mass Ratio	
			X	Z
16	2.268	0.441	0.778	0.000
34	5.158	0.194	0.161	0.000
47	6.200	0.161	0.047	0.002
69	9.318	0.107	0.005	0.001

(b)

mode	Natural Frequency (Hz)	Natural Period (s)	Effective Mass Ratio	
			X	RZ
1	2.278	0.439	0.756	0.096
2	5.172	0.193	0.240	0.354
3	11.52	0.087	0.000	0.003
4	13.40	0.075	0.000	0.206

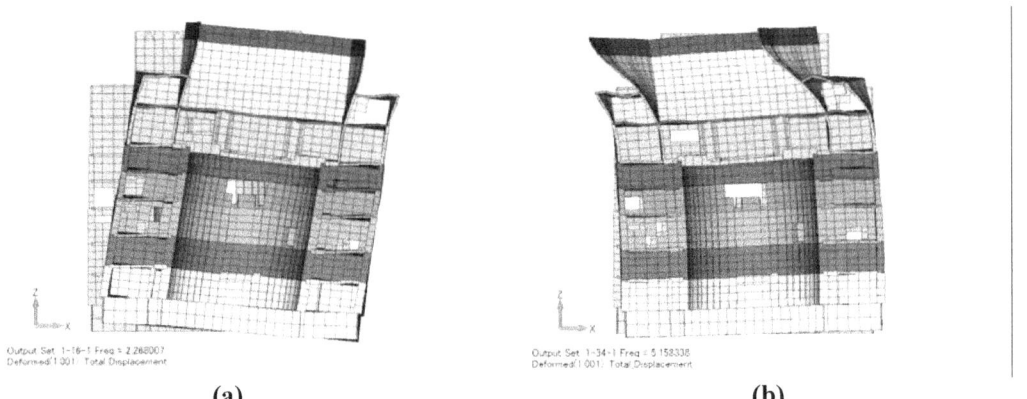

(a)   (b)

Figure 8: Natural Modes of FEM Model (NS Direction): Mode16 = 2.268 Hz (a) and Mode 34 = 5.158 Hz (b).

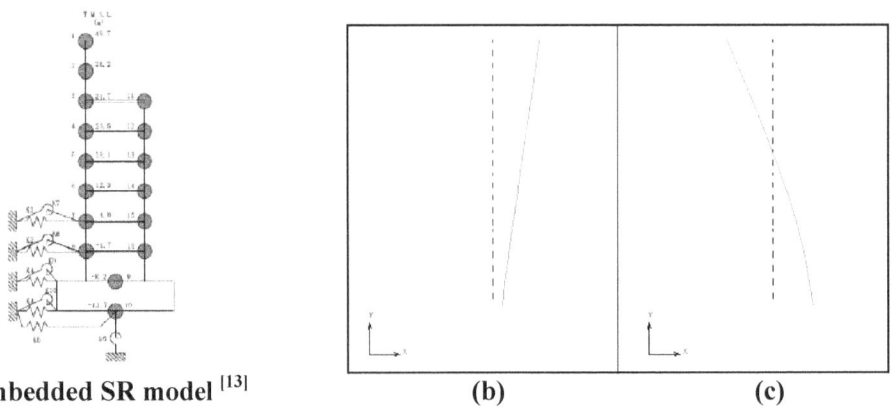

(a) Embedded SR model [13]   (b)   (c)

Figure 9: Embedded SR Model (a) and its Natural Modes (NS Direction): Mode 1 = 2.278 Hz (b) and Mode 2 = 5.172 Hz (c).

(2) Comparison with Observed Earthquake Data

To validate the constructed FEM model by another aspect, we compared the analytical results and earthquake data. The right-hand side of Figure 10 shows the positions where data were recorded, and the left-hand side of Figure 10 shows the output positions of analytical results.

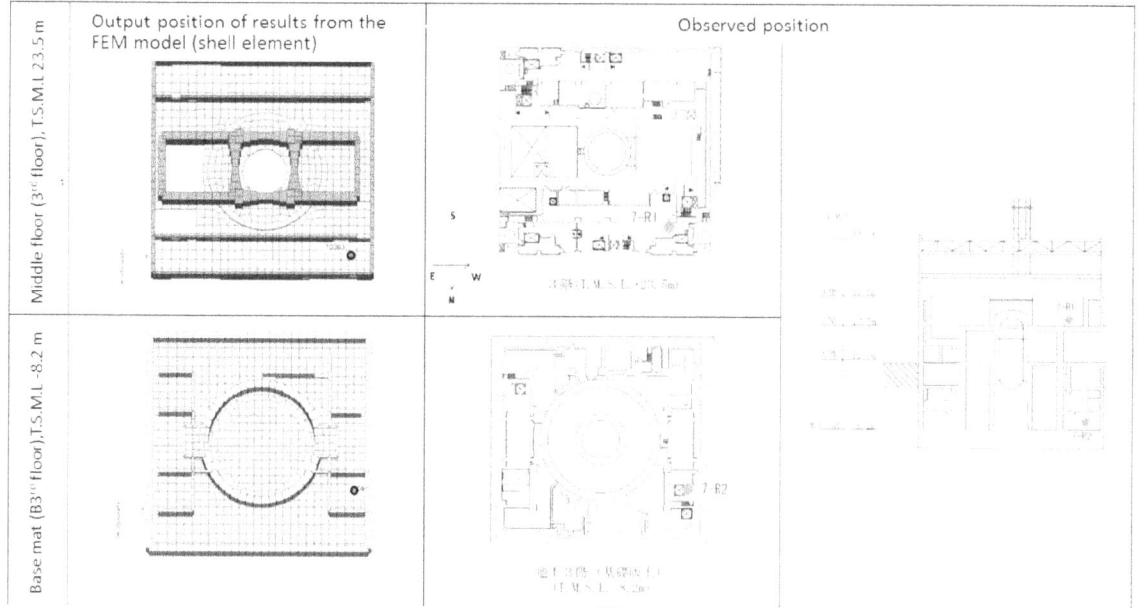

Figure 10: Positions Where Data Were Recorded [9] and Output Positions of Analytical Results.

Figure 11 shows observed- and analytical earthquake data for the SR model (up) and the FEM model (down), on the third (middle) floor. The horizontal axes show periods (in seconds) and the vertical axes show the acceleration response spectrum (in gals). The black and red lines indicate observed and analytical results, respectively.

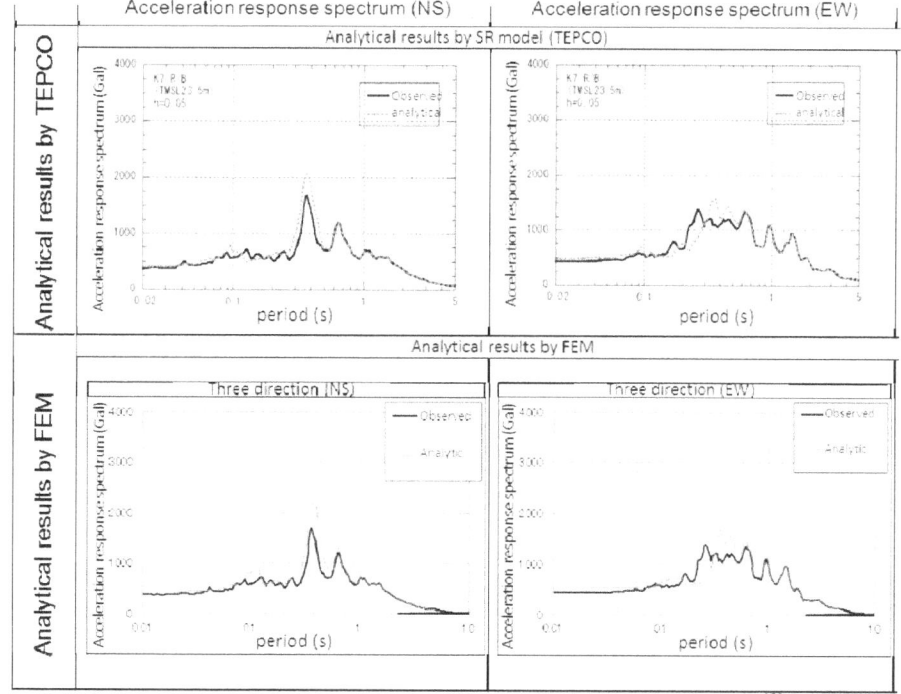

Figure 11: Observed- and analytical earthquake data for SR model (up)[9] and FEM (down) (third floor).

These results show that observed and analytical results are largely consistent. However, the observed spectrum in EW direction between 0.2 and 0.3 seconds is not captured by analytical results. Because a deliverable of this study is a selection guide of uncertainty factors- and ranges for response analysis, and not a simulation of observed data, we decided to use this model to perform sensitivity analysis and response analysis.

3.4. Pre-study for Sensitivity Analysis

Table 5 shows the uncertainty parameters used in the sensitivity analysis. The FEM model will be used for parameters one through seven. The SR model will be used for parameter eight because it is soil-property related. Literature survey will be performed for parameters nine and ten. Figure 12 shows the result of pre-study of sensitivity analysis to investigate the influence of the difference of equipment arrangement position on floor response. To explore relationships between input and responses, responses of three directions to analytical input are investigated. Results are shown as acceleration transfer functions. For example, the upper three figures in Figure 12 show the responses of a slab in EW, NS, and UD directions, to the input in EW direction. The horizontal axis shows frequency (in Hz) and the vertical axis shows the acceleration transfer functions (in dimensionless quantity). The three diagonal figures show the responses in EW, NS, and UD directions for the inputs in EW, NS, and UD directions, respectively. The lines of different color show responses to different points on the same floor. We can infer that differences in response amplitude occur on the same floor, and that differences become large in the high-frequency region. Furthermore, it is found that the response amplitude tends to become larger at the end of the opening compared to other points.

Table 5: Uncertainty Parameters Used in the Sensitivity Analysis.

S. no.	Parameters	Comments
1	Modeling area (quake-resistant wall, non quake-resistant wall, opening part of wall/slab, steps, non-structural member, etc.)	
2	Modeling errors (mesh size, error of center axis, etc.)	
3	Interaction between heavy equipment and building	
4	Influence of rigid/non-rigid floor assumption	
5	Influence of rigid/non-rigid base mat	
6	Influence of input wave (three directional input, one directional input, etc.)	
7	Soil-structure interaction (join condition of soil-base wall or soil-side wall, etc.)	
8	Consideration of non-linearity of soil, building structure, soil-building structure, etc.	SR model
9	Interaction between adjoining building	Literature survey
10	Material property (stiffness, strength, dumping of concrete, etc.)	Literature survey

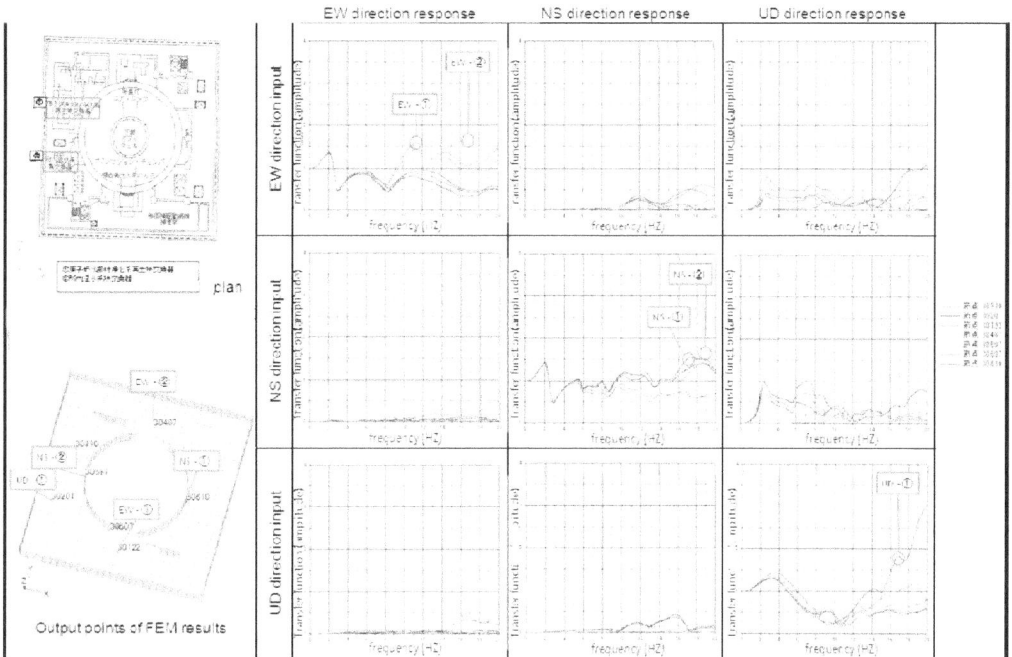

Figure 12: Amplitude of Acceleration Transfer Function for Each Direction [14].

4. CONCLUSIONS

Based on the one-dimensional wave theory essential to seismic response analysis of buildings and equipment, and the embedded SR model, we have attempted to extract opinions on high-impact factors from experts to reduce epistemic uncertainty through expert-opinion elicitation. Deep opinions were extracted and implemented by exchange between experts. Furthermore, we were able to confirm the relevance of the extracted factors in the sensitivity analysis, allowing for consideration of these important high-impact factors by policymakers. Since we found that seismic-safety evaluation is subdivided into many fields, and we identified this as an issue in this study, we are planning to strengthen the cooperation between these fields in the future. Furthermore, in preparation for the response- and sensitivity analyses of target buildings and ground of the model plant, a reactor-building model was created, the validity of which we confirmed by comparing observed to analytical records and by performing eigenvalue analysis. We will continue to perform sensitivity analyses related to the uncertainty evaluation of buildings and soil, for the purpose of fragility evaluation of equipment by using the present model. In the future, we also plan to carry out mapping of uncertainties using a logic tree.

Acknowledgements

This study is the result of "Reliability Enhancement of Seismic Risk Assessment of NPP as Risk Management Fundamentals" carried out under the Initiatives for Atomic Energy Basic and Generic Strategic Research by the Ministry of Education, Culture, Sports, Science and Technology of Japan.

References

[1] Implementation Standard Concerning the Seismic Probabilistic Risk Assessment of Nuclear Power Plants 2008, Atomic Energy Society standard, Japan Atomic Energy Society of Japan, 2007.
[2] USNRC, Recommendations for Probabilistic Seismic Hazard Analysis: Guidance on Uncertainty and Use of Experts , 1997
[3] USNRC, Practical Implementation: Guidelines for SSHAC Level 3 and 4 Hazard Studies, 2012

[4] Nuclear Standards Committee of Japan electric Association: seismic design technical regulations for nuclear power plant (JEAC4601-2008), 2009

[5] Seismic design standards subcommittee for LWR improvement:: research report on the standardization of seismic design (building system), 1981.

[6] Improvement and maintenance of FEM based analysis code (SANREF), Additional seismic safety function of 3D seismic input, JNES/SAE05-074, solutions, report of analysis evaluation unit-0074.

[7] IAEA, General specification for the KARISMA Benchmark, IAEA-EBP-SS-WA2-KARISMA-SP-002, (in press)

[8] Seismic safety evaluation report the results associated with the revision of the "seismic design review guidelines for nuclear power reactor facilities" (revision 1), Kashiwazaki-Kariwa nuclear power plant unit 7, TEPCO, 2009.

[9] Inspection and evaluation report in accordance with the equipment soundness after the Niigataken Chuetsu-Oki earthquake (buildings and structures, ed.) for Kashiwazaki-Kariwa nuclear power plant unit 7 (Rev. 1), 2008-0808, TEPCO, 2008.

[10] N. Fukuwa, Seismic performance of the reactor building (interaction analysis, etc.), "Niigataken Chuetsu-Oki earthquake and buildings and structures of the Kashiwazaki-Kariwa nuclear power plant," open lecture sponsored by the Science Council of Japan, 2009.

[11] K. Hijikata, R. Kikuchi, Y. Nukui, A. Imamura, F. Yagishita, T. Mase, H. Yoshida, T. Shiomi, K. Koyamada and K. Yoshida, "Dynamic response of unit No.7 reactor building during the Niigataken Chuetsu-oki Earthquake in 2007," J. Struct. Constr. Eng., AIJ, Vol. 76, No. 660, pp.319-327, 2011.

[12] The acceleration time history data in the Kashiwazaki-Kariwa nuclear power plant, 10/23/2004 17:56-12/31/2007 24:00, TEPCO (distributed by disaster prevention Association)

[13] Seismic safety evaluation report associated with the revision of the "seismic design review guidelines for nuclear power reactor facilities" for Kashiwazaki-Kariwa nuclear power plant unit 7 (Rev. 1), TEPCO, 2009.

[14] Application form for permission of establishment change for nuclear reactor, TEPCO, 1988.

Appendix: Questionnaire on Embedded SR Model

Q1-1   Uncertain factors of the embedded SR model
What factor among possible epistemic uncertainties relevant to the model is largest when the model is used to predict seismic input (FRS) to pipes and equipment?

Q1-2   Factors requiring sensitivity analysis
On what factors should the sensitivity analysis be performed?

Q1-3   Embedment effect
The model can reflect complex SSI (Soil-Structure Interaction) of a reactor building deeply embedded in the ground. Can the embedment effect be taken into consideration appropriately?

Analyses of Severe Accident Sequences During Shutdown and Caused by External Hazards

Michael Kowalik[a*] , **Horst Löffler** [a], **Oliver Mildenberger**[a], **Thomas Steinrötter**[a]

[a] Gesellschaft für Anlagen- und Reaktorsicherheit (GRS) mbH, Köln, Germany

Abstract: According to the German regulations for periodic safety reviews it is obligatory for each nuclear power plant to perform a Level 1 PSA for full power and shutdown operating conditions and for events caused by plant-external hazards. In contrary, a Level 2 PSA has to be performed only for full power operating conditions. The German regulatory body therefore supports a project with the objective of closing this gap of knowledge. First, a limited set of scenarios covering most of the relevant scenarios with respect to the time scale of the physical effects to be expected, the pressure buildup in the containment and the source term has to be identified. In order to calculate the set of scenarios by the computer code MELCOR the plant has been modeled in a plant-specific input deck and some scenario-specific settings need to be defined. Then the scenarios will be calculated by MELCOR and analyzed accurately regarding the relevant physical effects including core melting and the release of radionuclides. The results of the deterministic analyses will support development of a probabilistic event tree approach and recommendations for prevention and mitigation of such accidents.

Keywords: Level 2 PSA, shutdown modes, MELCOR, severe accident analyses

1. INTRODUCTION

The present obligation in Germany to perform Level 2 PSA with the only focus on full power operation modes is based on the assumption that pressure and decay heat are quite low at shutdown operational conditions. Despite these facts pertinent analyses have shown that shutdown modes represent a significant contribution to the overall core damage frequency, e.g. due to limited availability of safety systems. Furthermore, it is obvious that external hazards can cause damage of the relevant barriers.

Since the German PSA Guide [1] does not require performing Level 2 PSA for shutdown modes and external hazards, research and development (R&D) activities recently performed by GRS and depicted in this paper can be subdivided into the following five major parts:

1. Identification of the state of the art considering already performed studies in respect of shutdown operational modes;
2. Identification of relevant sequences covering all other sequences that lead to similar core damage states and determination of the initial and boundary conditions;
3. Analyses of the relevant accident sequences (PWR / BWR) using the integral code MELCOR;
4. Conclusions based on the above mentioned analyses concerning:
   a. Phenomena during accidents caused by external hazard or during shutdown operational modes;
   b. Behavior and release of fission products;
   c. Designated and possibly additional emergency procedures;
5. Quantitative assessment of the significance of accident sequences during low power and shutdown operational modes in comparison to sequences during full power operation and plant internal initial events including the influence of uncertainties on the results.

This paper will give an overview of the work done so far.

---

* Michael.Kowalik@grs.de

2. IDENTIFICATION OF RELEVANT SEQUENCES

2.1. Shutdown Operational Modes

To identify relevant sequences of accidents for Level 2 PSA it is self-evident to use the results of an appropriate Level 1 PSA that depicts sequences leading to so-called system damage states, which are defined as states that lead to core damage if preventive measures do not succeed.

There is a pre-defined interface between PSA Level 1 and Level 2 characterizing the physical and technical state of the facility. Whilst using the information delivered by this interface it is possible to continue these states to the scope of Level 2 PSA. This interface is defined in the technical document on PSA methods [2] supplementary to the German PSA Guide [1] intended to be used for full power operation. Hence before this interface can be used it has to be extended according to the characteristics of low power and shutdown operational modes or external hazards. The extensions, necessary in this project, concern conditions such as an open RPV (reactor pressure vessel), time after shutdown, state of the reactor protection system or the water level in the refueling cavity. This extended interface has been applied to Level 1 PSA [3] that had been performed for a PWR of KONVOI type, also in the context of a research project to evaluate sequences leading to core damage if no preventive measures are executed. Using this Level 1 PSA, it has been possible to systematically assign the entire system damage states to newly defined core damage states. The transition from system damage states to core damage states requires some assumptions for failures such as a not-initiated primary depressurization. If the primary pressure release is available, some further unavailabilities as those of the residual heat removal (RHR) systems has to be assumed because this system, if it is intact, it could inject or remove the decay heat in case of a successful primary pressure release. So the system damage states with the additional assumptions are summarized to a certain set of core damage states.

The next step is the choice of a set of relevant sequences leading to core damage, which is intended to be calculated and analyzed. The claim that has to be met by this set of sequences is to cover all other relevant sequences in respect of their frequency but also in respect of the expected severity of their consequences. This process is mainly based on the quantification of the system damage states, the obvious extrapolation of the sequences in consideration of the state of the facility and on expert judgment. Furthermore, system damage states considering phenomena of deborated primary coolant and sequences in the spent fuel pool are not considered within this selection. The reason for the exclusion of the deborating events is the absence of expected fuel element damages, which is described in [4]. The events related to the spent fuel pool are subject of another R&D project that GRS is working on.

In the case of BWR-type nuclear power plants the KWU-type BWR72 is chosen to be the reference object because it is the only facility representing a BWR in Germany that is in service. Screening the given documents (the most relevant is [5]) a set of four relevant initiators has emerged:

1. Loss of the modified heat removal during cool down,
2. Incorrect injection into RPV,
3. Leakage at the flood compensator,
4. Leakage at the bottom of the RPV due to dismounting a circulation pump.

The sequences leading to a set of six core damage states are selected according to the most evident system-technical states. The chosen sequences also cover a broad range of severe accident progression due to very different initial states including e.g. a filled and a dry RPV which may affect the access of atmospheric oxygen. Furthermore, the time since shutdown ranges between $16\,\text{h} \leq \Delta t_{\text{scram}}^{\text{shutdown mode}} \leq 200\,\text{h}$.

2.2. External Hazards

In the case of external hazards, literature research provided scenarios that comprise LOCAs and transients which should be controlled by the design features of the facility. Even in the case of aircraft crashes, earthquakes, floods or blasts caused by explosions no differences between these scenarios and known ones with internal initiators and induced additional unavailabilities have been identified. That means that no further phenomena emerged that should be studied. According to the objectives of this project to extend the knowledge in the context of shutdown operational modes and external hazards, nevertheless some scenarios are created disregarding the corresponding probability. This is the reason that the basic scenario of a station blackout is chosen because it is well known and thus comparable to former analyses using another input deck for MELCOR. Additional to this station blackout scenario, further scenarios are defined with some additional damages or unavailabilities respectively. The station blackout is as far as possible derived from the external hazard as well as the additional damage. The selected sequences comprise earthquakes and aircraft crashes onto the reactor building and the reactor auxiliary building. Damages are assumed at the primary circuit (LOCA) as well as at the reactor building and the containment respectively or reclusive at the venting system in case of a PWR. The scenarios related to BWR consider only earthquakes with additional postulated damages at a feed water line within the containment and low level leakage at the suppression chamber due to a rupture in the residual heat removal system.

3. DESCRIPTION OF THE MELCOR INPUT DECK

To perform severe accident analyses it is necessary to model the plant in an appropriate manner considering certain accuracy on the one hand and a certain simplification in order to limit the numerical effort on the other hand. Both input decks (PWR / BWR) will be presented in the following.

3.1. Pressurized Water Reactor (PWR)

The modelling of the reactor coolant system (RCS) is shown on the left side of Figure 1. The four loops of the real power plant are modelled using two loops. One of them comprises three real loops and the other one is a single loop with the pressurizer and the relief tank. The modelling consists of certain numbers of control volumes, heat structures and flow paths listed in Table 1. The systems attached to the primary side of the reactor coolant system incorporate reactor coolant pumps, safety injection pumps, residual heat removal system, accumulators, volume control system, and the extra borating system. The RPV itself is thermo-dynamically modelled (CVH package in MELCOR) by using only one control volume. Within the COR package it is modelled by using 5 core rings and 15 axial meshes whereof 12 are related to the active core region. Moreover, the two pressure relief valves and the blow-off control valve are modelled including the corresponding control from the reactor protection system.

At the secondary side, the steam generators are modelled including separator, main steam line and also the blow-off valve, the safety relief valve including the corresponding control by the reactor protection system, which implies the shutdown, the runback and the safety relief function. The conventional part of the nuclear power plant including the turbine, main condenser and the feed water heater line are modelled using one time-independent control volume for each the turbine and the feed water station. The nodalization of the containment is shown on the right side of Figure 1. The flow paths between the zones of the containment are partially equipped with doors that are assumed to be closed initially in most of the scenarios. The control of these doors implies the possibility of being opened by damage at the lock or the frame of the door depending from the direction of the pressure gradient. In the first case, the door can be re-closed again; in the latter case this is impossible. In addition to the doors, some rupture discs are modelled, too. Furthermore, some control volumes of the containment accommodate passive autocatalytic recombiners (PAR) – as realized in the reference plant – which keep the hydrogen concentration low in order to avoid large-scale hydrogen combustions. The hydrogen originates mainly from the reaction of zircaloy and steam but also from the reaction of steel and the corresponding alloy additions with steam. The areas in which the molten core concrete interaction (MCCI) takes place are modelled by MELCOR cavities. In this input deck, three cavities are defined, which stand for the reactor cavity, the zone between the biological shield and the support shield, in the following called "gap volume", and the reactor sump. The biological shield itself is quite thin ($\Delta r_{\text{biol. shield}} = 0.55$ m) and can be penetrated quickly by radial ablation.

Figure 1: MELCOR modelling of the reactor cooling system and the containment

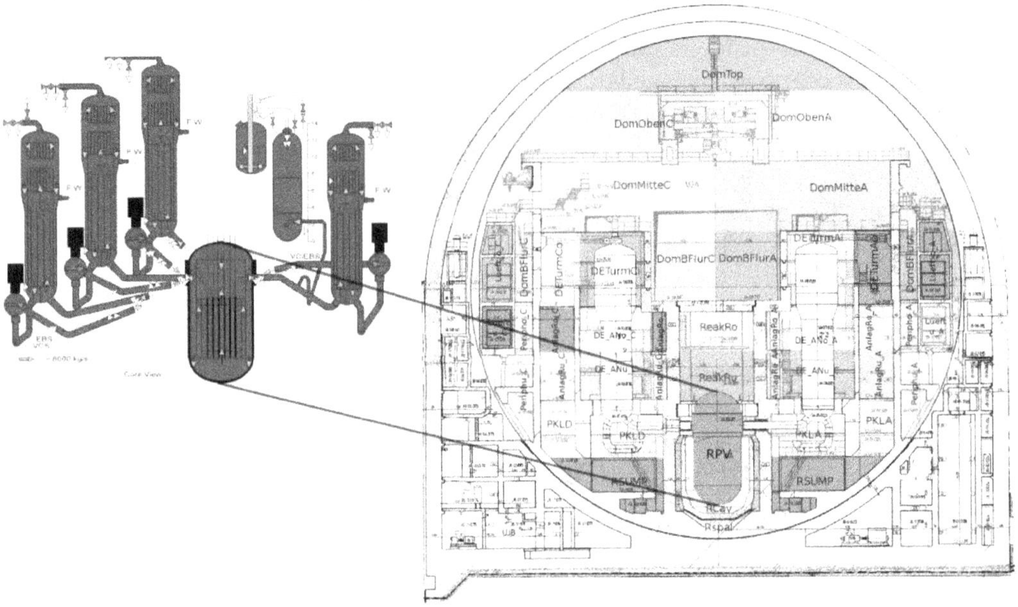

In addition, the ventilation channels below the surface of the reactor cavity bottom are considered as well. Thus, two modes of cavity rupture are possible to transfer molten material into the next cavity ("gap volume"). Between the latter one and the sump cavity there are dampers within the support shield whose lower edges are located very low above the bottom of this cavity. So there is only a certain small amount of corium necessary to trigger this rupture mode of the second cavity.

Table 1: Objects for the MELCOR modelling in the case of the PWR (pri: primary side of the RCS, sec: secondary side of the RCS, int: internal connections within area / object, ext: connections between the area / object and its environment)

Area / Object	n^{CV}	$n^{HS}_{\text{int/pri}}$	$n^{HS}_{\text{ext/sec}}$	n^{FL}_{internal}	n^{FL}_{external}
single loop (primary/secondary)	5/6	20	10	9	31
triple loop (primary/secondary)	5/6	36	15	9	10
RPV	6	30	10	10	7
containment	77	218	10	256	7
annulus	12	21	22	19	7
burst elements (door / disc)	82/56				
n^{CV} equipped with recombiners	37				

3.2. Boiling Water Reactor (BWR)

The modelling of the reactor cooling system and the containment is shown in Figure 2 and consists of a set of control volumes, heat structures and flow paths whose numbers are given in Table 2. In the case of a BWR, the reactor cooling system mainly consists of the RPV and the main steam lines that conduct the saturated steam which leaves the separator towards the turbine. The RPV of the BWR is nodalized in the CVH package using only one control volume. In the COR package it is divided into 5 rings and 19 axial meshes whereof 12 meshes belong to the scope of the active core region, the remaining ones are assigned to the lower plenum. The systems that are attached to the reactor cooling system comprise the circulation pumps, the residual heat removal system, the purging system, the safety relief valves and the corresponding control by the reactor protection system. The control of these valves implies the safety relief function to limit the pressure in the RPV and the automatic depressurization to decrease the pressure in the RPV in order to make the low pressure injection available. In reality there are 11 safety relief valves so there is a high level of redundancy but the valves are of the same kind. To handle common cause failures the facility provides 3 diverse pressure limiting valves which open at lower pressures in order to conserve the safety relief valves. In the case of the residual heat removal systems also several operational modes are possible. The system has high and low pressure pumps for the different pressure in the RPV. The high pressure pumps are intended to stabilize the water level in the RPV in the case of loss of feed water. The low pressure pumps are intended to flood the RPV in the case of LOCA events and remove the decay heat out of the RPV or the suppression chamber. The conventional part of the nuclear power plant is modelled like that of the PWR (see par. 3.1).

Table 2: Objects for the MELCOR modelling in the case of the PWR (pri: primary side of the RCS, sec: secondary side of the RCS, int: internal connections of the area / object, ext: connections between the area / object and its environment)

Area / Object	n^{CV}	$n^{HS}_{int/pri}$	$n^{HS}_{ext/sec}$	$n^{FL}_{internal}$	$n^{FL}_{external}$
RCS	14	32	14	18	3
containment	22	43	69	38	5
reactor building	196	859	212	602	21
burst elements (door / disc)	240/27				
n^{CV} equipped with recombiners	13				

In contrary to the modelling of the PWR an arrangement of only two cavities is considered here. The first one is related to the room where the control rod drives are located. There is also a cylinder symmetric wall on which an assembly machine is supported. The space that is surrounded by this wall represents the first cavity. Due to the purpose of this wall to support a device and not to retain a molten pool it has only a thickness of about 0.48 m, thus the radial ablation will quickly penetrate the wall. Then a part of the molten pool has access to an area of the basement of the containment which is separated from the reactor building by steel plates that are not able to cope with an attack of molten corium so they will immediately rupture in such a case. The molten corium will then flow into the basement of the reactor building. This area represents the second cavity but it is only that part of the basement that may be expected to cause the most severe consequence. This assumption is based on the circumstance that the boundary of this area provides a door which directs to the environment.

Figure 2: Reactor cooling system and the containment on the left side and the reactor building including the containment on the right side respectively (BWR72 type)

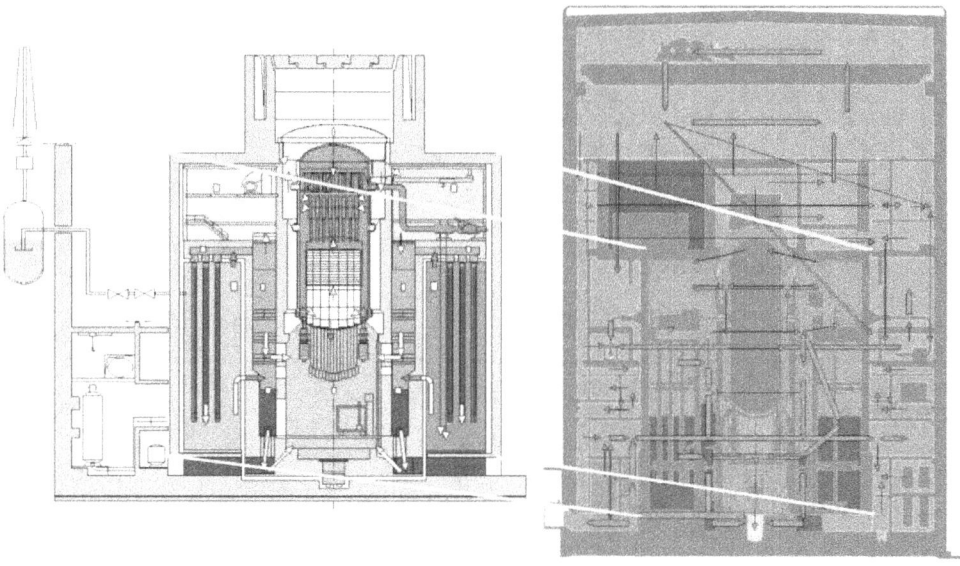

BWRs in Germany are also equipped with autocatalytic passive recombiners (PAR) which are located within the containment. To adapt the input values to the real devices in the facility, experimental data of the international THAI project [6] has been examined. So the values related to the PAR start concentration, dead time, relaxation time and the parameters for the simple dependence of the volume flow rate from the hydrogen concentration could be obtained. In the real facility there are 59 recombiners within the containment whereas 13 control volumes in the input deck are equipped with a PAR input. The doors and rupture disks are treated like that in the PWR (see par. 3.1).

4. EXEMPLARY ANALYSIS

One exemplary sequence related to a PWR in shutdown mode will be described in the following.

4.1. Initial Conditions

The PWR plant is in the shutdown mode called 1B2, in which the water level of the RCS is decreasing down to mid-loop (that means ¾ of the height of the reactor coolant line). Two trains of the residual heat removal system (RHR) assure the transportation of decay heat. One train is in maintenance and the last one is in standby. Under these initial conditions a postulated leakage occurs in one operating train of the RHR, for example due to thermal stress. This leak is located between the residual heat removal pump and the plunger check valve. It is assumed that the first shutoff valve and the plunger check valve (i.e. first and second isolation of the residual heat removal from the primary circuit) or the shutoff valve in the bypass line for the plunger check valve fail to close. The result is a permanent bypass from the RCS to the reactor building annulus. The residual heat removal is then lost by the drop of the water level below the suction point of the corresponding pumps. Due to reaching the threshold in the minimum flow line of these pumps they are shut down. According to the plant operating manual the accumulators inject to fill up the RCS. Moreover, some inventory of coolant is present in the flooding tanks that could be injected. This measure may delay core damage but it will not prevent it. This measure is not considered in the basic calculation shown here.

Table 3: Main events / phenomena during the sequence

Event / Phenomenon	Time	
shutdown	-23:00 h	
loss of all cooling systems due to leak in RHR system	0:00 h	
reach of the boiling point in the RPV (core related control volume)	0:33:20 h	
begin of the core uncovering at $L_{RPV} \leq 6.63$ m	6:09:18 h	$\Delta t_{uncovering}$
end of the core uncovering at $L_{RPV} \leq 2.73$ m	8:09:20 h	$= 2:00:02$ h
begin oft the production of hydrogen	6:21:40 h	
gap release (begin; core ring 2)	6:47:33 h	
begin of the core melt process (first relocation of core material)	**7:02:33 h**	
rupture of the lower core grid, core drop, quenching	**11:04:00 h**	
dry-out of the lower plenum	11:35:50 h	
rupture of the RPV, begin of relocation of the molten pool into the cavity	**12:42:22 h**	
contact of the molten pool with the ventilation channels (dry)	17:23:54 h	
reaching the design temperature of the containment ($T_{design}^{containment} = 418.15$ K)	58:32:50 h	
rupture of the burst disc of the relief tank	76:59:51 h	
maximum pressure in the containment	76:59:51 h ($= t_{burst}^{relief\,tank}$); $p_{max}^{cont} = 0.16$ MPa	
large combustion in the reactor building annulus	07:54:29 h	
end of the calculation	336:33:31 h	

4.2. Conditions in the RCS

At the beginning of the scenario at 00:00:00 hours, the leak occurs and at the same time (simplification) the accumulator injection takes place. This increases the pressure and decreases the temperature for a short time period. The further progression is characterized by local pressure maxima and minima that are based on the leakage rate and relocation of material of the uncovered part of the core which falls into the water pool and provides an increase of the evaporation for a certain time which increases also the pressure. The leakage rate at the beginning is about 13.88 kg/s for an interval of 03:18:00 hours. After that the leak becomes uncovered. From this time on only steam is discharged which means a significant decrease of the leak mass rate. This behavior can also be seen in Figure 3, where the whole mass of liquid water in the RCS is shown. At the time of the rupture of the RPV at 12:42:22 hours, only the water in the pump suctions resides in the RCS.

Figure 3: Water inventory of the RCS

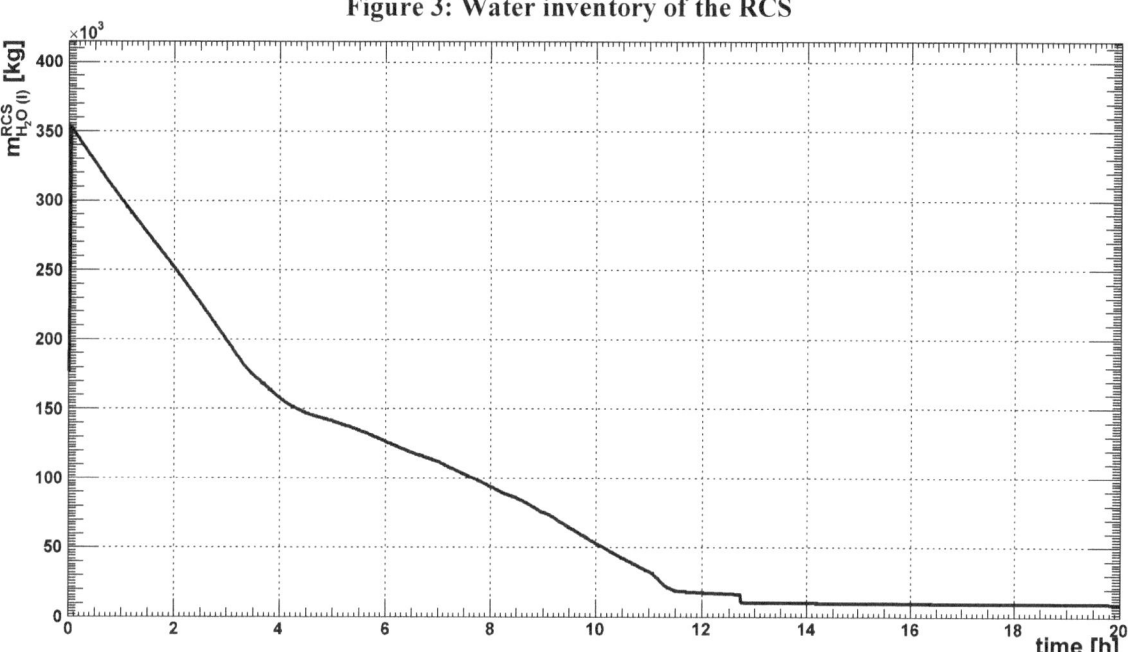

In Figure 4 the temperatures of the cladding of the inner ring of the MELCOR modelling can be seen. A significant increase from the boiling temperature ($T_{\text{boiling}}^{\text{RCS}} = 462.0$ K) can be realized at 06:09:10 hours. This time correlates with the beginning of uncovering the active region of the core at 6.63 m. The slope of the increase of the cladding temperatures rises at 06:50:00 hours significantly due to the beginning of the zircaloy steam reaction. This reaction ends if a certain oxide layer in the cladding is achieved. Thus the temperatures decrease through the loss of thermal radiation and conduction. Then the temperature increases once more due to the main heat up of the RCS and ends with culmination at 2500.0 K which is assumed as the melting point of the interacting ZrO_2 and UO_2. At 07:02:33 hours this temperature is achieved for the first time in the uppermost cell of the inner ring. The gap release of the five MELCOR rings occurs quite earlier at a temperature of 1173.0 K at the following times for the different MELCOR rings: (06: 47: 37 h, 06: 47: 33 h, 06: 48: 39 h, 06: 47: 39 h, 07: 07: 43 h).

The relocation of core material to the lower plenum begins at 11:04:00 hours and ends 00:09:20 hours later. This also means that the steel temperature of the lower head increases so it strains. Considering the stress and strain of the steel MELCOR assumes a rupture at 12:42:22 hours which results in an ejection of debris. This process ends in essence 00:26:39 hours later with a mass of $2.0797 \cdot 10^5$ kg which is ejected into the reactor cavity (cavity #1).

Figure 4: Cladding temperatures during the uncovering of the reactor core

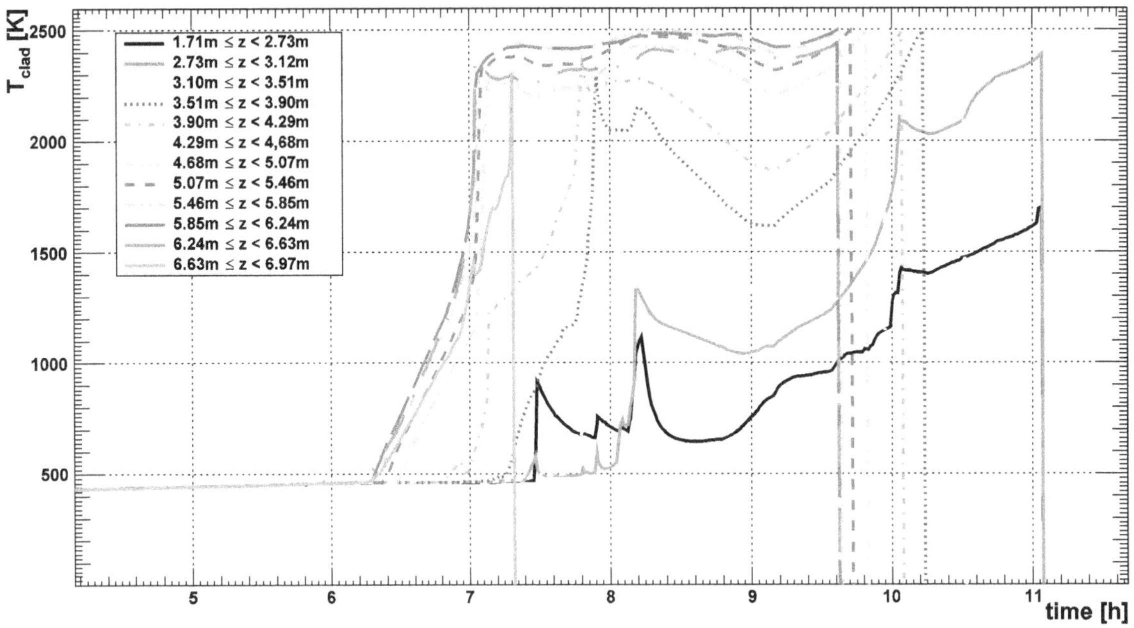

4.3. Conditions in the Containment and the Annulus

Due to the leak in the RHR connecting the RCS with the reactor building annulus, the essential barrier of the containment is bypassed, so no significant pressure built up is expected. There are local maxima ($p_{cont.} < 0.13$ MPa) of pressure within the containment due to effects like the rupture of the lower head and the rupture of the first cavity triggered by the radial rupture criterion at 17:23:54 hours which results in a discharge of molten pool into the second ("gap volume") and immediately into the third (reactor sump) cavity. This discharge means an increase of the production rate of hydrogen, which causes several combustions. Before these phenomena also a large combustion within the annulus occurs at 07:54:29 hours which results in a short increase of pressure and temperature which opens burst discs connecting the annulus with the environment directly and enlarges the existing connection through the reactor auxiliary building which has opened before due to the pressure built up resulting from the leakage. Hence no further significant pressure build-up is possible. The absolute maxima achieved in the containment correlates with the rupture of the burst discs of the relief tank at 76:59:51 hours, which is heated up by the containment atmosphere. The resulting maximum pressure is 0.16 MPa, which is not conserved due to condensation of the steam and the leak rate. This pressure is significantly below the design pressure of the containment, which is 0.63 MPa. In contrary, the design temperature, which is 418.15 K, is achieved at 58:32:50 hours and is increasing monotonically during the further devolution of the accident achieving a maximum value of 592.45 K at the end of the calculation. The conditions in the annulus also exceed the design in respect to the temperature, which reaches a value of 495.85 K at the end of the calculation. The design temperature of important safety devices in the case of a residual heat removal pump for example is 473.15 K according to [7].

In Figure 5, the production of hydrogen based on several chemical reactions during the in-vessel phase is shown. This phase begins at 06:21:40 hours, slightly before a significant increase of the cladding temperature can be observed. It ends with the rupture of the lower head of the RPV. The hydrogen production within the core continues until 18:22:31 hours with a generated total mass of 1082.4 kg of hydrogen. The main contribution with a mass of 874.0 kg is delivered by the reaction of zircaloy with steam. The remaining reactions with steam considering the steel and its alloy additions (chrome and nickel) deliver a mass of 208.4 kg of hydrogen. The hydrogen mass that is recombined by the passive autocatalytic recombiners which is also given in Figure 5 is exactly 0.0 kg during the in-vessel phase, because no hydrogen is delivered into the containment.

Figure 5: Hydrogen production during the in-vessel phase

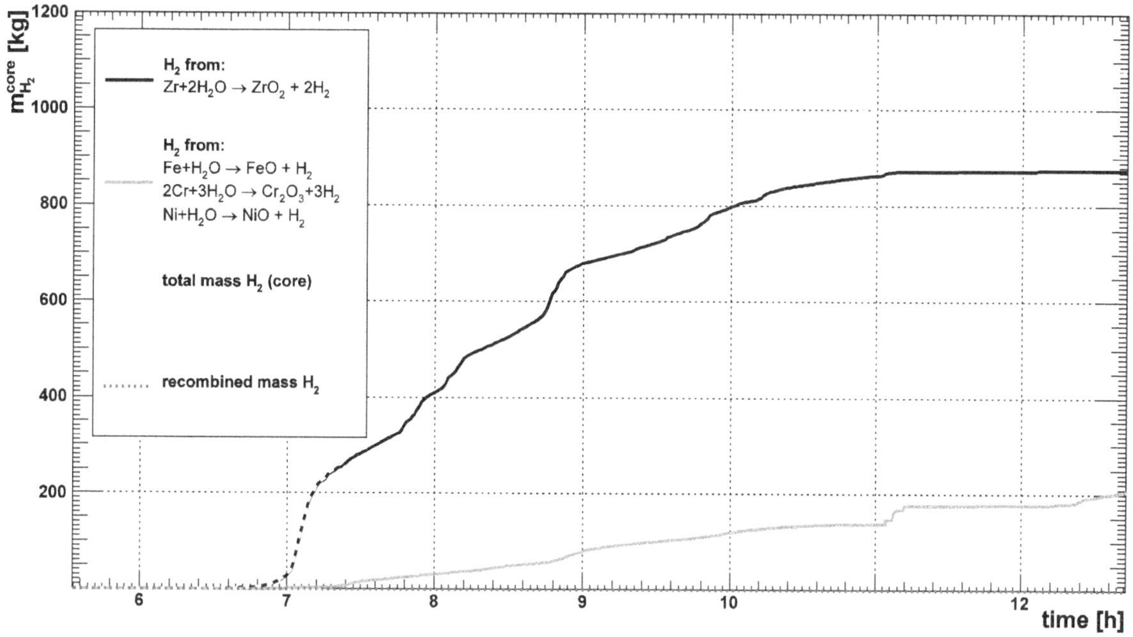

The whole hydrogen balance is shown in Figure 6 where the produced mass, the recombined mass and the consumed mass due to the combustions in the different areas of the reactor building are depicted. The total hydrogen mass produced in the core and the three cavities at the end of the calculation is 3460.9 kg, in which $\left(m^{H_2}_{cav00}, m^{H_2}_{cav01,\,sat}, m^{H_2}_{cav02,\,sat} \right)\big|_{t=t^{calculation}_{end}} = (2167.4\,kg, 5.3\,kg, 133.4\,kg)$ is related to the corresponding cavity. But only the first cavity produces hydrogen up to the end of the calculation. The productions originating from the other cavities go into saturations due to the cooling down of the thin pools. This is the case of the second cavity ("gap volume"), because the connecting path between the second and the third cavity (rector sump) is near to the floor of this cavity. Thus, the amount of the remaining molten pool mass is very limited. In the case of the third cavity the reason is based on the large area which is covered by the molten pool. In both cases the pool is cooled down below 1420 K which represents the solidus line of the concrete. Hence, no further liquefaction is possible and the gas production stops. As already mentioned the recombination of the PARs starts when the rupture of the RPV occurs at 12:42:22 hours. The total amount of catalytic recombined hydrogen is 1173.1 kg which also goes into saturation due to oxygen starvation within the containment at a molar fraction of $\frac{n_{O_2}}{\sum_i n_i} \approx 0.28\,\%$. The combustions are not dependent from such concise events such as the RPV rupture because the hydrogen is discharged continuously into the annulus since the zircaloy steam reaction takes place in a significant manner at about 06:21:40 hours. So here the first large scale combustion appears at 07:54:29 hours. Overall a mass of 485.03 kg recombine till the end of the calculation through combustions in which 246.07 kg are related to the reactor building annulus. An amount of 647.97 kg remains in the containment, RCS and annulus.

Figure 6: Balance of the hydrogen production and consumption during the accident devolution

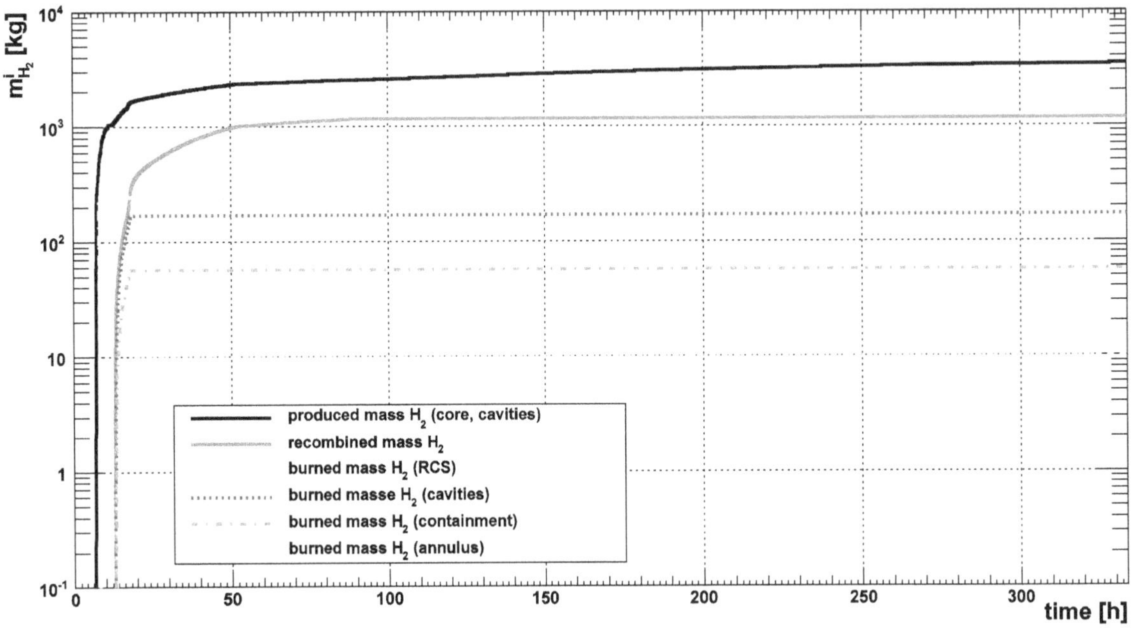

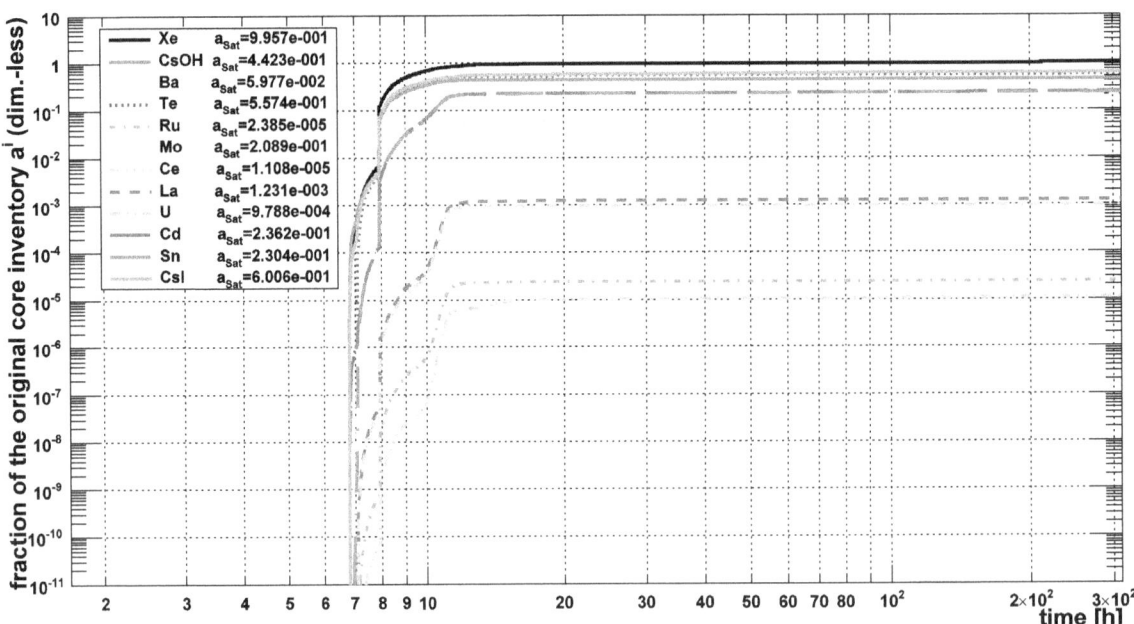

4.4. Release of Radionuclides

Figure 7 depicts the release into the environment of the plant. It begins with the burst of the fuel rod cladding at 06:47:33 hours (gap release). There is an early significant increase at 07:54:29 hours which is based on the hydrogen deflagration in the annulus that causes the activation of the release path into the environment of the facility by opening the door into the reactor auxiliary building.

Figure 7: Released fraction of the original core inventory of the MELCOR element classes

According to the explanation above the significant release into the environment starts with the large-scale deflagration in the annulus. The values at the end of the calculation are given in Table 4. For example, the fraction of the most volatile MELCOR group from the original core inventory is

$$\frac{m_{released}^{Xe}}{m_{core\ inventory}^{Xe}} = 99.57\,\%.$$

Table 4: Released fractions of the element classes from the original core inventories. The numbers in brackets indicate a hierarchy in relation on the volatility (fraction, total mass).

MELCOR element classes		Release fraction into the environment:	Original core inventory:	Released mass into the environment:
Xe	(1,1)	$9.957 \cdot 10^{-1}$	$7.111 \cdot 10^{+2}$ kg	$7.081 \cdot 10^{+2}$ kg
CsOH	(4,2)	$4.423 \cdot 10^{-1}$	$4.050 \cdot 10^{+2}$ kg	$1.791 \cdot 10^{+2}$ kg
Ba	(8,7)	$5.978 \cdot 10^{-2}$	$2.969 \cdot 10^{+2}$ kg	$1.775 \cdot 10^{+1}$ kg
Te	(3,5)	$5.574 \cdot 10^{-1}$	$6.501 \cdot 10^{+1}$ kg	$3.624 \cdot 10^{+1}$ kg
Ru	(11,12)	$2.385 \cdot 10^{-5}$	$5.327 \cdot 10^{+2}$ kg	$1.271 \cdot 10^{-2}$ kg
Mo	(7,3)	$2.089 \cdot 10^{-1}$	$5.070 \cdot 10^{+2}$ kg	$1.059 \cdot 10^{+2}$ kg
Ce	(12,11)	$1.107 \cdot 10^{-5}$	$2.095 \cdot 10^{+3}$ kg	$2.320 \cdot 10^{-2}$ kg
La	(9,10)	$1.232 \cdot 10^{-3}$	$1.009 \cdot 10^{+3}$ kg	$1.243 \cdot 10^{+0}$ kg
U	(10,4)	$9.788 \cdot 10^{-4}$	$9.750 \cdot 10^{+4}$ kg	$9.544 \cdot 10^{+1}$ kg
Cd	(5,9)	$2.362 \cdot 10^{-1}$	$1.310 \cdot 10^{+1}$ kg	$3.093 \cdot 10^{+0}$ kg
Sn	(6,8)	$2.304 \cdot 10^{-1}$	$1.764 \cdot 10^{+1}$ kg	$4.064 \cdot 10^{+0}$ kg
CsI	(2,6)	$6.006 \cdot 10^{-1}$	$5.806 \cdot 10^{+1}$ kg	$3.487 \cdot 10^{+1}$ kg

Figure 8 depicts the total masses of xenon as a representative class for the release through the corresponding release paths is shown. Most of the gas mass ($m_{\text{release, BD}}^{\text{Xe}} = 647.04$ kg) is released through a burst disc which connects the annulus (lower cv) and the atmosphere. This disc is opened at 07:54:29 hours when the pressure in the annulus reaches 20.0 kPa for a short time due to the hydrogen burn. This pressure peak opens also other release paths.

Figure 8: Total mass released through different paths into the environment

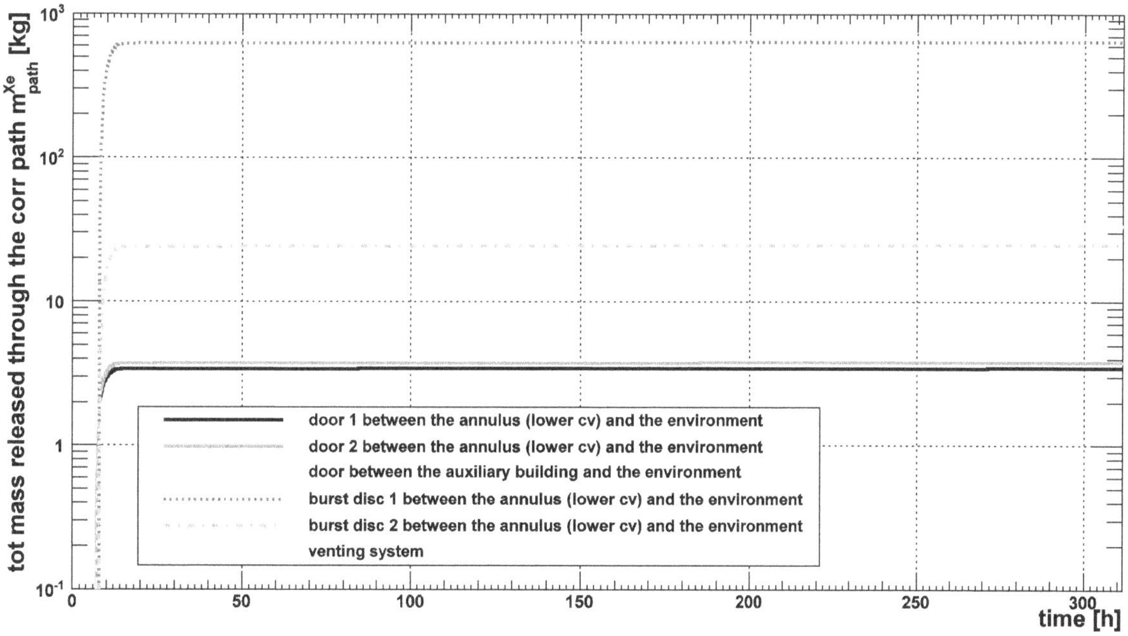

5. CONCLUSIONS

As described above, the selection of relevant scenarios with respect to shutdown operational modes and external hazards in order to deepen the knowledge in this context have been accomplished for PWR and BWR. Furthermore, several severe accident sequences have already been modelled in the input deck of MELCOR and calculated. This shows that MELCOR is capable to be applied also for sequences different from full power operation.

The exemplary analysis shown in this paper is initiated by a leak in the RHR system at mid-loop operation of a PWR, bypassing the containment. It shows a hydrogen mass of 1082.4 kg during the in-vessel phase lasting up to about 12:42:22 hours until the RPV rupture occurs. In comparison to the full power operation analyses (in [8] it is about ~600 kg ... 800 kg) that means a quite high amount of hydrogen mass during a relatively long in-vessel phase.

Beside this, the analysis has shown that the calculated temperature in the containment as well as in the annulus significantly exceed the design temperature whereas no significant pressure built up occurs. Furthermore, early hydrogen combustion in the reactor building before RPV failure opens direct release paths to the environment. No noteworthy retention inside the plant is possible; therefore a very large release into the environment will occur.

Further conclusions will be drawn in the next steps of this ongoing R&D project. In those steps the remaining calculations will be performed and analyzed in respect of discovering weak points in the design of the facility and they will contribute to a deepened comprehensive assessment of such kinds of sequences. Furthermore, they will be assessed in respect to their probabilities and considered in an event tree in order to evaluate the relevance of such severe accident sequences also in comparison to full power operation.

Acknowledgements

The authors thank the Federal Ministry for the Environment, Nature Conservation, Building and Nuclear Safety (BMUB) for funding this R&D project.

References

[1]   Federal Ministry of the Environment, Nature Conservation and Reactor Safety (BMU), *"Safety Review for Nuclear Power Plants pursuant to § 19a of the Atomic Energy Act - Guide Probabilistic Safety Analysis - of 30 August 2005*, Federal Bulletin No. 207a, (2005), http://www.bfs.de/de/bfs/recht/rsh/volltext/A1_Englisch/A1_08_05.pdf.

[2]   Facharbeitskreis (FAK) probabilistische Sicherheitsanalysen. *"Methoden zur probabilistischen Sicherheitsanalyse für Kernkraftwerke"*, Stand: August 2005, Bundesamt für Strahlenschutz (BfS), BfS-SCHR-37/05, Salzgitter, Germany, (2005).

[3]   D. Müller-Ecker et al. *"Sicherheitstechnische Bedeutung von Zuständen bei Nichtleistungsbetrieb eines DWR"*, Gesellschaft für Anlagen- und Reaktorsicherheit (GRS) mbH, GRS-A-3114, Köln, Germany, (2003).

[4]   S. Kliem et al.: *"Core response of a PWR to a slug of underborated water"*, Forschungszentrum Rossendorf e.V., Institute of Safety Research, Dresden, Nuclear Engineering and Design 230 (2004) 121-132, (2003).

[5]   D. Müller-Ecker et al. *"Untersuchung von Ereignissen außerhalb des Leistungsbetriebes"*, Gesellschaft für Anlagen- und Reaktorsicherheit (GRS) mbH, SWR-Sicherheitsanalyse, Phase II, Abschlussbericht, Band 2, GRS-A-2713, Köln, Germany, (1999).

[6]   T. Kanzleiter et al. *"Hydrogen Recombiner Tests HR-14 to HR-16 (Tests using an NIS-PAR), AREVA, AECL and NIS Comparison"*, Technical Report (Quick Look Report), Rector Safety Research Project 150 1326, OECD/NEA THAI Project (contract 18 July 2007), Becker Technologies GmbH, Eschborn, Germany, (2007).

[7]   EnBW Kernkraft GmbH. *"Systembeschreibung Not- und Nachkühlsystem und Beckenkühlsystem JN, FAK"*, Gemeinschaftskraftwerk Neckar, Block 2 (GKN-2), (2002).

[8]   M. Sonnenkalb et al. *"Unfallanalysen für DWR vom Typ KONVOI (GKN-2) mit dem Integralcode MELCOR 1.8.4"*, Gesellschaft für Anlagen- und Reaktorsicherheit (GRS) mbH, GRS-A-2954, Köln, Germany, (2001).

Lessons Learned from the New Fire PRA Methodology (NUREG/CR-6850) Application in Korea under Fire Ignition Frequency Perspectives

Sung-Hyun Kim[a], Kwang-Nam Lee[b], and Hak-Kyu Lim[a]

[a] KEPCO-E&C, Integrated Engineering Department, Korea, chiz@kepco-enc.com
[a] KEPCO-E&C, Integrated Engineering Department, Korea, hklim@kepco-enc.com
[b] KEPCO-E&C, Power Engineering Research Institute, Korea, knlee@kepco-enc.com

Abstract: The objectives of the Fire Probabilistic Risk Assessment (PRA) are to estimate the contribution of in-plant fires to overall plant Core Damage Frequency (CDF) and Large Early Release Frequency (LERF), to identify its vulnerabilities, and to provide recommendations for reducing fire-induced plant risk. Risk due to internal fire has been one of the major concerns in design and for operation of nuclear power plants. So far, Korea has applied Fire PRA Implementation Guide (EPRI TR-105928: FPRAIG) to conduct Fire PRA. In the meantime, NUREG/CR-6850 was issued as a current state-of-the-art method, which was studied by joint activity between Electrical Power Research Institute (EPRI) and U.S Nuclear Regulatory Commission (NRC) office of Nuclear Regulatory Research (RES), in August 2005. This paper covers comparison results for fire ignition frequency analysis separately conducted by FPRAIG and NUREG/CR-6850 and lessons learned from outcomes performed by newly developed Fire PRA methodology, NUREG/CR-6850, from fire ignition frequency perspectives. As a result, when applying new Fire PRA methodology, NUREG/CR-6850, compared to the previous Fire PRA methodology, FPRAIG, fire frequency for fixed ignition source has been decreased, while fire frequency for transient ignition source has been increased.

Keywords: PRA, Fire PRA, NUREG/CR-6850, Fire, CDF, LERF, FPRAIG, CCDP, CLRP

1. INTRODUCTION

The objectives of the Fire Probabilistic Risk Assessment (PRA) are to estimate the contribution of in-plant fires to overall plant Core Damage Frequency (CDF) and Large Early Release Frequency (LERF) to identify vulnerabilities and to provide recommendations for reducing fire-induced plant risk. Risk due to internal fire has been one of the major concerns for design and operation of nuclear power plants.

That's why Korea has performed Fire PRA for all plants, considering its 23 operating plants and 5 plants under construction. And Korea also has applied Fire PRA Implementation Guide (EPRI TR-105928: FPRAIG) to implementation of Fire PRA so far. In the meantime, NUREG/CR-6850 was issued as a current state-of-the-art Fire PRA method, which was studied by joint activity between Electrical Power Research Institute (EPRI) and U.S Nuclear Regulatory Commission (NRC) office of Nuclear Regulatory Research (RES), in August 2005.

Especially, NUREG/CR-6850 consists of 16 tasks and 2 support tasks, and it shows substantially details and deep approach task by task and suggests more realistic values for ones which are assumed a little conservatively compared to FPRAIG especially in part of Circuit Analysis, Human Reliability Analysis (HRA), etc.

Also, the new Fire PRA methodology, NUREG/CR-6850 needs relatively much more time and efforts than FPRAIG in many areas in order to perform in-deep and detailed analysis. But, in approach perspectives, both FPRAIG and NUREG/CR-6850 have similar approaches, except that NUREG/CR-6850 uses more realistic and recent fire ignition frequency, detailed fire scenarios, detailed cable failure probability, detailed human error probability, etc. than FPRAIG.

This paper tries to find out the difference from the results between new Fire PRA methodology, NUREG/CR-6850 and old Fire PRA methodology, FPRAIG, in same fire compartments for plants with almost similar design from Fire PRA perspectives.

Especially, comparison of fire frequency results conducted by FPRAIG and NUREG/CR-6850 separately and lessons learned from outcomes performed by newly developed Fire PRA methodology, NUREG/CR-6850, will be covered.

2. THE APPLICATION OF NUREG/CR-6850 IN KOREA

Korea has applied new Fire PRA methodology, NUREG/CR-6850, to an advanced nuclear power plant under design whose reference plant has already conducted Fire PRA in accordance with FPRAIG, EPRI TR-105928 before. One of the prime design characteristics of both plants analyzed is to adapt quadrant arrangement concept as shown in Figure 1, where most cables and equipment are located in each quadrant (A/B/C/D).

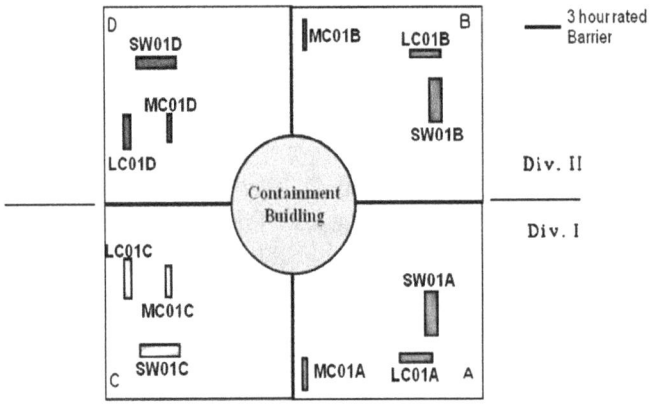

Figure 1. Quadrant arrangement concept in Auxiliary Building

And Main Control Room (MCR) has employed fully digitalized control system including Large Display Panel which is one of the differences from the conventional plants.

2.1. New Fire PRA Methodology

New Fire PRA methodology, NUREG/CR-6850, is composed of 16 tasks as below and shows task flow like Figure 2.

Task 1: Plant Boundary Definition and Partitioning is to define the Global Plant Analysis Boundary, and to divide the Global Plant Analysis Boundary into discrete physical analysis units (fire compartments).

Task 2: Fire PRA Component Selection is to select the plant equipment that will be included and/or credited in the Fire PRA.

Task 3: Fire PRA Cable Selection is to identify the cables associated with all Fire PRA components, and their physical routing throughout the plant.

Task 4: Qualitative Screening is to identify physical analysis units whose potential fire risk contribution can be judged negligible without quantitative analysis.

Task 5: Plant Fire-Induced Risk Model is to create the Fire PRA model that will be used in estimating the fire risk.

Task 6: Fire Ignition Frequency is to determine the fire ignition frequencies for fixed and transient ignition sources on a fire compartment basis.

Task 7A/7B: Quantitative Screening is to screen physical analysis units located within the Global Plant Analysis Boundary from further consideration based on preliminary conservative estimates of fire risk contribution using established quantitative screening criteria.

Task 8: Scoping Fire Modeling is to eliminate or reduce the frequency of those fixed ignition sources in a fire compartment that do not pose a threat to any Fire PRA target.

Task 9: Detailed Circuit Failure Analysis is to conduct a more detailed analysis of circuit operation and functionality to determine equipment responses to specific fire-induced cable failure modes. This information is then used to screen out cables that cannot prevent a component from completing its credited function.

Task 10: Circuit Failure Mode Likelihood Analysis is to quantify the probabilities for fire-induced hot short circuit failures that lead to component failure modes of interest. The failure mode probabilities are estimated for the cables of risk-significant components.

Task 11: Detailed Fire Modeling - In prior tasks, the analyses assumed that a fire would have widespread impact within the fire compartment. In this task, for those fire compartments found to be potentially risk-significant (i.e., unscreened compartments), a detailed analysis approach is provided. As part of the detailed analysis, fire growth and propagation may be modeled. Furthermore, the possibility of fire suppression before damage to a specific target set is analyzed. This task is composed of the following three sub-tasks:

   a. Detailed fire modeling of single fire compartments
   b. MCR fire analysis
   c. Multi-compartment fire analysis.

Task 12A/12B: Post-Fire HRA - In this task, human failure events (HFEs) associated with the fire scenarios are identified, and associated human error probabilities (HEPs) are estimated. Operator actions after fire ignition are assumed to be affected by the fire unless it can be clearly shown otherwise.

Task 13: Seismic Fire Interactions is to identify and correct any weaknesses in the fire protection systems and vulnerabilities in the ignition sources due to seismic events. This is the qualitative evaluation of the potential for: 1) seismically induced fires, 2) degradation of fire suppression systems and features, 3) spurious actuation of fire suppression and/or detection systems, and 4) degradation of manual fire fighting effectiveness. No risks are computed.

Task 14: Fire Risk Quantification - In this task of the analysis process, the Fire PRA model is quantified for each final fire scenario, the associated risk values (i.e., CDF and LERF) are computed and risk contributors are identified.

Task 15: Uncertainty and Sensitivity Analyses are to determine, characterize and assess the impact of uncertainty on the CDF and LRF estimates. In addition, sensitivity analyses are used to identify and understand the impact of risk significant modeling assumptions.

Task 16: Fire PRA Documentation is to ensure that the previous analyses are documented in a manner which facilitates review and update.

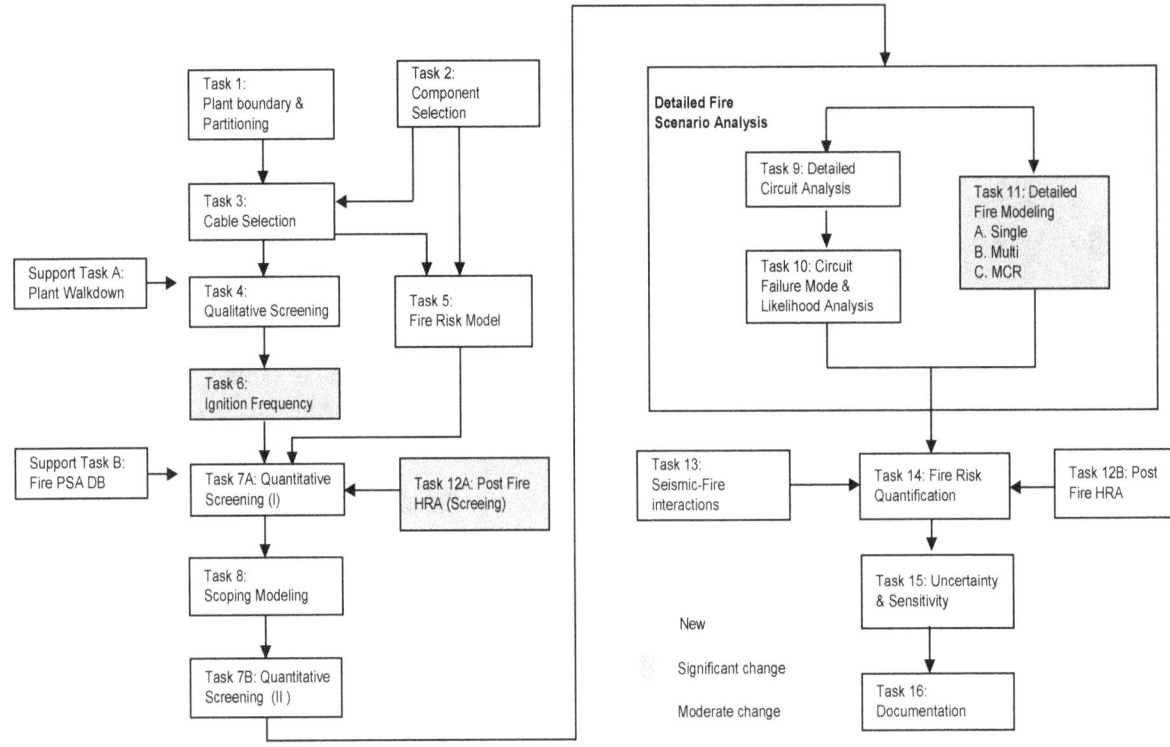

Figure 2. Overview of the Fire PRA Process in NUREG/CR-6850

2.2. Comparison of Frequency Results from NUREG/CR-6850 and FPRAIG Methodology

According to the result for fire frequency comparison between plant that NUREG/CR-6850-based fire frequency has been applied to and plant that FPRAIG based fire frequency has been applied to and those two plants have same design concept and are almost similar except several design changes such as from 2 diesel generators to 4 diesel generators, battery room's location, Essential Service Water system configuration, etc.

2.2.1 Electrical Equipment Room

For major electrical equipment rooms, comparison results for fire frequency are given in Table 1, which shows that fire frequency for fire compartment with NUREG/CR-6850 methodology has a tendency to be lower than that with FPRAIG methodology.

Table 1. Electrical Equipment related Fire Compartment Fire Frequency

Fire Compartment	Description	NUREG/CR-6850	FPRAIG
F078-A25A	Class 1E SWGR01A Room	3.21E-04	8.05E-04
F078-AEEB	Class 1E SWGR01B Room	3.66E-04	1.04E-03
F157-A01D	I &C Equipment Room	1.75E-04	2.06E-04
F157-A19C	I &C Equipment Room	1.94E-04	2.38E-04
F157-A19D	I &C Equipment Room	1.96E-04	2.61E-04
F157-A20C	I &C Equipment Room	1.65E-04	1.91E-04
F157-A20D	I &C Equipment Room	1.67E-04	1.76E-04

Figure 3 shows more explicitly difference for fire frequency results between NUREG/CR-6850 methodology and FPRAIG methodology for same fire compartments.

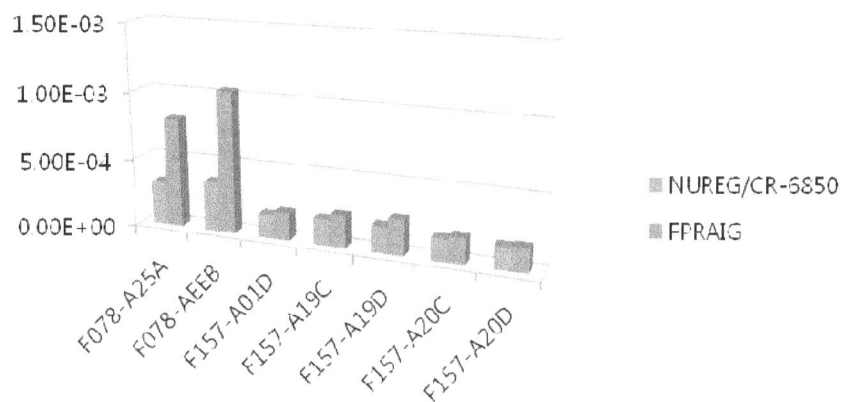

Figure 3. Electrical Equipment Room

2.2.2 Pump Room

For major pump rooms, comparison results for fire frequency are given in Table 2, which shows that fire frequency for fire compartment with NUREG/CR-6850 methodology has a tendency to be lower than that with FPRAIG methodology.

Table 2. Pump Room Fire Frequency

Fire Compartment	Description	NUREG/CR-6850	FPRAIG
F050-A03A	SI Pump A Room	8.29E-05	1.83E-04
F050-A03B	SI Pump B Room	8.28E-05	1.83E-04
F050-A02C	SI Pump C Room	8.33E-05	1.83E-04
F050-A02D	SI Pump D Room	8.33E-05	1.83E-04
F055-A02A	CCW Pump A Room	8.71E-05	1.83E-04
F055-A02B	CCW Pump B Room	8.51E-05	1.83E-04
F055-A02C	CCW Pump C Room	9.41E-05	1.83E-04
F055-A02D	CCW Pump D Room	9.10E-05	1.83E-04

Figure 4 illustrates more explicitly difference for fire frequency results between NUREG/CR-6850 methodology and FPRAIG methodology for same fire compartments.

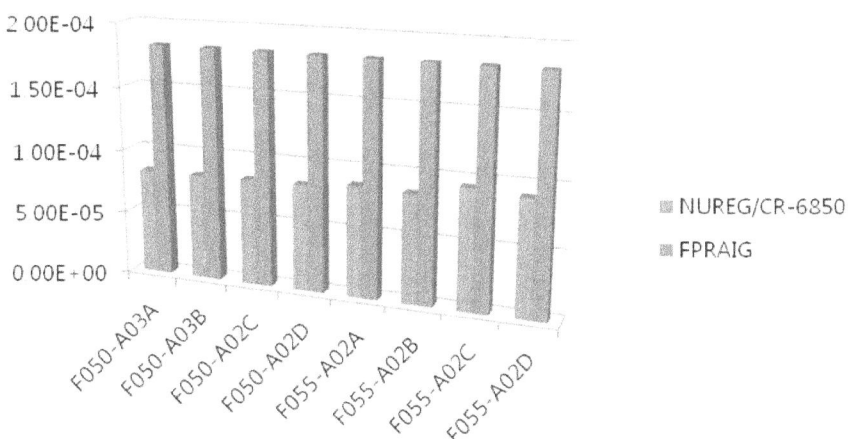

Figure 4. Pump Room

2.2.3 Transients

For transient fire, comparison results for fire frequency are given in Table 3, which shows that fire frequency for fire compartment with NUREG/CR-6850 methodology has a tendency to be higher than that with FPRAIG methodology for almost same fire compartments unlike fixed ignition sources. This is because in FPRAIG methodology, transient fire includes cigarette smoking, extension cord, heater, candle, overheating and hot pipe, which can be considered the impact to be ignored when procedurally prohibited or not existing. On the other hand, in NUREG/CR-6850, when calculating transient-relevant fire frequency, it is assumed that transient fires may occur at all areas of a plant unless precluded by design or operation, and also administrative controls don't preclude their occurrence in light of industry evidence. This is one of the differences between FPRAIG based frequency and NUREG/CR-6850 based frequency application.

Table 3. Transient Fire Frequency

Fire Compartment	Description	NUREG/CR-6850	FPRAIG
F073-T08	Stair	8.68E-05	0.00E+00
F073-T10	Stair	8.68E-05	0.00E+00
F079-P01	Access Area	7.89E-05	0.00E+00

Figure 5 represents more explicitly difference for fire frequency results between NUREG/CR-6850 methodology and FPRAIG methodology for same fire compartments.

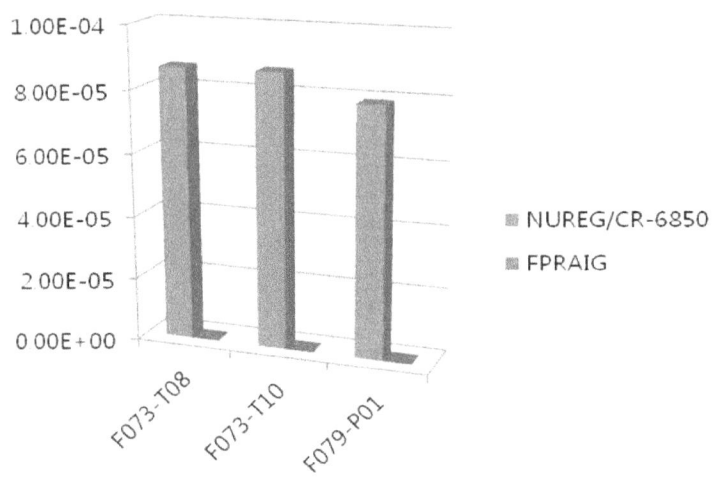

Figure 5. Transient Fire

2.2.4 Main Control Room

For main control room fire, fire frequency comparison was performed like Table 4, which shows that fire frequency for fire compartment with NUREG/CR-6850 methodology has a tendency to be much lower than that with FPRAIG methodology for same fire compartment.

Table 4. Main Control Room Fire Frequency

Fire Compartment	Description	NUREG/CR -6850	FPRAIG
F157-AMCR	Main Control Room	1.22E-04	7.94E-03

Figure 6 depicts more explicitly difference for fire frequency results between NUREG/CR-6850 methodology and FPRAIG methodology for same main control room.

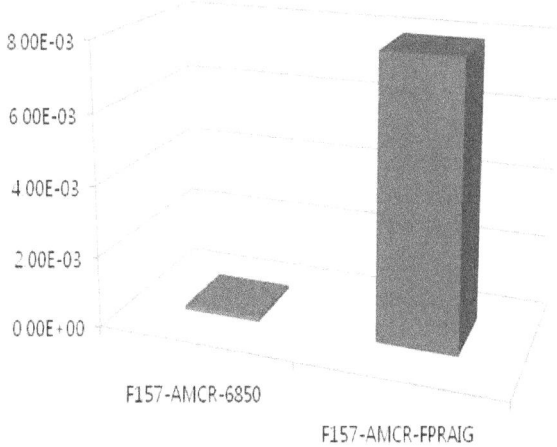

Figure 6. Main Control Room

2.2.5 Total Plant Fire Frequency

For plant total fire, fire frequency comparison was performed like Table 5, which shows that plant total fire frequency with NUREG/CR-6850 methodology has a tendency to be much lower than that with FPRAIG methodology. Especially, in terms of fixed ignition source, fire frequency with NUREG/CR-6850 methodology is lower than that with FPRAIG methodology. On the other hand, in case of transient fire frequency, fire frequency with NUREG/CR-6850 methodology is higher than that with FPRAIG methodology for reason mentioned in 2.2.3.

Table 5. Total Plant Fire Frequency

Ignition Source	NUREG/CR-6850	FPRAIG
Fixed Ignition Source	1.02E-01	2.92E-01
Transients	3.39E-02	1.19E-03
Total Frequency	1.36E-01	2.93E-01

Figure 7 shows more explicitly difference for fire frequency results between NUREG/CR-6850 methodology and FPRAIG methodology for same fire compartment.

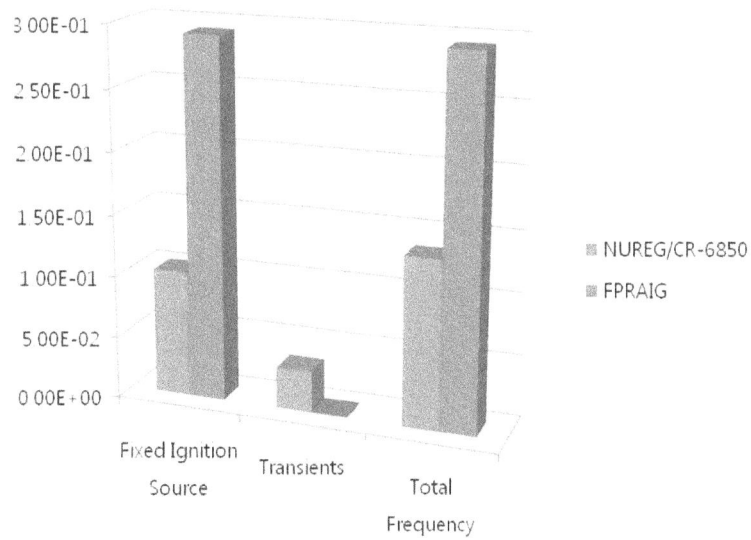

Figure 7. Total Plant Fire Frequency

3. CONCLUSIONS

When applying new Fire PRA methodology, NUREG/CR-6850, compared to the previous Fire PRA methodology, FPRAIG, fire frequency for fixed ignition source has been decreased, while fire frequency for transient ignition source has been increased. However, increased transient fire frequency is judged to be fully able to get lowered through transient-relevant procedure improvement and ignition source management. Consequently, NUREG/CR-6850 methodology leads that individual compartment fire frequency has been decreased compared to FPRAIG methodology.

Additionally, one of things that we should take care of is that NUREG/CR-6850 has the high chance to increase Conditional Core Damage Probability (CCDP)/Conditional Large Release Probability (CLRP) because Fire PRA equipment in NUREG/CR-6850 methodology should be incorporated into analysis more than that in the previous methodology. Therefore, CDF calculated from fire frequency multiplied by CCDP is not always reduced in proportion to the decrease in fire frequency based on NUREG/CR-6850 methodology.

References

[1] NUREG/CR-6850, "EPRI/NRC-RES Fire PRA Methodology for Nuclear Power Facilities Volume 1: Summary & Overview," USNRC, Washington, DC, August 2005.

[2] NUREG/CR-6850, "EPRI/NRC-RES Fire PRA Methodology for Nuclear Power Facilities Volume 2: Detailed Methodology," USNRC, Washington, DC, August 2005.

[3] EPRI 1019259, "Fire Probabilistic Risk Assessment Methods Enhancements: Supplement 1 to NUREG/CR-6850 and EPRI 1011989," EPRI, September 2010.

[4] EPRI TR-105928, "Fire PRA Implementation Guide," EPRI, December 1995.

Crisis organization and severe accident management: Contribution of ergonomic considerations in the definition of Severe Accident Management Guidelines (SAMG)

Violaine Bringaud[a*], Jean-Paul Labarthe[a]

[a]Industrial Risk Management Department, EDF R&D, Lab Clamart, France

Abstract: This document presents an ergonomic action led in the framework of the definition of the organizational reaction to a crisis and the associated technical and documentary supports, defined for the design of a new nuclear plant.

In the event of an incidental or accidental situation occurring in a functioning plant, the risk management approach is based on a local and corporate crisis management organization, the objective being to ensure that the situation at the plant is under control and to protect people endangered by the situation. The more specific case of a severe accident is defined as a state of functioning with deterioration and possible loss of the plant and the probability of this type of accident occurring is extremely low. Nevertheless, these cases are taken into account in the design stage and emergency simulation drills are organized to help prepare the staff to manage these situations.

The first part of this document presents the crisis organization and the associated documentary supports (the SAMG) designed to manage a severe accident. The second part describes the ergonomic approach to the design of the SAMG and concludes with the value of such an approach in preparing the teams to manage a crisis in a complex and high risk social and technical system.

Keywords: design, ergonomics, crisis organization, severe accident management guidelines,

1. GUIDES TO REACTING TO A SEVERE ACCIDENT IN A CRISIS SITUATION

1.1. What is a crisis situation?

A severe accident, leading to the loss of the plant and potential external impacts (on people, the environment, etc.), would place the entire company in a crisis situation. Consulting existing literature on this subject, mainly in the fields of risk science and psychology, provides elements that can help to define this type of situation [1], [2], [3], [4].

- A crisis is a threat which can lead to catastrophes or disasters. A crisis is defined by the enormity of the original fault and its multiple consequences. It creates severe constraints, difficult environmental conditions and deteriorated working conditions. It is a step into the unknown resulting in a clear difference between 'before' and 'after' the crisis. It involves many risks and huge organizational, economic, environmental and political stakes.
- A crisis creates a very specific temporal reality characterised by a continued situation of instability, unpredictability and surprise. Crisis situations evolve at different speeds but often in a non-linear manner. Another characteristic of a crisis is absolute urgency demanding the implementation of pertinent actions in very short spaces of time. In this context, teams have very little time to establish a diagnosis and to predict how the situation will evolve, in order to coordinate, decide upon and implement the chosen actions. Moreover, a crisis situation can last for a very long time.
- From a psychological standpoint, a crisis is characterised by exceeding (cognitive) resources [5]. In these situations, events are hard to apprehend and understand. A crisis arises "*when a system is confronted with an event, generally unforeseen, the consequences of which develop rapidly producing significant risks, and the management of which exceeds the pre-existing resources in terms of actions and people*" [4].

1

-   Lastly, a crisis is a stressful situation for those responsible for managing it due to the following contradiction: they must be capable of bringing rapid and pertinent responses to a complex and fast-moving situation which at least partially escapes their understanding. Furthermore, a context of uncertainty and extreme conditions of intervention add to the stress felt by the teams.

This brief list of characteristics give an idea of the situation that the teams would find themselves in if a severe accident were to occur (which nevertheless remains very hypothetical). To prepare ourselves to manage crises in general, and specific cases in particular, the company has designed a local and corporate organization which is presented in the following paragraph.

1.2. A crisis organization to lead the response to a severe accident

At the new plant, the management of a severe accident is part of the company's general crisis organization defined to manage other facilities, with adaptations at the local and corporate levels. This organization is consistent with the standards of the mechanisms designed to manage these situations [2, 6]. The functions typically attributed to crisis organization concern: expertise, decision-making, logistics and communication. The implementation of these functions relies on competent and trained teams in the different work groups at the local and centralised level. Furthermore, technical, documentary and communication supports help in the management of these situations.
The following diagram shows the four functions of the crisis organization defined for the company.

Figure 1: Overall crisis organization diagram

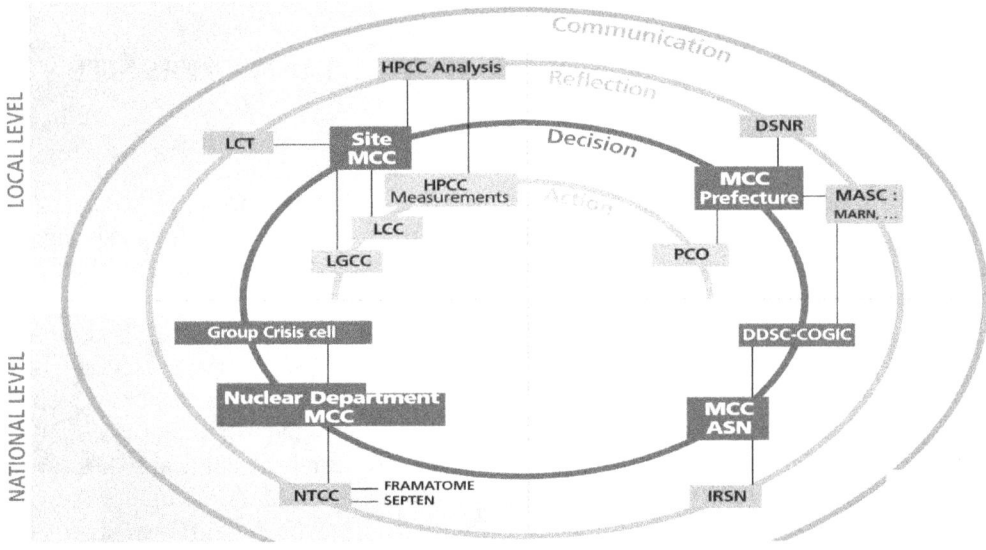

This organization is implemented in order to control the situation at the plant, minimize the environmental consequences, and protect people.

Specialists in crisis management [1, 6] insist upon certain points to distinguish crisis organization from organization in a normal situation. In particular, they identify:

2

- The scope of the players concerned by the crisis which goes beyond the scope encountered in a normal operating situation (public authorities, surrounding populations, media, etc.),
- The need to call upon internal experts and external experts who are not available within the company,
- Requirements in terms of extra equipment,
- The stakes linked to in-house and external communication.

To manage a crisis situation as defined in paragraph 1.1 and for which we point out the similarity with a severe accident situation, the organization proposed, and concisely presented in this paragraph, must allow the teams to bring adapted responses to the situation encountered. On this subject, the approaches proposed in HRO (High Reliability Organizations) and resilience engineering [7, 8, 9] are interesting because they have considered how best to organize the activity of the teams so that the organization as a whole is capable of dealing with unexpected critical or threatening events [3]. Concerning how to manage a severe accident, several points caught our attention.

Firstly, crisis situations demand the combination of a process of anticipation and a process of adaptation. The process of anticipation refers to the preparation of methods of coordination and action within the teams, backed by instructive supports and preparative drills. The process of adaptation concerns the way in which the methods of coordination and action adjust according to events.

Secondly, we retain the fact that crisis organization must allow for changes in the configurations of the plant over time. Therefore it must be sufficiently flexible to give the teams optimal understanding and control of events. There must also be appropriate communication and information exchange capacities to allow the emergency response team, comprising people dispersed in different places, to have a clear overview of the state of the plant and the actions led. In these conditions, the emergency response team is able to adapt its actions at every stage of the evolution of the crisis. This clearly requires an organized method of coordination even if the deteriorated situation may cause communication difficulties.

Following on from the organizational approaches mentioned above, ergonomic psychology, which is mainly focused on the capacities of individuals to deal with situations, also insists upon the notion of adaptation [3, 4]. Adaptation appears as a response to the extraordinary nature of the situation which exceeds the ordinary (cognitive)[1] resources called upon by individuals. When circumstances go beyond a certain point, individuals and groups need extra resources to prevent them from losing control of the situation and to allow them to try to regain control. Extra resources are needed to allow the (short-term) adaptation process to be implemented in a pertinent manner. "*Short-term adaptation refers to the processes by which an individual (or group or organization) modifies or regulates its activity to respond to variations in its external or internal environment with the aim of maximising the appropriateness of its responses to the difficulties encountered*" [4].

This concise presentation of elements taken from studies on the subject of crisis organization gives a good insight into what is required to manage these often complex situations, known for their deteriorated contexts and the unpredictability of their evolution, requiring actions to be implemented rapidly and obviously stressful for those in charge of managing them. The crisis organization put in place must allow the teams to bring adapted responses to the rapidly-changing situations encountered. In this context, one of the main aims of the organization is to allow anticipation and adaptation. This means offering teams a precise framework for analysing and exchanging information about the situations, and making it possible to periodically assess these situations and implement the associated adjustments. All this requires organized modalities of coordination between the members of the emergency response team.

In the framework of its crisis organization, to manage severe accidents, which, to reiterate, lead to the loss of the plant and the probability of which is very low, the company has created a specific

---

[1] In the domain of psychology, these are the cognitive resources called upon by an individual, as opposed to resources in the sense of 'human resources' (i.e. staff).

3

organizational mechanism reinforced by documents called Severe Accident Management Guidelines (SAMG). These documents are presented in the following paragraph.

1.3. SAMG as supports for managing a severe accident

Management of a severe accident begins only once a technical threshold of plant deterioration has been reached. At that moment, and only after validation from the site's senior management, the approach changes from 'incidental/accidental' management to the status of a 'severe accident', the consequence of which is the loss of the plant and the beginning of a crisis requiring long-term management. The decision to use the SAMG changes the configuration of the crisis organization already in place to manage the 'incidental/accidental' situation. The SAMG provide the different people concerned with an operational support in the form of Operating Strategies for Severe Accident (OSSA). One of the particularities of the SAMG is that they transfer the responsibility of operating the plant from the control room to the site's senior management division, which decides upon the action to take with the help of a network of local and centralised experts both within and outside of the company.

Seven SAMG documents are made available to each of the members of the emergency response team in charge of operating the plant in a severe accident situation and can be found in four areas on the site (Control Room, Emergency Technical Centre, Local Management Command Centre) and in the company's centralised department (Corporate Technical Emergency Response Team). The SAMG cover three circles of the overall crisis organization diagram (figure 1): expertise, decision and action. The SAMG offer precise indications concerning the actions of each user, according to their mission in the crisis organization, but avoids directing the team members down excessively restrictive paths of action which could prove to be inappropriate in the long-term. This enables them to be adapted to the situations encountered in order to allow users to react to the unforeseen events and surprises mentioned in paragraph 1.1. The SAMG also contain periodic written messages concerning the state of the plant and documents guiding the emergency response teams in their analysis, diagnosis and predictions concerning the situation.

The SAMG designed for the new nuclear plant contain two major innovations for the experts in the emergency response teams, compared to those existing for the other plants:
1. A 'looped' approach which regularly poses the question of the state of the plant to provoke a diagnosis of the three main safety functions which are: control of radioactivity release, the confinement of the plant, and the cooling of the core.
2. A decisional aid matrix which defines, depending on the degree of deterioration of the three safety functions, the priority mitigation actions to conduct on the plant.

As soon as the plant's senior management division has decided that the situation has reached 'severe accident' status, the teams are faced with an emergency situation which necessitates immediate actions which do not require prior analysis. The operating team works autonomously in the control room to implement these initial actions which do not require direct external expertise. Guided by the SAMG and thanks to the organized communication points in the guide as well as specific messages about the state of the plant, the other work groups involved in the crisis organization can follow the performance of these immediate actions.

After performing the immediate actions, application of the SAMG allows the local and corporate emergency response teams to continue to supervise the state of the plant (by monitoring certain parameters) and to perform periodic diagnoses of the damaged unit; to suggest, decide and implement appropriate actions according to the situation in order to limit the consequences of the severe accident. The SAMG also provides guidance for predicting the future state of the plant. At the crisis organization level, the circle of experts is particularly solicited at this point. The second phase of the response to a severe accident is surveillance and potentially the implementation of pre-established counter measures (amongst other mechanisms) according to the faults encountered during the evolution of the plant, hour after hour and day after day. This phase is characterised by successive

4

adaptations by the emergency response teams to take actions adapted to the rapidly-changing and uncertain situations encountered. These adaptations are made possible thanks to communication between the members of the emergency response team, with each individual sharing their understanding of the situation and the actions suggested. The SAMG, in their capacity as documentary supports, have been designed to facilitate and optimise the reliability of coordination and synchronisation between the different work groups involved in managing a severe accident (consistency of vocabulary, formalisation of important coordination measures, traceability of the actions led, etc.).

Finally, the temporal specificities of a severe accident have been taken into account in the design of the plant. These guides are designed for long-term management of the plant and integrate the temporal aspect of severe accident management by offering the possibility of suggesting actions to be led over time. They also take into consideration the fact that, for this type of accident, the dynamics of the plant are not entirely foreseeable at the moment of designing the process (by a regular diagnosis of the state of the plant and an associated diagnosis, potentially completed by suggestions of actions validated by the plant' Senior Management).

In this first part of the present document, based on the characteristics of crisis situations, we have addressed the modalities that shape crisis organization, the objective of which is to allow the teams to come up with responses that are adapted to the complex and often unforeseeable situations that they encounter. Bearing this in mind, the management of a severe accident is guided by documentary supports designed so that their use will allow teams to adapt to the situations encountered throughout the crisis period. Following on from this, the second part of this document presents the ergonomic approach to the design of the SAMG, created from 2009 to assist in the management of a severe accident at the new nuclear plant.

2. CONTRIBUTION OF ERGONOMICS IN THE DESIGN OF THE SAMG

2.1. Characteristics of the ergonomic[2] approach in documentary design

An ergonomic approach integrated into the SAMG design project

The ergonomic approach applied to the design of the SAMG is just one aspect of an overall industrial project to design a new nuclear plant. This large-scale project depends on many different spheres of competence found in the company's different divisions (engineering, future operator, R&D). A 'Human Factor' engineering program structures the actions led in this design project. The human factor contribution of Research & Development consists in providing ergonomic markers for the design of the management means and associated organizations, and for evaluation campaigns.

The project to design the SAMG documents, launched in 2009, has been underway for several years and will end before the industrial start-up of the new plant. This project depends on different players: the authors of the SAMG, engineers specialising in severe accidents, the future operators, the company's experts in ergonomics from Research and Development, etc. Experts in the 'human factor' have been involved from the very start of the project and intervene throughout the process, the aim being to act while there is still sufficient room for manoeuver in order to detect as early as possible (and therefore resolve at a low cost) any use and performance problems in the mechanism being designed [10].

In this project, the ergonomists' role is to help the authors of the SAMG to compile documentary supports that will, in the event of a severe accident, facilitate and maximise the reliability of the emergency response teams' different missions comprised in the circles of expertise, decision and action of the crisis organization presented in paragraph 1.2. To reiterate what has been said in part 1, the ergonomists' contribution is to help the authors of the SAMG to design documentary supports

---

[2] In the company, the ergonomics and 'human factor' engineering approaches cover the same scope.

5

which will reinforce the approaches of anticipation and adaptation allowing the team to come up with responses adapted to the complex, uncertain and stressful situations encountered.

<u>The priority is to design documents that are easy to use and that reinforce the coordination and synchronisation of the emergency response team members</u>

The ergonomists focus on two main aspects. Firstly, they offer their expertise to <u>ensure that the documents can be used without difficulty</u>. In this framework, the documentary ergonomic requirements are suggested for the project and ergonomic evaluations are conducted. Several ergonomic requirements need to be considered. They concern [10]:
- Usability: the SAMG documents must be easy to use from the very first time they are followed;
- Guidance: the SAMG should offer clear guidance for the users;
- Coherence/uniformity of the documentary design elements (graphic design and presentation of the information) ;
- The comprehensibility of the information (syntax, vocabulary, and use of abbreviations);
- The quantity and density of information;
- The coherence between the SAMG and the other supports used when following them (Man-Machine interface, other documentary supports, communication supports, and technical supports).

Following on from this, ergonomic evaluations [12, 13, 14] ensure the quality of the documents by considering, depending on the experts, two or three fundamental aspects. Some experts [15] focus above all on the pertinence of the document, i.e. *"its capacity to respond to the user's needs in terms of information"* and its usability i.e. *"its capacity to be used and understood easily and to give access to the pertinent information that it contains"*. Other experts [15] focus more on a third aspect which is its acceptability, in other words the extent to which the design of the documents encourages people to use them. Given the stakes, complexity and infrequent use the SAMG, which are designed to be used solely in the event of a severe accident, the acceptability aspect is very limited in these evaluations.

Secondly, the ergonomists help the authors to produce SAMG documents which will be <u>supports that facilitate coordination and synchronisation</u> between the team members applying these guides. In addition to the elements presented in paragraphs 1.2 and 1.3, temporal coordination and synchronisation between the team members help them to better understand and control situations. Thanks to the communication and information-sharing means, the teams should be able to get an overview of the state of the plant and the actions led. In the SAMG design project, the ergonomists are very careful to integrate this organizational aspect into the documentation.

In addition to evaluations concerning the usability of the documentation, the aspect relative to the possibilities for coordination and synchronisation offered by the SAMG is also evaluated. The evaluation procedure is an important part of the ergonomic documentary design approach and deserves some explanation here.

<u>Successive ergonomic evaluations throughout the SAMG design process</u>

The ergonomists conduct several documentary evaluations throughout the SAMG design process. These evaluations involve a representative panel of future users of the SAMG. They are participative evaluations [15, 17, 18]. They are evolutive; in other words they contain specificities determined according to the degree of completion of the documents (evaluation by an expert, 'static' evaluation with representatives of the future users, 'dynamic' overall evaluation in the framework of mini crisis drills). The evaluations are iterative: they encourage teamwork between the designers of the SAMG, the representatives of future users and the ergonomists. However, the final analysis of the results of the evaluation is performed independently by the ergonomists.

Each ergonomic evaluation is subject to an analysis of user difficulties, the results of which are then presented to the designers of the SAMG as a series of recommendations. In compliance with the

6

method of this documentary design process, these recommendations are used when compiling updated versions of the SAMG through until the industrial start-up of the plant.

In the following paragraph, the presentation of key points concerning the modalities of the last 'dynamic' evaluation conducted shows the contribution of the ergonomic approach in the SAMG design process. It also reveals findings concerning training in the management of crisis situations in the context of the implementation of a specially-dedicated organization.

2.2. Evaluating the SAMG by simulating the management of crisis situations with maximum realism

Framework and objective of the 'dynamic' evaluation of the SAMG

The ergonomic evaluation approach is part of the SAMG design project and is therefore a participative and iterative process between designers, evaluators (ergonomists) and representatives of future users. These players find themselves in an environment which is in the process of being designed, as close as possible to the design stage considered but different from the framework of the use of the documentation in a real situation (at several levels: technical, documentary, organisational, in terms of team skills, etc.). Despite the differences between the situation during the design phase and a real operating situation, this evaluation method, because it is a key aspect of the design process, helps to identify the main difficulties that the users could encounter in real situations (in terms of the usability of the documents, comprehensibility of the messages, and effectiveness of the document as a support for analysis, action, and coordination between the different work groups involved in managing a severe accident). This is the framework of the 'dynamic' evaluation process presented in this paragraph.

Its objective was to analyse the usability of the SAMG documentation in the context of the most realistic simulation possible, at this stage of the design of the crisis organization and the associated human and logistic resources. To satisfy this objective, the evaluation involved the different team members contributing to the management of a severe accident. They applied the SAMG in a coordinated and synchronised manner, in the different work groups at the future plant (operating simulator, local emergency response team, management command post) and at the centralised emergency response centre (corporate emergency response team). For this, 'mini' crisis drills were organized based on severe accident scenarios. The three selected scenarios were chosen to ensure that the SAMG could be used in different ways and very varied cases. The simulator was specially prepared to improve the representativeness of the management and use of the SAMG in the control room. Finally estimated performance times of local actions, taking into account the availability of field operators, were integrated into the scenarios.

The evaluation lasted three days with a different scenario each day from 8-12am. The afternoons were devoted to debriefings, firstly in separate work groups and then with all four participating work groups.
Based on the analysis of the usability of the SAMG, the aim was to make the designers of the documents aware of changes to integrate into the next series of documents.

The ergonomic evaluation method

The people involved in the evaluation:
- Representatives of the future users: The ergonomic approach to documentary evaluation is a participative approach. Therefore this type of evaluation is conducted on a representative panel of future users. In the present case, the user panel covered all the work stations applying the SAMG. In order to be able to apply the SAMG, the representatives of the future users and all those participating in the evaluation followed a short training session to familiarize themselves with the SAMG, the associated supports and the crisis organization. We should add that the participative evaluation concerns the ergonomics of the document and its

7

usability. It does not concern the competence of the users or their technical knowledge of severe accidents.

- The independent Human Performances (HP) evaluation team: The evaluation team was present in each work group comprising users of the SAMG and gathered data during the simulation drills and the debriefings that followed the drills. The HP team then analysed the data. A technical support team assisted the HP evaluation team, notably by giving their opinions concerning the achievement of the technical objectives.

- The instructor-pilot of the simulator (from the training department) and the severe accident operating engineer: During the several months spent preparing the evaluation and until it was launched, this duo defined the drill scenarios, made the necessary technical preparations for the scenarios and compiled the technical information to be communicated to the users. During the evaluation, these two individuals supervised the progress of the scenarios and operated the simulator.

NB: The SAMG designers are not involved in the evaluation. Their role was nonetheless important because they had to produce all the documentation required for the evaluation in time.

Representativeness and limits of the evaluation:

It is important to address the representativeness and the limits of the evaluation. This allows us to analyse the risks of the evaluation to identify potential counter measures to ensure the pertinence of the data gathered.

In the case discussed, the representativeness had been improved compared to the previous 'static' evaluation (improvement of the representativeness of the work stations thanks to specific use of the operating simulator and the national emergency response resources; the scenarios lasted several hours which is a more realistic simulation of the temporal conditions of a severe accident, particularly in terms of coordination and synchronisation between the different work groups; the documents were all at a more advanced state than for the previous evaluation).

Various limits in terms of representativeness have been identified and taken into account in the analysis. The fact that the same team participated in the scenarios for three days, despite the differences in the scenarios, produced an effect of repetition that caused the individuals to become relatively 'practised' in the exercise. For logistic reasons, it was not possible to simulate the entire crisis organization and only the users of the SAMG participated in the evaluation; this undoubtedly had an impact insofar as the real management of a severe accident would be part of an overall crisis organization involving more interfaces to deal with. In the end, the limits identified did not prevent the collection of instructive and pertinent information.

Study hypotheses to satisfy the aim of the evaluation and guide the analysis:

In order to evaluate the usability and the capacity of the SAMG to facilitate coordination and synchronisation between team members, as defined in paragraph 2.1, the following general hypotheses, developed into detailed hypotheses, have been suggested;

1. The structure of the SAMG and the associated documentary ergonomics allow each of the team member-users of the SAMG to diagnose/appraise, decide, act and monitor what their mission demands of them;

2. The supports associated with the SAMG (operating resources, traceability supports and communication means) enable each user to apply the guidelines relative to their mission.

3. The SAMG and associated supports reinforce coordination and synchronisation between the team members called upon to manage a severe accident.

Analysis of the data and main results

Using methodological data collection and analysis tools, the data analysis is performed as follows:

- Firstly, the data analysis provided insights into study hypotheses based on:
  - The opinion of technical support specialists on the achievement of the safety objectives of the crisis simulation drills;
  - The data gathered during the scenarios and debriefings ;
  - The overall reconstitution of how the users reacted during the scenarios;

8

- Individual feedback from the users of the SAMG.
- Secondly, the data analysis according to the study hypotheses led to the formulation of HF recommendations for the SAMG and the associated supports.

In terms of results, the main safety objectives associated with the different scenarios were achieved. The analysis produced recommendations for optimisation concerning the documentary supports, the operating resources, the crisis management activities and the missions of certain posts of responsibility in the control room. These recommendations will be taken into account in the compilation of the future version of the SAMG which will in turn be subject to a final evaluation before the industrial start-up of the plant.

The presentation of this documentary ergonomics evaluation describes how the evaluation and data analysis was conducted, primarily in order to produce HF recommendations for the authors of the SAMG. From the perspective of the industrial start-up of the plant, the aim is to propose SAMG documents that are easy to use and that optimise coordination and synchronisation between the team members applying the SAMG.

For this new plant, the performance of this first 'dynamic' documentary evaluation simulating crisis situations and activating the associated organization has also helped to train the different players at the new plant in the management of complex, uncertain and stressful crisis situations [6, 18].

This type of evaluation can also encourage experience feedback about the capacities of the organizational set-up and the teams to manage these situations and about the ways of conducting this type of crisis simulation drill.

CONCLUSION

The definition of elements characterising crisis situations, and the concise presentation of elements taken from studies on the subject of crisis organization, provide a good insight into what is required to manage these often complex situations, known for their deteriorated contexts and the unpredictability of their evolution, requiring actions to be implemented rapidly and obviously stressful for those in charge of managing them. This type of organization must enable the teams to bring adapted responses to the rapidly-changing situations encountered. In this context, one of the main aims of the organization is to allow anticipation and adaptation. This means offering teams a precise framework for analysing and discussing the situations, and making it possible to periodically assess these situations and to support the associated adjustments. All this requires organized modalities of coordination between the members of the emergency response team. At the new plant, the management of a severe accident has been taken into account as of the design phase. The documentary supports (the SAMG) have been designed and evaluated to ensure that those called upon to use them can adapt to the situations encountered throughout the duration of the crisis.

The ergonomic approach to the documentary design applied to the SAMG, and initiated from the outset of their design, guides the designers to help them to create documents that are easy to use and that optimise coordination and synchronisation between emergency response team members. Along the same lines, the modalities of documentary evaluation proposed by the ergonomic approach, thanks notably to the simulations of crisis situations, help to prepare the teams at the new plant to manage crisis situations and encourage experience feedback on this subject at the corporate level.

9

References

[1] P. Lagadec. *"La gestion des crises. Outils de réflexion à l'usage des décideurs"*, McGRAW-HILL, 1991, Paris.

[2] L. Combalbert. *"Le management des situations de crise. Anticiper les risques et gérer les crises"*, ESF Editeur, 2005, Paris.

[3] M. Bourgy. *"L'adaptation cognitive et l'improvisation dans les environnements dynamiques : pour une intégration de l'expérience sensible dans les modèles de l'activité experte"*. Doctorate thesis, 2012, Université de Paris 8.

[4] C. De La Garza, F. Darses, M. Bourgy. *"Caractéristiques des situations de crise dans les environnements à hauts risques et conséquences sur les opérateurs et les équipes"*, EDF R&D internal document

[5] J. Rogalski. *"La gestion des crises"*, in P. Falzon, Ergonomie, pp. 531-544, PUF, 2004, Paris.

[6] L. Combalbert, E. Delbecque. *"La gestion de crise"*, PUF Que sais-je ?, 2012, Paris.

[7] K. Weick, K.M. Sutcliffe. *"Managing the Unexpected: Assuring Performance in an Age of Uncertainty"*, CA: Jossey-Bass, 2001, San Francisco.

[8] E. Hollnagel, D.D. Woods, N. Levenson. *"Resilience Engineering: Concepts and Precepts"*. UK: Ashgate, 2006, Aldershot.

[9] P. Le Bot. *"The Model of Resilience in Situation (MRS) as an Idealistic Organization of At-Risks Systems to be Ultra Safe"*, PSAM 10 Seattle, 2010, Washington.

[10] L. Graglia, J.P. Labarthe. *"Facteurs humains et systèmes complexes à risques : principes directeurs et réponses apportées pour la conception des moyens de conduite d'un nouveau réacteur"*. Tutorial presented at the "Lambda Mu" Congress on risk management and operational safety, 2012, Tours.

[11] V. Bringaud, J.P. Labarthe. *"Une évaluation ergonomique documentaire dans un projet de conception de moyens de conduite d'un process continu – Le cas de l'évaluation des guides d'intervention en accident grave"*, communication at the 7[the] EPIQUE conference, Arpege, 2013, Brussels.

[12] J. Nielsen. *"Utilisability Engineering"*. Academic Press, 1993, Boston.

[13] J. Nielsen. *"Heuristic evaluation"*, In J. Nielsen, R.L. Mack (Ed.), Utilisability inspection methods (pp.25-65), Wiley, 1994, New York.

[14] International Organisation for Standardization, ISO 9241-110. *"Ergonomie de l'interaction homme-système – Partie 110 : principes et dialogues"*, AFNOR, 2006, Paris.

[15] C. Bastien, A. Tricot. *"L'évaluation ergonomique des documents électroniques"*, in A. Le Chevalier, A. Tricot (Ed.), Ergonomie des documents électroniques, PUF, 2008, Paris.

[16] A. Dillon, M. Morris. *"User acceptance of new information technology – theories and models."* In M. Williams (Ed.), Annual Review of Information Science and Technology (pp. 3-32), Information Today, 1996, Medford.

[17] F. Darses, F. Reuzeau. *"Participation des utilisateurs à la conception des systèmes et dispositifs de travail"*. In P. Falzon (Ed.), Ergonomie (pp. 405-420), PUF, 2004, Paris.

[18] F. Ganier. *"Comprendre la documentation technique"*, PUF, 2013, Paris.

[19] S. Gaultier-Gaillard, M. Persin, B. Vraie. « *Gestion de crise – Les exercices de simulation : de l'apprentissage à l'alerte* », Afnor Editions, 2012, Paris.

10

Disaster Context Modeling for the Creation of Exercise Scenarios

Taro KANNO[a*], Wataru ONO[a], Shengxin HONG[a], and Kazuo FURUTA[a]

[a]The University of Tokyo, Tokyo, Japan

Abstract: Disaster training and exercises are widely employed to improve preparedness for and the ability to respond to unprecedented natural and man-made disasters. While various types of drills and exercises such as Serious Game, Disaster Imagination Game, and Cross Road have been proposed, less attention has been paid to how to make effective exercise scenarios efficiently. This study develops a disaster context model that provides a foundation for creating new exercise scenarios and describing what happened in actual past disasters. The study also develops a method of creating semi-automatically a new imaginary disaster context that will be used as an assumption in an exercise scenario.

Keywords: Disaster Response Exercise, Exercise Scenario, Context Modeling

1. INTRODUCTION

In many sectors of society, such as governmental agencies, medical institutions, commercial companies, and local communities, disaster drills and exercises play an important role in improving preparedness for and the ability to respond to unprecedented natural and man-made disasters. While various types of drills and exercises such as Serious Game [1], Disaster Imagination Game [2], and Cross Road [3] have been proposed, less attention has been paid to how to create effective scenarios efficiently for such drills and exercises. In fact, the quality of the exercise scenario is heavily dependent on expertise, and the design of scenario requires much cost and many man-hours. This becomes a big obstacle in preparing effective drills and exercises in many organizations that do not have enough know-how and human resources. Another serious problem that hinders the preparation of a disaster exercise is that it is rare, at least in Japan, for many organizations to share the knowledge and experiences gained through the preparation and execution of an exercise, which is partly because there is no easy-to-use technology that supports and facilitates collaboration and knowledge sharing among different organizations and communities.

This paper proposes a disaster context model that provides a framework with which to describe both actual and imaginary disaster contexts and provides a foundation for creating new exercise scenarios. Employing this context model, a method of interactively creating a new imaginary disaster context, which is used as an assumption for the exercise scenario, is proposed.

Section 2 of the paper explains the modeling of disaster context. In this study, the example of disaster medicine and nursing was used for the modeling and development of the proposed methodology. Section 3 introduces a model-based method of creating a new imaginary disaster context and a method of converting a disaster context into a narrative/textual description. In Section 4, a preliminary evaluation of the proposed methodology is described. A conclusion and discussion of future application in collaborative scenario design are presented in Section 5.

2. MODELNG DISASTER CONTEXT

A disaster context in this paper refers to the settings of physical and functional parts of a disaster situation that is related to response activities and the relationships among them. According to reviews and qualitative analyses of existing disaster exercise scenarios, reports on past disasters, and many personal notes on actual disaster experiences, three major components constituting a disaster context—the situation, tasks, and constraints—were summarized. The components were extended further employing existing models and frameworks, which are explained in the following subsections.

A disaster context also refers to the temporal changes in the setting of each component. Figure 1 is a schematic of the framework of disaster context. The snapshot of a context at any point in time is called a scene. A disaster context is thus represented as a series of scenes that contain the conditions of the situation, tasks, and constraints. The details of each component are explained in the following.

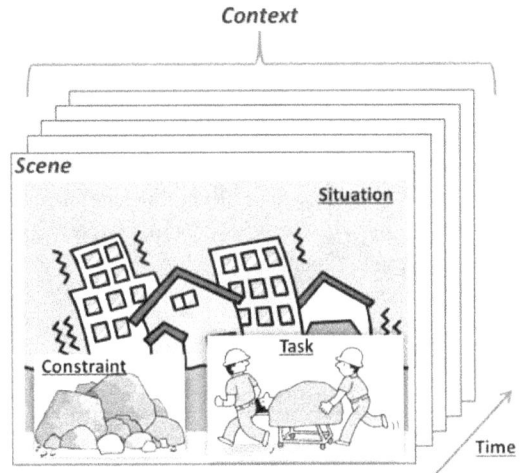

Figure 1: Framework of a disaster context model

2.1. Situation

A situation describes the conditions and statuses of objects and actors in relation to a disaster scene. We further detailed the situation using a general model describing service contexts [4]; the model comprises providers, recipients, tools, the environment, and interactions among them. In the context of disaster medicine and nursing, providers are medical staff such as medical doctors and nurses, recipients are inpatients, outpatients, and their family members, tools are medical tools and devices such as medicines, injectors, wheelchairs, and dialyzers, and the environment is that surrounding these three elements such as hospital rooms and wards.

Figure 2 is a schematic representation of the service context model. The four major model elements were used as the top-level ontology of the situation and were taken as the starting point from which to extend sub-level concepts. Table 1 presents the extended patient model, which has the form of a frame with slots and values. All slots and values were extracted from reports on past disasters, and many personal notes on actual disaster experiences such as those in [5]. The combination of values in Table 1 generates, for example, various types of patient instance such as an injured person, an old person, and a person with chronic disease. Other model elements such as the environment and tools were implemented using the frame model in the same way and were stored in a database of situation elements. A situation is represented by a set of model elements with different values.

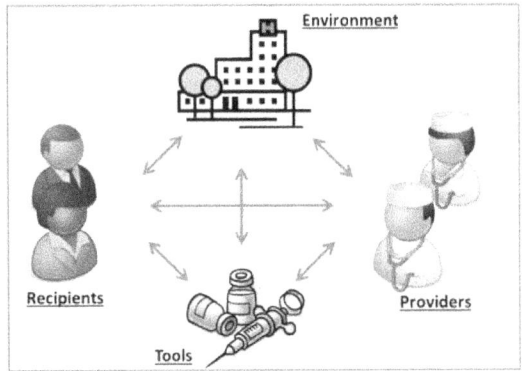

Figure 2: Service context model

Table 1: Patient model

Slot	Values
Age	0, 1–4, 5–9, 10–19, 20s, 30s, 40s, 50s, …
Sex	Male, female
Level of care	1–5
Disease	Cerebral infarction, terminal cancer, COPD, influenza, etc.
Injury	Minor, bruising, laceration, broken bones, etc.
Special conditions	Pregnant, etc.
Ability to walk	Cannot walk, requires a walking stick, requires a wheel chair, etc.
Mental condition	Anxiety, fear, panic, depression, insomnia, etc.
Requirement for an accompanying person	Yes, no

2.2. Task

A task is an action or series of actions carried out to achieve a goal. Tasks are usually hierarchical and situation dependent, and therefore, different tasks need to be performed in different situations and domains. In the example domain, tasks need to be performed in response to the disaster situations at medical intuitions; e.g., confirming and securing the safety of patients and evacuation. The disaster medicine and nursing tasks are listed according to CSCATTT, which covers the major response tasks of command, safety, communication, assessment, triage, treatment, and transportation. Table 2 gives several examples of the tasks categorized by CSCATTT. For the prototype system, 64 tasks were extracted and implemented in the task database.

Table 2: Typical tasks that need to be performed in an emergency

Category	Task	Subtask
Command	Initiate command system	Contact a commander
		Declare an emergency
		Call up appropriate staff
		Confirm the command chain
Safety	Secure own safety	Step away from shelves and windows
		Stand by a thick pillar or get under a table
		Wear a helmet
	Secure safety	Find a search light
		Confirm the safety of patients and visitors
		Prevent fire
Communication	Establish communication	Check communication devices
		Contact headquarters
		Establish communication with other institutions
Assessment	Assess damage	Check for building damage
		Check the availability of the power supply
		Check the availability of dialyzers

2.3. Constraints

Constraints prevent someone from accomplishing tasks as planned or wished. Constraints are situation dependent and emerge in specific disaster situations; e.g., wall collapses, water leaks, and fire. In other words, a constraint can be identified by the relationship between a task and its background situation. Constraints were extracted from reports on past disasters and many personal notes on disaster experiences, categorized according to the concept of performance-shaping factors used in human reliability assessment, and stored in a relational database. In the prototype system, 52 different constraints were implemented in the relational database.

3. CREATING A NEW CONTEXT

The databases of each component were constructed using the above component models of disaster context. The databases store data extracted from the same materials used in the modeling. We also developed prototype database software to support the creation of a new imaginary disaster context. This section describes how a new imaginary disaster context is created using the software. An overview of the procedure is shown in Figure 3. The interactive and tailor-made process is expected to allow users to create a new disaster context for an exercise to be held at his/her hospital with less time and effort. The detail of each step in Figure 3 and the supporting functions are explained in the following subsections.

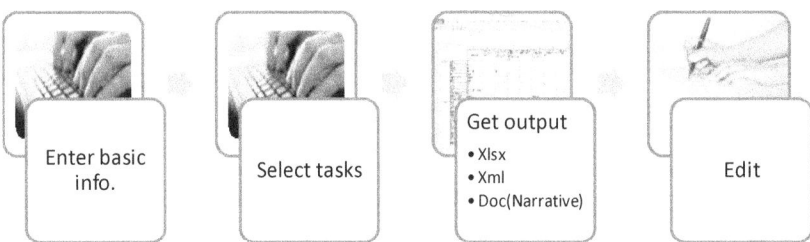

Figure 3: Procedure for creating a new context

3.1. Entering Basic Information

The first step for the user is to enter basic information on the situation assumed for an exercise such as the season, weather, time of occurrence, and numbers of staff and patients. The user can also specify a specific set of values for some slots of the situation model such as a large fire in a patient room. The software then randomly assigns values to the remaining slots of the situation model and creates a complete initial situation. Subsequently, the software creates a changing situation from the initial situation. There are four basic modes of change of a situation: static, escalation, de-escalation, and cyclic. Unless the user specifies one of these modes, one is randomly applied to each variable to create a continuous situation comprising plural situation instances.

3.2. Selecting Tasks

The next step is to select training tasks from the task list stored in the database and to specify the starting time for each task in the context. Once a task is selected, candidate constraints are searched for in the relational database considering the background situation of the task. For example, if windows of the patient room are broken and there are inpatients who need support to walk, while the required task is evacuation, constraints such as the requirements for more supporting staff and special attention to be paid to patients in the evacuation are retrieved from the database.

3.3. Outputting and Editing

When a disaster context is completed through the instantiation of situations, tasks, and constraints, the context is output in xml, xlsx, or doc format. The contents of the output data are shown in Figure 4.

The software has a limited function to maintain consistency among the variables of the situation model when it automatically assigns values, for it is technically difficult to maintain consistency among all values perfectly. For example, if the value of the season is "summer", then a value "snowing" should be excluded from the candidates for the slot of "weather". Such simple exception handlings were partly implemented in the prototype but exception rules that are more complicated were not. Instead, we expect users to find and modify such inconsistencies in the output context. Additionally, the user can edit the details of the contents at any step in the procedure.

3.3.1 Narrative Output

One of the useful functions of the software is the conversion of the output into a narrative/textual format. A story structure was designed according to the situation/service model, which provides a framework with which to arrange the order of the description of each model element. For each model element, several "fill-in-the-blank" sentence formats were prepared. Once a disaster context is specified, the textual explanations for the model elements are generated using the sentence formats, and these sentences are then arranged according to the story structure and a whole story describing the context is completed. This function also generates tables that complement the textual explanation. Figure 5 shows part of a story generated by the function.

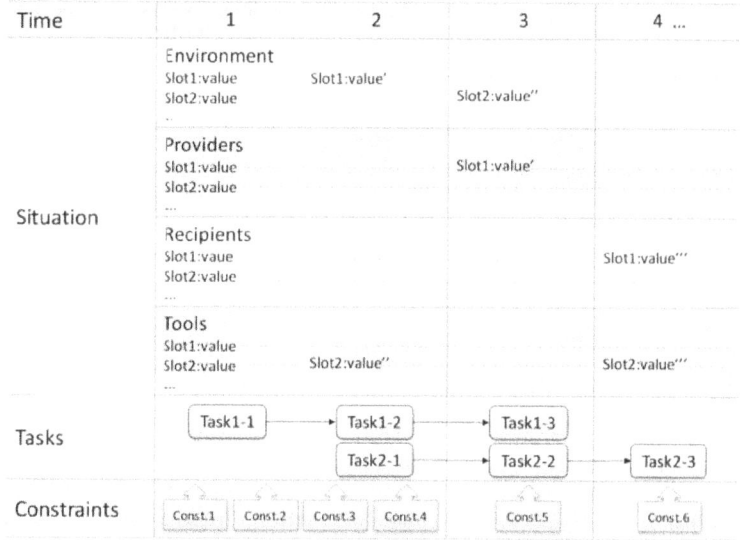

Figure 4: Contents of the output file

On a day in autumn, a level-seven large earthquake strikes near Shizuoka Pref. It occurs about 6pm, while many people are going home. The weather is fine with a moderate wind, and the weather forecast says it will rain tomorrow. There are tens of fires in the area, but because the wind is only moderate, the possibility of the spread of fire is low. Because the earthquake is a subduction earthquake, a tsunami having height of 1 m reaches the coast after 57 minutes. There is a nuclear power plant in the area and already a report of a slight leak of radioactive materials. There is no report of liquefaction. Regarding public transportation, all highways and four main roads are closed. 80% of railway services are stopped or delayed. It will take about 21 hours to reopen highways, but they will only be made available to emergency vehicles. Full restoration of freeways will take 3 days. Full railway services will restart tomorrow afternoon....

季節は秋、平日の帰宅時に震度7の海溝型地震が発生した。地震発生時の天気は晴れで弱風、翌日の天気は雨であった。火災は数十件発生したが、弱風のため、延焼の可能性は少ない。海溝型地震であり、1m程度の津波が57分後に到着した。原子力発電所があり、地震発生直後にすでに放射線漏れが確認されている。また、液状化が発生する可能性はない。交通機関に関しては、高速道路は通行止め、一般道路は4区間が通行止め、鉄道は8割以上が運転見合わせ、もしくは遅延となっている。高速道路が緊急車両のみ通行可となるのは21時間後、一般道路が全線通行可能になるのは30日後、鉄道は翌日の午後には通常通りの運転となった。

Figure 5: Textual description of a disaster context

4. QUALITATIVE EVALUATION

We conducted a preliminary qualitative evaluation of the proposals by interviewing with subject matter experts (SMEs). We asked four nurses about the validity of the context model and practical utility of the prototype. Before starting the group interview, we demonstrated how to create a new imaginary disaster context using the software and showed several output examples, and we then asked the nurses to make comments. Table 3 gives some of the comments made in the group interview. In general, many positive comments on both the model and output contexts were obtained. Several valuable suggestions, such as allowing the creation of scenarios with multimedia and extending the software to an online environment, were also made.

Table 3: Comments from SMEs

Context model
- It is helpful that we can view the entire picture of a disaster context.
- It is necessary to consider hospital staff other than medical doctors and nurses.
- The model does not have sufficiently comprehensive information on medicine and medical devices.
- Constraints are better to be divided into two finer categories: comprehension and projection.
- The remaining amounts of water and food should be given in units of "day·person" and not "days".
Output
- It is better to present a disaster through textual descriptions, while numerical information should be summarized in a table.
- Pictures and/or video clips should be presented with the textual descriptions.
- We can easily make a new context by integrating experiences and expectations.
General comments
- We can consider characteristics of the target region and facility in simulation design.
- We would like to see if it would be better to focus on a more specific topic such as triage training.
- It would be nice to use the system on the Internet and share the contents with other hospitals.

5. CONCLUSION

This paper presented a disaster context model that aims to provide comprehensive and detailed assumptions for disaster exercise scenarios. The model has three basic elements, which are the situation, tasks, and constraints. A disaster context is represented by a series of snapshot scenes that consist of the instances of these model elements. The paper also presented a method of generating disaster context instances in an interactive and semi-automatic manner. The example context of disaster medicine and nursing was used in developing prototype database software. The software generates imaginary disaster contexts in a hospital by combining data pieces stored in the database and outputs them in xml, xlsx, and textual format. We demonstrated the prototype to four nurses and asked them to evaluate it qualitatively in a group interview. Most comments were positive and there were high expectations of further development.

In this paper, we focused on how to use the context model to create imaginary disaster contexts. However, the model is also expected to be used as a common framework for describing and recording both virtual and actual disaster experiences. The model is therefore expected to provide a foundation for sharing (1) knowledge on past experiences, (2) information on how to design good exercise scenarios, and (3) good examples of disaster preparedness, among different organizations.

References

[1]   Susi T., Johannesson M., Backlund P. *Serious Games-An Overview*, The Journal, volume, pp. 110-120, (2000).
[2]   Kobayashi K., et al. *DIGTable:A Tabletop Simulation System for Disaster Education*, Proc. Int'l Conf. on Pervasive Computing, pp.57-60, (2008).
[3]   Yamori K., *Using Games in Community Disaster Prevention Exercises*, Group Decision and Negotiation, Vol. 21, Iss. 4, pp.571-583, (2012).
[4]   Kanno T., and Furuta K., *Service Cognition for Service Systems Design*, Proc. INFORMS, pp. 160-164, (2010).
[5]   Japan Red Cross Ishinomaki Hospital, 石巻赤十字病院の100日間 ("100days of Japan Red Cross Ishinoaki Hospital）, Shogaku-kan, (2011) (in Japanese).

Bayesian networks as a decision making tool to plan and assess maritime safety management indicators.

Osiris A. Valdez Banda[a*], Maria Hänninen[b], Floris Goerlandt[b] and Pentti Kujala[b]

[a] Aalto University, Department of Applied Mechanics, Kotka Maritime Research Centre, Heikinkatu 7, FI-48100 Kotka, Finland.
[b] Aalto University, Department of Applied Mechanics, P.O. Box 15300, FI-00076 Espoo, Finland.

Abstract: Today, maritime safety management norms, self-assessment guides and frameworks demand and/or recommend the collection, report, and analysis of indicators to measure the safety performance of shipping companies. However, the characteristic of classic indicators only provide information about the specific evaluated activity. In this paper, a new quantitative and qualitative option to jointly analyze the performance of individual and collective indicators of a maritime safety management system is proposed. For this purpose, the dependencies between the quality of the most representative components of maritime safety management and their designated indicators levels are probabilistically estimated using a Bayesian network model and two expert views. Each component has one or more designated indicators which aim to identify practical values for the performance of those components. Based on the findings of this study, the implementation of the Bayesian network model seem to provide a unique decision support tool to plan and set indicators, and also to evaluate the indicators' performance and the effect on their designated components. Furthermore, the use of the indicators in the model enable detecting their repercussion on other components of an evaluated safety management system, even when those components do not seem to be directly related.

Keywords: Maritime safety management, Safety management systems, Indicators, Bayesian networks.

1. INTRODUCTION

There are several definitions regarding to the concept of Key Performance Indicators (KPIs), these definitions vary according to the field of indicators' application (e.g. financial, management business, operational). From a general and simple perspective, indicators are quantitatively and/or qualitatively references used to measure how processes perform to obtain planned goals [1]. In maritime safety management, indicators are described as discrete measures which track organization's effectiveness in meeting its aims and objectives [2].

Commonly, KPIs have four defined phases before, during, and after assessing the performance level of certain activity. These phases are related with the initial step of clearly defining the indicator with an accurate setting of the aim of the indicator and target values. A second phase includes the continual monitoring of the indicator, even when there is no need to report it. Then, a third phase is performed for collecting and reporting the indicator. And finally, a phase for a posterior development of actions based on the information reported [3]. In the maritime industry, shipping companies have established safety management systems (SMS) which constantly need to be evaluated in order to evidence if companies are gradually obtaining their planned safety performance. Furthermore, the mentioned evaluations may also evidence if the safety management planning phase is being realistic [3].

The setting of safety management indicators is normally supported by the integration of expert knowledge, the evaluation of organization's available resources, and the available company's historical data of each analyzed safety management aspect (see section 2.1). The monitoring and reporting of the established indicators constantly demand an analysis of the current situation in a performed activity [4]. However, this traditional process of monitoring and reporting indicators allow only analyzing the indicator's influence on a single evaluated component of a safety management system (SMS), without evidencing its affectation in some other different components. In this paper, a new proposal to plan, monitor, and evaluate maritime safety management indicators through the implementation of Bayesian networks is provided. The aim is to quantitatively analyze the performance of the main components of the SMSs in two different shipping companies based on the experts' estimations on practical indicators of those components. The properties of the resulting model is then demonstrated with a hypothetical evidence which could have been derived from periodic reports of a shipping company's SMS.

The paper is organized as follows. Section 2 describes the different material and methods utilized in this study. The main results and findings are presented in Section 3. The section 4 discusses the results further. And finally, conclusions are drawn in Section 5.

2. MATERIAL AND METHODS
2.1. Maritime safety management indicators framework

Today, a common organizational approach when setting, monitoring and evaluating indicators is composed of three key features. The first one is the knowledge of the experts regarding the analyzed safety management area and/or component. This knowledge includes adapting the indicators to the safety management strategy, targets and priorities of the organization, and also the expert knowledge regarding the experience of previous performance of the area/component analyzed by the indicator [5]. The second feature is the designated resources to manage the area evaluated by the indicator, including monetary aspects and the available personnel and technology [6]. The last feature includes the utilization of all available historical data which provide evidence on previous performance of the safety management component or area analyzed by the indicator [7]. Figure 1 present a general perspective of the mentioned components utilized for setting key performance.

In this research, the proposed framework and its key features have been used as an initial guidance for the consulted experts regarding the main aspects to consider when selecting their personal estimation of the presented indicators.

Figure 1. Key features on the setting, monitoring and evaluating of indicators

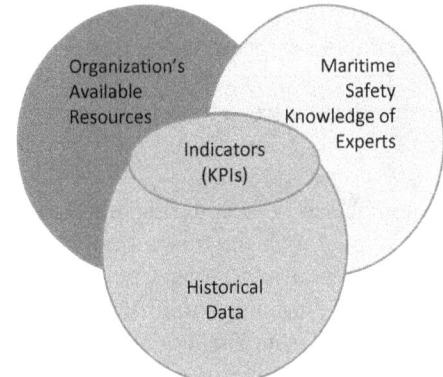

2.2. Bayesian networks

Bayesian networks (BNs) is a technique that can depict relatively complex, possibly but not necessarily causal dependencies and confront with uncertain and unobserved variables while also having a graphical volume [8]. Basically, a BN is a graphical model that encodes probabilistic relationships among variables of interest [9]. Each variable consists of a finite number of mutually exclusive states. And each state has a probability of event and it may also depends on the states of the variable's parent nodes, i.e., the variables with a straight link to the variable under analysis. The utilization of Bayesian networks have become more popular in the last twenty years because their application has been benefited from the development of new computational algorithms and software tools [10]. In the maritime domain, Bayesian networks have already been applied in several maritime traffic safety related models for accident analysis, accidents occurrence estimations, and vessels' potential oil spills e.g. [11; 12; 13; 14; 15; 16]. In this paper, the dependencies between safety management indicators and SMS components are also modeled with BNs. The probability distributions of these safety management indicators are modeled with triangular distributions, whose parameters (min, max, and mode) are based on expert opinion. The constitution of the analysis of indicators presented in this study is explained more in details in the following subsections.

2.3. Analyzing safety management indicators through Bayesian networks

2.3.1 The Safety management model.

The structure and components of maritime safety management utilized in this study are based in the components of maritime safety management proposed in [17], and a Bayesian network model of maritime safety management proposed in [18]. In those studies, 23 components of maritime safety management were extracted from the contents of three documents including maritime safety regulations and frameworks: the International Safety Management (ISM) Code [19], the Tanker Management Self-Assessment (TMSA) [21], and the analysis of the safety management framework proposed in [21]. The 23 components are: accident and incident reporting and analysis, communication, company responsibilities and authority, designated persons, documentation, emergency preparedness, external audit, feedback, internal audit, IT system for the safety management, maintenance of the ship and equipment, management commitment, management review, master's responsibilities and authority, no-blame culture, personnel awareness and involvement, planning, resources and personnel, safety and environmental protection policy, shipboard operations, status of the corrective actions, status of the preventive actions, and training. All these variables were allocated within a Bayesian network model of maritime safety management (see [19]), and three mutually exclusive states (good, average, and poor) were designated to each variable. The links between the network variables were determined with expert opinion.

The conditional probability tables of the safety management variables are based on expert elicitation. Two Safety Designated Persons Ashore (DPAs) in two different shipping companies have contributed in this task:

- Expert 1: A safety DPA of a Finnish shipping company operating ro-ro and ro-pax vessels and providing port operations.
- Expert 2: A safety DPA of a Finnish shipping company operating ro-ro vessels and general cargo ships.

2.3.2 The estimation of the indicators' values

53 indicators extracted and proposed in [18], were allocated to each safety management variable, having components with a minimum of 1 indicator and maximum of 7 indicators. Table 1 contains an example of 10 indicators designated to the maritime safety management variables. For the indicators' estimation by the experts, a structured questionnaire was implemented in order to extract numerical indicators given each of the three mutually exclusive states of the parent variables. In this questionnaire, the experts had to assess the parameters of the triangular distribution of all the indicators. Thus, for each safety management variable the experts were required to provide three values per three established states (poor, average, and good) of the analyzed variable, providing a total of 477 values. These values should provide the minimum, maximum and mode number or percentage of every indicator. Table 2 presents a simple example of the questions proposed to the experts.

Table 1: An example of indicators designated to some of the maritime safety management variables

Variable	Indicators
Communication	Average grade on the annual internal communication evaluation (e.g. from staff satisfaction survey)
Ship operations	Number of blackouts reported by ships per year Number of fires reported during ships operations per year Number of navigational errors reported in a year? Percentage of the ships reaching destination on time (plan vs. real) Safety Percentage of the ships reaching destination on time (plan vs. real) technical
Maintenance	Total out of service time due to a failure in the Maintenance Management System.
IT SM system	Percentage of organization's personnel satisfied with the IT SM System
Acc. and Inc. rep.	Number of accidents reported per year?
Training	Average grade on the internal training provided to the organization's staff?

Table 2: An example of the utilized questionnaire in the experts' estimation of the indicators

Element	Indicator	Amount that represents:	Max	Mod	Min
Ship operations	Number of blackouts reported by ships per year	Good ship operations	3	2	0
		Average ship operations	7	5	3
		Poor ship operations	15	12	7

3. RESULTS

3.1 The network model

Figure 2 presents an extract from the model, showing the network structure with some of the safety management variables (based on [19] and their respective indicators.

Figure 2: A Bayesian network with safety management variables and their indicators

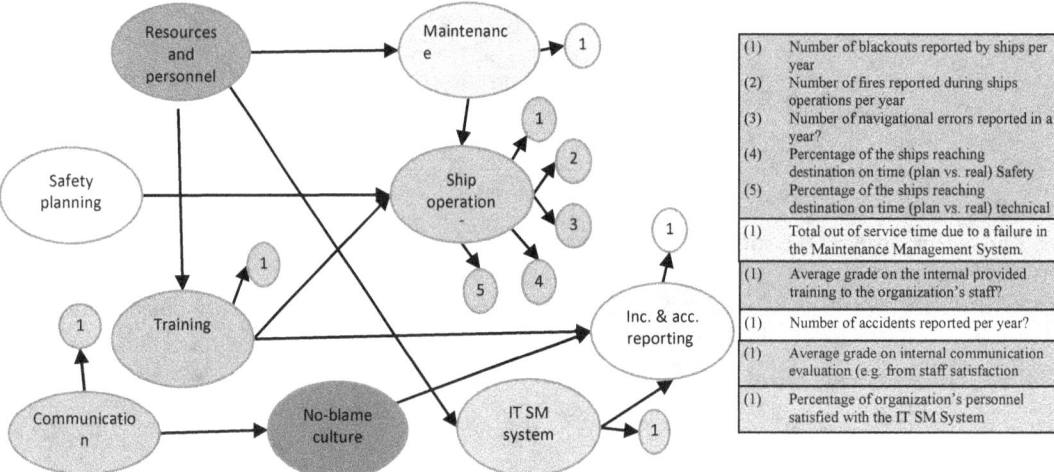

The conditional probability distributions of the safety management variables presented in the Figure 2 have been adopted from [19].

3.2 The estimated indicators

Table 3 presents some of the estimated indicators by the experts. At this point, it is important to remember that such estimations are based on the components introduced in Section 2.1. Thus, the estimations are provided by two experts from two different organizations with different: strategies, objectives, resources and organizational structure.

Table 3: Experts' estimation of the indicators

Element	Indicator	State	Expert 1			Expert 2		
			Max	Mod	Min	Max	Mod	Min
Ship operations	Number of blackouts reported by ships per year	Good	0	3	5	0	0	0
		Average	6	10	15	1	1	1
		Poor	15	25	35	1	1	1
	Number of fires reported during ships operations per year	Good	0	2	4	0	0	0
		Average	5	7	10	0	0	0
		Poor	11	15	20	1	1	1
	Number of navigational errors reported in a year?	Good	0	3	5	3	2	0
		Average	6	10	15	5	4	4
		Poor	15	25	35	10	7	6
	Percentage of the ships reaching destination on time (plan vs. real) Safety	Good	100	92	90	100	97	95
		Average	90	85	80	94	87	80
		Poor	80	75	70	79	67	50
	Percentage of the ships reaching destination on time (plan vs. real) technical	Good	100	99	97	100	100	100
		Average	96	94	92	99	98	98
		Poor	92	90	88	97	90	80
Training	Average grade on the internal provided training to the organization's staff?	Good	5	4	4	5	4.5	4
		Average	4	3.5	3	4	3	3
		Poor	2.5	2	1	2.9	2	2
Maintenance	Total out of service time (in days) due to a failure in the Maintenance Management System.	Good	2	1	0	2	1	0
		Average	7	5	3	5	4	3
		Poor	15	11	8	15	9	6
Communication	Average grade on internal communication evaluation (e.g. from staff satisfaction survey)	Good	5	4.2	4	5	4.5	4
		Average	4	3	2	4	3.5	3.5
		Poor	2	1	0	3	2.5	2
Accident & incident reporting and analysing	Number of accidents reported per year?	Good	2	1	0	0	0	0
		Average	15	7	3	5	3	1
		Poor	30	23	16	10	7	6

The resulting model can be used in evaluating the indicative properties of the chosen indicators not only for its parent variable, that is, the safety management component to be estimated through the indicator in question, but also for the other safety management components and their indicators.

3.3 Indicators in action

3.3.1 Clear specified values and "targets"

Figure 3 graphically presents the marginal distributions for the indicators of the component ship operations. In this figure, the experts are able to visualize the different values for each state of the practical measured area and component based on their probability estimations of these components, and the designated values for the indicators. In the figure, the same scale is utilized in order to compare how different values can be represented in the indicators according to the needs, targets, and general structure of the companies.

Figure 3: Marginal distributions of the indicators for the component "ship operations"

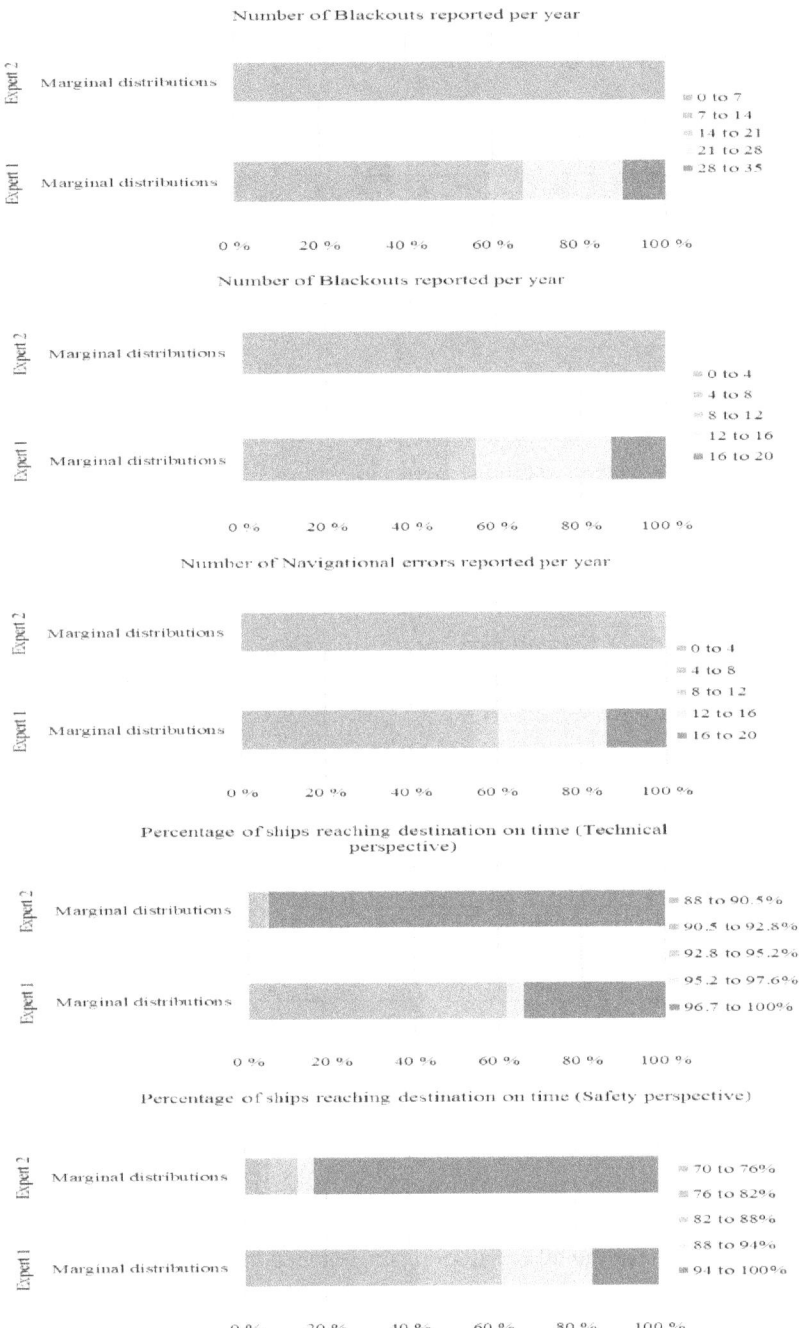

3.3.2 Hypothetical scenarios

Figure 4-5 presents the functioning of the Bayesian network when evidence is available for some of the indicators. In the hypothetical example of Figure 5, the shipping company 1 has had 3 navigational errors reported, 2 blackouts experienced, 0 fires reported, and 92% of ships reaching destination on time. The status observed in the component ship operations are: 1% probability of having good ship operations, 99% probability of having average ship operations and 0% probability of having poor ship operations. And for the expert 2 (Figure 5), having 1 navigational error reported, 0 blackouts experienced, 0 fires suffered, and the 99.2% of ships reaching destination on time, represents: 100% probability of having good ship operations.

Figure 4. Introducing the real data of the indicators of ship operations (Expert 1)

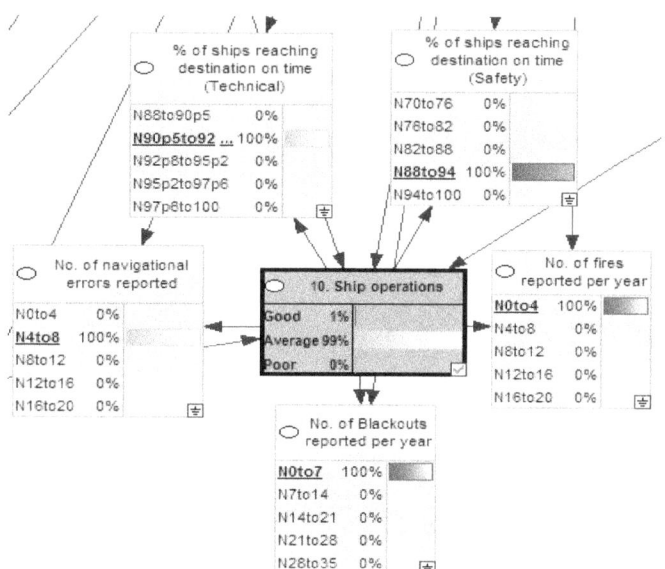

Figure 5. Introducing the real data of the indicators of ship operations (Expert 2)

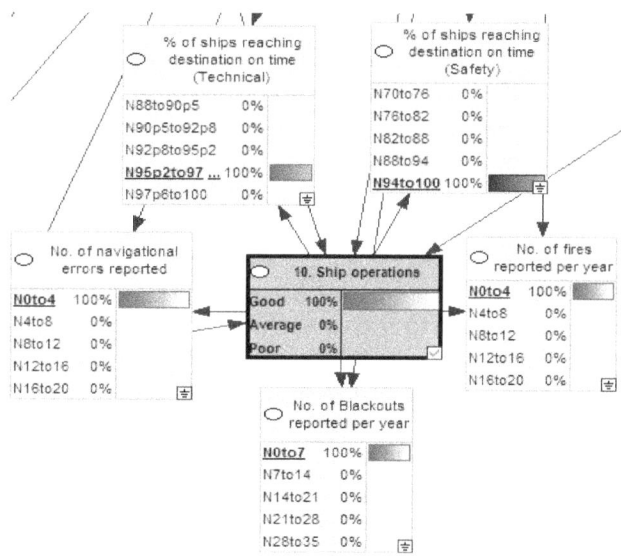

Figure 6 presents the effects of hypothetical knowledge on 5 reported days of out of service time in the fleet and/or machinery belonging to the organization of expert 1. Thus, this amount of service time directly represents 100% probability of having average maintenance of the ship and equipment. However, observing this indicator value has also updated the knowledge on other variables, and also on the other indicators. For example, this observation has updated the variable ship operations to 10, 50 and 40% probabilities of having good, average and poor ship operations respectively. The status of the IT SM system has also been updated by this hypothetical, the component presents: 27, 25 and 48% probabilities of having a good, average and poor levels, respectively. Furthermore, the observed probabilities in other indicators have also been also affected by the inclusion of the mentioned observation. For example, the observation of the mentioned indicator yields a 10% probability of having reported 0 to 4 navigational errors per year, 11 % of having reported 0 to 7 blackouts yearly, and 9% probability of having 97 to 100% of the vessels reaching destination on time (from a technical perspective).

Figure 6. The influence of one indicator on its parent variable, and other network variables and indicators.

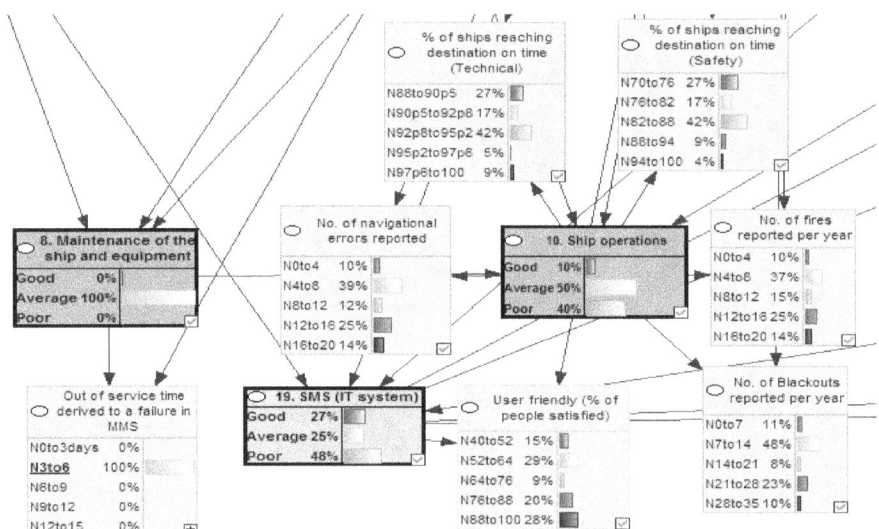

4. DISCUSSION

The implementation of Bayesian networks to set and assess key performance indicators seen to provide a feasible methodology option to set the adequate amounts to all the state values of the maritime safety management components and its indicators based on company's budgets, strategies and objectives, the knowledge of the experts, and data collected from past experiences. The network structure and the designated probability estimations of those variables have represented the initial step to consider the current situation of the SMS installed within experts' organization. The incorporation of the complementing methodology presented in this paper has attempted to provide an adequate initial approach to set the respective practical values of the components' indicators. By implementing a triangular distribution in the estimation of the indicator's values, the experts have had the opportunity to designate specific amounts to the different states included in the indicators. These amounts provided to each indicators seen to follow a linear scale where the values designated to the states good, average and poor are not commonly overlapped. Thus, this seem to reflect that experts have a clear idea of what can be tolerated as good or average levels and what can be not. Figures 3 presented the marginal distributions of the indicators designated to one safety management component based on the knowledge of two experts from different shipping companies. Through this figure, the experts can visualize the actual numerical values of each indicator based on their previous probability estimation for the components contained within the utilized network model.

The hypothetical scenarios attempt to represent how a real assessment of the indicator can be implemented when comparing actual obtained information (e.g. generated from periodic reports of the SMS) against the estimated during the planning and setting phase. Figures 5 and 6 present the effect of including hypothetical data as an observation in the indicators designated to the variables. Thus, including "actual information" can provided numerical estimations of the states in the analyzed variable, which automatically represent an indicator of the current status of safety management component (in this case: ship operations). This option of including data can be exploded e.g. during the assessment of the complete SMS, and/or any individual components of the system.

The advantage provided by the adopted Bayesian network to complete the process of establishing and assessing key performance indicators is clearly reflected in the results provided in Figure 6. Having an assessed structure of the components integrated within maritime safety management, provided the opportunity of identifying affectations in also different variables and/or indicators which do not necessary need to be directly related to the indicator in which data is included. This effect enable detecting the possible effect of indicator(s) on the complete SMS, and of course, its effect on a single component of the system. Thus, using Bayesian networks for the mentioned purpose provided a new method to analyze the performance of a specific and/or several components of maritime safety management, and their repercussion in different areas. This particular aspect represents a differentiate advantage to jump from the classical methods to establish and assess key performance indicators, to a new method to jointly analyze indicators and the general performance of a SMS.

The resulted inclusion of the indicators on each maritime safety management component included in the original "BN model of safety management" [11] could present unlimited number of different types of queries on the current state of the maritime safety management. Unfortunately, this paper cannot present the results to all possible queries due to the limited amount of information allowed to present here. For this reason, any query regarding this proposal can be provided to some extent by contacting the authors of this paper.

5. CONCLUSIONS

This study has presented how Bayesian networks can be implemented to set and assess maritime safety management indicators. The adopted Bayesian network model of maritime safety management proposed by [11] have represented the initial step where experts have considered the current status of their SMS, and the influence among all the variables. With the inclusion of the indicators to each variable, the two consulted experts were able to represent an example of how organizations with different characteristics may follow a similar process to accurately plan, set and evaluate safety management indicators. The work produce from this research provided the opportunity to analyze the management of safety within one shipping company, or also for comparing targets and results from companies with similar characteristics or characteristics of any particular interest.

The results presented in this paper represent how two different organizations may set and evaluate maritime safety management indicators. The main outcome of this research and its results is represented by the option of having a new methodology which allow not only establishing and evaluating key performance indicators in a traditional way, but also to detect affectations derived from the performance of a specific indicator belonging to a single safety management component on any other component and indicator integrated in the network. It can be concluded that while there is still room for improvement and further validation of the methodology proposed, the option presented here seems to adequately represent several planned targets for performance of different areas or components of maritime safety management which can be easily serve as a starting point for building a detailed maritime safety management decision support tool.

Acknowledgements

The work presented in this paper has been carried out as part of the research project 'Competitive Advantage by SaFEty' (CAFE project). A project funded by the European Union and the financing comes from the European Regional Development Fund, the ERDF programme for Southern Finland, the City of Kotka, Varustamosäätiö, Kotka Maritime Research Centre corporate group: Aker Arctic Technology Inc., HaminaKotka port, the Port of Helsinki, Kristina Cruises Ltd, Meriaura Ltd. Project partners: the Kotka Maritime Research Centre, the Centre for Maritime Studies at the University of Turku, Kymenlaakso University of Applied Sciences , Turku University of Applied Sciences and Aalto University. The authors wish to express their gratitude to the maritime safety management experts from the two companies who have participated in this research.

References

[1] Pourshahid, A., Amyot, D., Chen, P., Weiss, M., & Forster, A. (2007). Business Process Monitoring and Alignment: An Approach Based on the User Requirements Notation and Business Intelligence Tools. Ottawa: SITE, University of Ottawa.

[2] OCIMF. (2008). Tanker Management Self-Assessment (TMSA) (2008 ed.). London: Oil Companies International Maritime Forum.

[3] Alwaer, H., & Clements-Croome, D. (2010). *Key performance indicators (KPIs) and priority setting in using the multi-attribute approach for assessing sustainable intelligent buildings.* Building and Environment, 45, 799-807.

[4] Øien, K. (2001). *A framework for the establishment of organizational risk indicators.* Reliability Engineering and System Safety, 74, 147-167.

[5] McDonald, M., Mors, T., and Phillips A. (2003). *Management System Integration: Can It Be Done?* American Society for Quality (ASQ), Quality Systems Bulleting.

[6] Koehn E., P.E., ASCE, F. and Datta, N. (2003). *Quality, Environmental, and Health and Safety Management Systems for Construction Engineering.* Journal of Construction Engineering and Management. DOI: 10.1061/(ASCE)0733-9364(2003)129:5(562)

[7] Choudhry, R., Fang, D. and Mohamed, S. (200t). *The nature of safety culture: A survey of the state-of-art.* Safety Science, 45, 993-1012.

[8] Pearl, J. (1988). *Probabilistic reasoning in intelligent systems: networks of plausible inference.* Morgan Kaufmann.

[9] Heckerman, D. (1995). *Learning Bayesian networks: The combination of knowledge and statistical data.* Machine Learning, 20(3), 197-243.

[10] Song, L., Yang, L., Han, J., & Li, J. (2011). *A Bayesian Belief Net Model for Evaluating Organizational Risks.* Journal of Computers, 6(9), 1842-1846.

[11] Hänninen, M., & Kujala, P. (2014). *Bayesian network modeling of port state control inspection findings and ship accident involvement.* Expert Systems with Applications, 41, 1632–1646.

[12] Hänninen, M., Mazaheri, A., Kujala, P., Montewka, J., Laaksonen, P., Salmiovirta, M., & Klang, M. (2014). *Expert elicitation of a navigation service implementation effects on ship groundings and collisions in the Gulf of Finland.* Proceedings of the Institution of Mechanical Engineers, Part O: Journal of Risk and Reliability, 228, 19–28.

[13] Montewka, J., Ehlers, S., Goerlandt, F., Hinz, T., Tabri, K., & Kujala, P. (2014). *A framework for risk assessment for maritime transportation systems - a case study for open sea collisions involving ropax vessels.* Reliability Engineering & System Safety, 124, 142–157.

[14] Trucco, P., Cagno, E., Ruggeri, F., & Grande, O. (2008). *A bayesian belief network modelling of organisational factors in risk analysis: A case study in maritime transportation.* Reliability Engineering & System Safety, 93, 845–856.

[15] Mazaheri, A., Montewka, J. and Kujala, P. (2013). *Modeling the risk ship grounding – a literature review from a risk management perspective.* Journal of the WMU Maritime Affairs, DOI 10.1007/s13437-013-0056-3.

[16] F. Goerlandt and J. Montewka, *"A probabilistic model for accidental cargo oil outflow from product tankers in a ship-ship collision,"* Mar. Pollut. Bull., vol. 79, pp. 130–144, 2014.

[17] Valdez Banda, O., Hänninen, M., Lappalainen, J., and Kujala, P. (2013). *Analysis of maritime safety management norms for identifying safety management key components and their indicators.* Submitted to Maritime Policy and Management.

[18] Hänninen, M., Valdez Banda, O., and Kujala, P. (2014). *Bayesian Network Model of Maritime Safety Management.* Submitted to Expert Systems with Applications.

[19] IMO (2013). Ism code and guidelines on implementation of the ism code 2010 [online]. http://www.imo.org/OurWork/HumanElement/ SafetyManagement/Pages/ISMCode.aspx.

[20] OCIMF (2008). Tanker Management Self-Assessment (TMSA) (2008 ed.).

[21] Grote, G. (2012). *Safety management in different high-risk domains–all the same?* Safety Science, 50, 1983–1992.

Addressing Off-site Consequence Criteria Using PSA Level 3 - Enhanced Scoping Study

Anders Olsson[*a], **Andrew Caldwell**[a], **Malin Nordqvist**[a], **Gunnar Johansson**[b], **Carl Sunde**[c], **Jan-Erik Holmberg**[c], **and Ilkka Karanta**[d]

[a] Lloyd's Register Consulting, Stockholm, Sweden
[b] ES-Konsult, Stockholm, Sweden
[c] Risk Pilot, Stockholm, Sweden
[d] VTT, Helsinki, Finland

Abstract: Based on an inquiry from the Nordic PSA Group (NPSAG) and the Nordic Nuclear Safety Research group (NKS), a consortium of Swedish nuclear risk consultancies (Lloyd's Register Consulting, ES-Konsult and Risk Pilot) and the Finnish research institute VTT has begun a multi-year study of Probabilistic Off-site Consequences Analysis, commonly referred to as Level 3 Probabilistic Safety Assessment (Level 3 PSA). Level 3 PSA is infrequently performed and generally regarded as a less developed analysis when compared to Level 1 and Level 2 PSA. Interest in the Nordic countries has been spurred based on new nuclear construction projects and plans. These activities have raised interest in objective, risk-based siting analyses for new nuclear reactors in order to better understand the risks of off-site consequences in the wake of the multi-unit disaster at the Fukushima Daiichi site. The objective of this study is to further develop understanding within the Nordic countries in the field of Level 3 PSA, in order to determine the scope of its application, its limitations, the appropriate risk metrics, and the overall need and requirements for performing a Level 3 PSA. The project's first year focused on the development and analysis of an industrial survey about Level 3 PSA, which included several workshops and meetings with Nordic utilities, regulators, and safety experts. Level 3 PSA risk metrics including health, environmental, and economic effects have been researched and discussed in the first year's project report. The project has generated significant interest internationally and has interfaced with international organizations including the IAEA and the American Nuclear Society. The long term objective of the work is to set the foundation for performing a "state-of-the art" Level 3 PSA for Nordic conditions.

Keywords: PSA, PRA, Level 3

1. INTRODUCTION

Level 3 PSA (Probabilistic Safety Assessment) provides a probabilistic assessment of off-site consequences from a radioactive release. The input to a standard Level 3 PSA is derived from several sources. The results from the identification and assessment of the accident sequences leading to core damages, which are provided by Level 1 PSA, and the severe accidents and radioactive source term analyses, which are provided by Level 2 PSA, are combined with meteorological, population and agricultural data to estimate the off-site societal, environmental, and economic risks posed by a nuclear facility.

The field of Level 3 PSA is generally weakly understood, but has been receiving significant attention by the risk community. Many diverse groups stand to benefit from the proposed activities. Those in particular are utilities with operating plants, utilities pursuing new construction, regulatory bodies, public health organizations, and emergency preparedness networks. For the utilities, both the ones with operating plants and those that pursue construction of new plants, it has been noted that the risk

_____
[*] anders.olsson@lr.org

Probabilistic Safety Assessment and Management PSAM 12, June 2014, Honolulu, Hawaii

presented in a Level 3 PSA is of particular interest for the owners of the utilities (shareholders) and the insurance companies.

1.1. Purpose

The project was initiated because due to the Fukushima accidents and the continuing interest in new reactors, interest in Level 3 Off-site consequence PSA has risen within the Nordic region, and around the world. This interest has been reflected in the volume of recent activity in the area of Level 3 PSA at the International Atomic Energy Agency (IAEA) and the large scale projects in the United States and elsewhere.

The goal of this study is to further develop the Nordic understanding of the potential for Level 3 PSA to determine the influences and impacts of off-site consequences, the effectiveness of off-site emergency response, and the potential contributions of improved upstream Level 1 and Level 2 PSAs. Level 3 PSA provides a tool to assess the risks to society posed by a nuclear plant, and could be integral in making objective decisions related to the off-site risks of nuclear facilities.

1.2. Scope of project

The project will develop guidance on several significant topics. The reports and seminars will include guidance on the following topics:

1.   A summary on industrial purpose for performing Level 3 PSA
2.   Recommended risk metrics for Level 3 PSA
3.   Requirements on existing Level 1 & Level 2 studies set by the Level 3 analysis.
4.   Insights on abilities of existing Level 3 PSA tools/codes and possible needs for further development.
5.   Collection of current regulations, guides and standards toward Level 3 PSA
6.   Methodology guidance document

1.3. Project organization

The project includes several separate tasks that are being conducted in parallel. Several of these tasks started during 2013, while others will start up in 2014 and be finalized in 2015. The project tasks address the following topics:

- Industry and Literature Survey,
- Appropriate Risk Metrics,
- Regulation, guides and standards,
- Development of a Guidance document
- Pilot Application including tools for dispersion and consequence analysis

1.4. Project interfaces

The project has had significant interaction with Nordic utilities and regulatory authorities. These include a Stakeholder Meeting where the project financiers provided input on the scope and direction of the project and the industry survey. These stakeholders also responded to the questionnaire that was developed in Task 0, and then assisted in drawing conclusions from the questionnaire during a "Questionnaire Response Workshop". Finally the working group held a seminar on January 21st, 2014 to summarize the progress during the first year of the project and to receive input on the pathway forward for the project.

The project has aroused interest in many international organizations and has fostered Nordic participation in several international Level 3 PSA activities. Currently, the IAEA is developing Level 3 PSA guidance through the drafting of a TECDOC. This project has allowed the working group to contribute to this effort through member participation in IAEA Technical Meeting & Consultant Meetings as well as an expert lecturer an IAEA Regional Workshop on Level 3 PSA. The project has also interfaced with groups such as OECD/NEA Working Group RISK and the ANS/ASME Level 3 PSA standard writing committee.

2. INDUSTRY AND LITERATURE SURVEY

2.1. Background

The first step in this task included the formation of the industrial questionnaire and for this a literature study was performed. The questionnaire was based on earlier similar studies and discussions between the working group and project stakeholders. The purpose of the questionnaire was to collect base information about current international practices and the motivations of utilities and regulators for Level 3 PSA. Even though Level 3 PSA is required only in a few countries, the interest is broader. The results from the questionnaire will therefore contribute to the ultimate objective and outcome of the project in total, a guiding document to provide clear and applied guidance towards regulators, utilities and Level 3 practitioners.

2.2. Questionnaire summary – Risk comparison and development of Level 3 PSA

Risk comparisons for society made risks are possible to do in theory; however, this might not be possible in practice. One reason is the difficulty in finding comparable units, based on risk. If risk comparisons are to be done this must be done carefully. The survey respondents were not in agreement whether or not risk comparisons between different nuclear power plants or between nuclear power and other energy sources are needed. From the questionnaire, it can be concluded that risk comparison in itself is not a strong driver for performing Level 3 PSA.

2.3. Questionnaire summary – Needs for Level 3 PSA

The scope for Level 3 PSA and the use of the results from this type of analysis need to be established before the "need" or value for Level 3 PSA can be fully defined. The main expected motivations for performing a Level 3 PSA are, however, to use the analysis as an objective guidance tool for decision making, e.g. regarding costs for rebuilds and emergency preparedness work.

The respondents attempted to define "unacceptable effects" of a nuclear accident. This was viewed differently between the nuclear expert respondents and the respondents from insurance companies. This indicated the needs for more clearly defining the scope for Level 3 PSA and the use of results.

This difference in opinions can be exemplified by looking at how a risk metric defining unacceptable health effects could be defined:

- Unacceptable health effects, from a <u>nuclear expert's</u> point of view, could be defined from national and international safety standards, e.g. no immediate deaths caused by radiation. Possible, unacceptable, health effects in long term could be compared to other health risks, for example background radiation. There is also the possibility of defining unacceptable health effects by setting dose criteria.
- An example of unacceptable health effects, from an <u>insurance company's</u> point of view, could be: all kinds of health effects that require a visit to hospital and would not exist if the accident would not have happened, the general public should not have any adverse effect from the operation of a nuclear power plant.

Another example of this difference in opinion is related to definition of economic impacts, which can be difficult to define in general. From the utility point of view, an unacceptable economic impact can be when the "bills" are higher than the economic preparedness. From an insurance company's point of view, however, it could be defined as costs related to third parties in terms of compensation to third party. The taxpayers should not be called upon to pay for the damages.

2.4. Questionnaire summary – Advantages of using Level 3 PSA and risk communication

If the use of Level 3 PSA could lead to defining the risk with nuclear power and expressing the risks in terms that are possible to compare, discuss and calculate (e.g. in monetary values) with other societal risks then the results would be communicable. Making the risks communicable could help to improve the communication between the nuclear industry, authorities, insurance companies and the community. The most important communication path consists of two parts. One consists of the communication from experts to authorities and the other one is from authorities to the community (e.g. private persons, non-governmental organizations, and media). However, the authorities are in a double role because they are both experts and authorities. Communication by authorities is more important than communication by experts.

The way of how to grade important communication paths may differ between different groups and persons. Different communications paths may vary in importance for different parts of the society. For example, media may rely on information coming from the government while information between two private persons can be of equal importance for an individual person. Within the questionnaire the respondents were asked to grade the most important communication paths and the result is displayed in the matrices below where the color coding is:

Red=Important, Yellow=Medium, Green=Not so important

From → To↓	Experts	Authorities and Government	Media	Health and Enviorinment	Private person
Private person	3	4	3	3	2
Health and Enviorinment	3	4	1	2	
Media	3	4	2		
Authorities and Government	4	3			
Experts	3				

Figure 1 – Communication path importance graded by nuclear experts.

From → To↓	Experts	Authorities and Government	Media	Health and Envoirinment	Private person
Private person		5	3		
Health and Envoirinment	5				
Media	3	3	1		
Authorities and Government	2	3			
Experts	3				

Figure 2 – Communication path importance graded by insurance companies.

One of the main conclusions from this is that a Level 3 PSA can be an efficient tool when it comes to communication of risk between different stakeholders. The most important communication path consists of two parts:

1. From experts to authorities
2. From authorities to everybody else (private persons, non-governmental organizations, media).

2.5. Questionnaire summary – Challenges with Level 3 PSA

There are several possible uncertainties involved in Level 3 PSA, e.g. those that are related to the dispersion and consequence assessment, those that are related to the chosen risk metrics to be used (health, environmental and/or economic) and the uncertainties that stem from the Level 1 and 2 PSA. Other challenges are related to the fact that the Level 3 PSA might be expensive to perform and require a lot of work and therefore there is a risk for a large gap in time between performing Level 3 PSA studies which leads to problems with knowledge transfer. On the other hand, there are many possible advantages of performing a Level 3 PSA. One of the (unique) advantages that Level 3 PSA can provide is the possibility to compare negative impacts from different technologies. There is also a possibility to see the uncertainties with Level 3 PSA to be, in fact, one of the reasons why the analysis is needed.

One conclusion from the "Questionnaire Response Workshop" was that *"The challenges are also the reasons for performing a Level 3 PSA"*, i.e. the challenges themselves do not motivate not performing a Level 3 PSA. To be able to work uniformly with Level 3 PSA suitable risk metrics should be defined, together with safety criteria that shall be met. There is also a need for specifying guidelines on how to perform the analysis.

3. APPROPRIATE RISK METRICS

3.1. Introduction

The main goal of this task was to discuss appropriate risk metrics for Level 3 PSA. The results from the task will contribute to the ultimate objective and outcome of the project in total, a guiding document to provide clear and applied guidance towards regulators, utilities and Level 3 practitioners. No safety goals, i.e., no numerical criteria, were explicitly connected to the risk metrics presented. However, safety goals were touched upon as a reference to which risk metrics that could be used. In the previous performed work in the NKS/NPSAG Safety Goals project [1], information can be found on what safety goals are being used in different countries and industries, together with arguments and historical background on why different criteria are being used in these countries. Some of the safety goals are related to Level 3 PSA.

There are a number of countries worldwide which have more or less clear safety goals or off-site consequence criterion connected to Level 3 PSA or risks with hazardous industries, see references [1], [2], [3], [4] and [5] for examples. Most of the off-site consequence criteria used in different countries are related to health effects, both to individuals and to the society at large. For numerical criteria, see e.g. [4].

3.2. Risk metrics for Level 3 PSA

Risk metrics of Level 3 PSA have two components: 1) probability metric and 2) consequence (or impact) metric. Regarding the probability metric, it is a matter of choosing the normalization unit for risk comparison purposes. The consequence metric is associated with the impacts which are quantified in the consequence assessment part of a Level 3 PSA. The following main group of consequence metrics has been identified:

- Health effects - Dose
- Environmental impact
- Economic impact (can include every other risk metric).

3.3. Probability units

The results of a PSA, at any level (1, 2 and 3), are typically presented as probabilities of the unwanted events (core damage, large release, offsite impact) per year, and, hence, it can be interpreted as a frequency. The interpretation of the probability per year is that it represents the average risk for a certain nuclear plant that has been analyzed. "Probability per year" is the unit which is used in the regulatory framework and it is almost always associated with a single reactor, since operating licenses are reactor specific. However, in some countries a "probability per year per site" is used (see [6]).

Since "probability per year per reactor" is the probability unit applied in the regulatory context, the probability metric is mainly considered here. Probability units "per lifetime" and "per produced energy over the complete fuel life cycle" can be considered for risk comparison purposes

3.3. Health effects — Dose

Both individual dose and collective dose are of interest for both short-term and long-term effects. From the individual short-term and collective long-term dose both prompt fatalities and cancer fatalities can be calculated.

The following metrics related to health effects are identified:
- Collective dose/individual dose (short- and long-term) [mSv]
- Prompt fatalities (short term)
- Cancer fatalities (long term).

The advantage with the dose related metric is that it is rather straight forward to calculate from the release of radioactive material following a nuclear accident. The dose metric can also be connected to fatalities both in short and long term. It should also be relatively straightforward to define consequence criterion to the dose risk metric. Both the individual and societal consequence can be estimated using dose risk metric (or fatality risk metric). The dose metric can also be used to improve plant design and emergency preparedness. The disadvantage with the dose related metric is that it does not cover the complete consequences of a nuclear accident. The impact to the biosphere is not captured with the dose related risk metric, e.g. contamination/restrictions (evacuation) on land and sea, impact on wildlife is not covered by the dose related metric.

The uncertainties connected to dose and fatalities are the general uncertainties with respect to dispersion calculations (which also affect all other risk metrics). Once the release and dispersion of radioactive material is calculated it is rather straight forward to calculate the dose exposure both on an individual and collective level if population densities are available. From the dose exposure it is easy to estimate fatalities. There is, however, uncertainties related to the validity of the linear, no threshold hypothesis used in the proposed way of calculating cancer deaths. It is being debated whether cancer risk is linearly proportional to dose, when doses are small. Some claim, small doses do not cause cancer or are even healthy, some claim the opposite, the model assume linear relation. This assumption can have a significant impact on the cancer risk estimate, since in many potential studies a major part of the population would get just small doses in case of an accident

3.4. Environmental impact

Different levels of contamination can be used. One level of contamination could result in a restriction for living within a certain area and another level of contamination could result in restrictions from farming and harvest within a certain area.

The following metrics related to environmental impact are identified:
- Ground contamination level due to Cs-134 and Cs-137 [Bq/m^2] or [mSv/year]
- Non-usable areal of land and sea [km^2]

Similar to dose related metrics, it should be relatively straightforward to calculate the environmental metric at least in terms of affected land area (sea may be more challenging). This metric can be further refined from the time perspective point of view (temporary land use restrictions and long term restrictions) and the type of land point of view. Environmental metric is in many respect closely related to the health based metric and these two metrics could be evaluated in an integrated manner. Environmental metric thus compensates part of the disadvantages of health impact metric. The disadvantage is that there is not yet any commonly agreed approach to evaluate different environmental impacts. A single number measuring the area of restricted land use does not reflect the differences between site locations. Type of land and time period of impact are relevant factors to be accounted, but then conversion factors need to be defined if the results are to be compared. This leads to the definition of economic metric.

The uncertainties connected to environmental impact are the general uncertainties with respect to dispersion calculations as well as the estimation of the long term impact on environment. The first issue is common to all other impact metrics, and the second one depends on the quality of environmental impact models. In practice, there should be sufficient input data for environmental impact estimation but the models include uncertainties, e.g., given that the release and dispersion can be calculated and given that the characteristics of the contaminated land area are known, it may be difficult to predict the time periods for land use restrictions and the significance for biosphere. Release to sea or river is even more complex to quantify but the air pathway is usually much more important than the sea pathway. Uncertainties are thus related to the definitions of the surrogate environmental impact metric that need to be applied.

3.5. Economic impact

The following metric related to economic impact is identified:
- Total cost of accident, EUR

Economic impact has the obvious theoretical advantage that all impacts of an accident can be converted into a single metric, which allows consistent risk comparisons and cost-benefit analyses. In principle, this kind of metric should be applied in decision making, while the other impact metrics are surrogates to it. In practice, it can be difficult to agree on what should be included in the quantification of economic impact and how to convert different impacts in a monetary scale. This is a general problem for risk decision making and not specific to nuclear power plant risk analysis, although nuclear accidents have specific complicating aspects such as multitude of impacts and involved stakeholders and the low probability of an accident. Despite the difficulties to evaluate economic impact, it should be sufficient to estimate the order of magnitude of different kinds of accidents, e.g., the Three Mile Island type of core damage accident with practically no external release would mean certain economic impact. Depending on the order of magnitude of release and direction of dispersion some other orders of magnitude of economic impact could be assumed. Knowledge from costs of other natural or industrial catastrophes could be also used as references to estimate the order of magnitude of a nuclear accident.

Despite possible difficulties to convert non-monetary impacts to monetary scale, it might nevertheless be useful to do this exercise, i.e., to try find some commonly agreed conversion factors. This process should lead to increased understanding of risks and facilitate risk communication. Given an economic impact assessment with explicit (parameterized) conversion factors, it is always possible to do sensitivity studies to determine the items that would be most critical to the economic impacts – even with the presence of uncertainties. Example for a multi-criteria decision analysis related to health, environmental, economic and societal impacts, see [7].

Since the economic impact assessment includes any consequences, the range of uncertainties is large and covers all kinds of uncertainties from the incompleteness issues, modeling uncertainties to parametric uncertainties.

3.6. Risk metrics for different stakeholders

Different stakeholders may need different risk metrics. Health effect and environmental impact metrics should be relevant to all stakeholders, but the way economic impact is assessed is more stakeholder dependent. The issue in selecting risk metrics for different stakeholders is thus mainly the question which costs are taken into account and in which way they are weighted. For instance, the safety authority may not necessarily want to take any position on the economic impact, while the utility and the insurance company may look at the economic impact on different risk perspectives.

It may be assumed that the Level 3 PSA is primarily done by the licensee and it would be advisable to consider a wide range of risk metrics (health effects, environmental and economic impact). The aggregation of different risk metrics into single risk metrics should be done explicitly with parametric models, which allows different weightings. The issue of selecting risk metrics can be reduced to the discussion on weightings of risk metrics.

3.7. Comparison with level 1 and 2 PSA risk metrics

The risk metrics related to Level 1 (core damage frequency) and 2 (unacceptable release frequency) PSA are to a large extent independent of the siting (location) of the plant. The only impact from the location of the site in Level 1 and 2 PSA is from the determination of external events which to some extent are dependent on the location. In Level 3 PSA the location of the site is of paramount importance since e.g. metrological data and distance to population and agriculture areas are affecting the output. Hence, Level 3 PSA can give useful information about siting issues. Basically, Level 1 PSA analyze the plant systems which are designed to prevent core damage and Level 2 analyze the plant systems design to prevent and mitigate the consequences of a severe accident. Level 3 PSA will give useful information about both off-site emergency response or preparedness and plant safety systems.

Risk metrics for Level 2 PSA can be applicable as surrogates for Level 3 PSA risk metrics. There is a strong correlation between the release magnitude/timing metric and the health effect/environmental impact risk metrics. The correlation is site-specific. In practice, at certain site it is only the effect of dispersion and evacuation which give variation in the consequence scale given certain release category. Core damage risk metric of Level 1 PSA is not a sufficient surrogate risk metric for Level 3 PSA purposes. On the other hand, if economic impact will be considered in Level 3 PSA, it would be consistent to consider economic impacts event at Level 1 PSA, i.e., to expand the consequence categories of Level 1 PSA to include even major economic losses (without a core damage). From the risk comparison point of view, there may be economically significant consequences without external release or even without core/fuel damage.

4. REGULATIONS, GUIDES, AND STANDARDS

4.1. Introduction

The probabilistic assessment of off-site consequences, often referred to as Level 3 PSA, was the subject of many large studies and international interest in the late 1980s, Organizations such as the IAEA, NEA, European Commission, and US NRC published reports or funded Level 3 PSA programs and studies. It was observed that relatively little has been done in the field since that time, but activities have started within some of these same organizations [2]. The purpose of this task is to provide the ability to observe and influence the development of Level 3 PSA regulations, guides, and standards. This task has also provided input to the other tasks within the project, as well as, provided feedback to external organizations based on the findings of the working group's activities.

Activity in the field of probabilistic off-site consequence analysis has had many peaks and valleys over the years. Internationally and within the Nordic countries, there was a large effort in the field of Level 3 PSA in the late 1980s, which included significant Probabilistic Consequence Analysis (PCA) methods work, large scope studies, and IAEA meetings and publications. Several countries have been performing Level 3 PSA consistently for many years (e.g. the Netherlands, South Africa). However, generally speaking, there was a significant drop-off in the work performed on Level 3 PSA methods and the number of studies performed since the work of the late 1980s and early 1990s. The interest in Level 3 PSA has risen in the last several years. This is based on several reasons, the fact that many of the large-scope well known studies are aging, the development and construction of new reactor units, and perhaps most significantly, the disasters at Fukushima. These reasons have prompted many in the nuclear safety community to re-investigate Level 3 PSA.

The primary focus of this task has been to follow the ongoing work regarding the peer review standards ANS/ASME 58.24 (Level 2 PSA) and ANS/ASME 58.25 (Level 3 PSA) and Level 3 PSA activities at the IAEA. The ANS/ASME 58.24 (Level 2 PSA) and ANS/ASME 58.25 (Level 3 PSA) standards have been under development in writing committee over the past several years. It is anticipated that it will take another several years until these standards will be published. It was envisioned that this task will allow the project to influence and report on the progress of these standards. The work performed under this task has also include monitoring and if possible participation in the development of international guides and regulations. This includes the developments made by the IAEA, the United State Nuclear Regulatory Commission, and similar organizations. Finally, any additional, applicable regulations, and standards will be included in this task, particularly those identified in the work performed for Task 0 and Task 1.

4.2. ANS/ASME Level 3 PSA standard 58.25

The ANS Standards 58.24 and 58.25 regarding Level 2 PSA and Level 3 PSA respectively have been under active development for several years. During this time a member of the working group has been actively involved in the 58.25 writing committee. This project will be integral in providing the resources to continue to engage in the ongoing work and report on the progress of these standards.
Since the work is relatively modest over the past year a large majority of the work to date in the area of the ANS/ASME 28.25 standard was provided in the thesis work provided in reference [2].

The standard is being written by a committee of American Nuclear Society (ANS) and American Society of Mechanical Engineers (ASME) members. The committee was first funded and assembled in the early 2004. Since that time, a draft standard has been completed and released for review. To date, approximately 800 responses have been collected critiquing the draft version of the standard. The ANS/ASME-58.25 standard provides requirements for application of risk-informed decisions related to the consequences of accidents involving release of radioactive materials to the environment. The consequences to be addressed include health effects (early and late) and longer term environmental impacts. These requirements are articulated for a range of technical Level 3 PSA areas in a specific structure. This structure is consistent with previously published ANS/ASME risk standards.

4.3. IAEA activities in Level 3 PSA

The IAEA issued a procedure guide on Level 3 PSA in 1996, IAEA Safety Series No. 50-P-12, "Procedures for Conducting Probabilistic Safety Assessments of Nuclear Power Plants (Level 3)," following significant work performed in the US, Europe, and Japan in the field of Level 3 PSA methods. The IAEA has recently reopened the issue of Level 3 PSA with an IAEA Technical Meeting on Level 3 PSA, which took place in July of 2012. The meeting was the first activity specifically discussing Level 3 PSA since the publication of the IAEA Safety Series No. 50-P-12.

Following the IAEA Technical Meeting, two further IAEA activities have taken place. The first was an Eastern European Regional Workshop on Level 3 PSA, and the second was a Consultant Meeting on Level 3 PSA. The funding provided by the project allowed the working group to participate in both activities. The Consultant Meeting on Level 3 PSA took place in Vienna Austria from November 25-28, 2013. The meeting included several individuals from countries with active Level 3 PSA projects. The recommendation from the attendees of the Technical Meeting was that the IAEA should provide further guidance on Level 3 PSA. The purpose of the IAEA Consultant's meeting was to determine in what form the IAEAs guidance on Level 3 PSA should take.

Following the Technical Meeting a Regional Workshop (Eastern Europe) on the topic "Level 3 PSA development and related issues" took place in July of 2012, which was the first activity specifically discussing Level 3 PSA since the publication of the IAEA Safety Series No. 50-P-12. The motivation for the meeting was due to the relative difficulty in finding information on Level 3 PSA. Due to this difficulty and many open questions in the Region, a 3-day workshop could provide significant insight into the basic constituents, uses, and scope of a Level 3 PSA.

During the course of the IAEA Technical Meeting it became apparent that widely varying approaches and opinions surround Level 3 PSA were held among the group of participating member states. As a result of this, the IAEA decided to pursue further guidance through the development of a TECDOC (IAEA Technical Document).

The objectives of the TECDOC are the following:
- Outline the methodology and indicate the techniques most widely used to date.
- Provide general guidance for conducting a Level 3 PSA with description of major technical elements (e.g. interface between Level 2 and Level 3 PSA, atmospheric dispersion, countermeasures, consequence results interpretation).
- Survey of current practices and computer codes available for consequence assessment (real difficulties learned by Level 3 PSA analysts).
- Provide information on the use of Level 3 PSA and applications, and effective presentation of the results.
- Identify areas of further research.
- Update previous (now outdated) IAEA of the previous IAEA Level 3 PSA publication.

The general scope of the TECDOC should not be completely different from the scope outlined in the IAEA Safety Series No. 50-P-12, publication. As today the TECDOC is planned to have the following scope:
- Level 3 PSA for nuclear power plants considering all facilities at the nuclear power plant (NPP) site is the primary focus of the document.
- Since the general methodology may be also applicable for other parts of the nuclear fuel cycle, (e.g. reprocessing plants, spent fuel storage installations, and research reactors), the document should not exclude these types of facilities, but should maintain the focus on NPP applications.
- The document shall provide guidance but refrain from being prescriptive in its guidance.

4.4. Task 2 Continuing Activities

The work in the area of regulation and standards will continue through 2014. The focus on the continuation of these activities will be the development of the IAEA Level 3 PSA TECDOC, which will have several Consultant meetings over the coming years. Progress on the Level 3 PSA standard has been modest over the past year and it is anticipated that there are several years before completion. IAEA work is poised to continue through the next several years. The IAEA TECDOC is in the very early stages of development, and several more Consultant Meetings will be required to continue and eventually complete it. Internationally there is significantly more work being done in Level 3 PSA. Countries such as the Netherlands and South Africa continue to maintain Level 3 PSA models as it is part of their regulatory requirements. A large scale US NRC study is underway and preliminary results will begin to be discussed and later published in the coming years. Development of a possible replacement to the COYSMA program "PACE" is underway and being discussed. There is also significant interest in this NPSAG / NKS project on Level 3 PSA and the next year seminar shall be planned at least 6 months in advance to accommodate the international participants.

5. PILOT APPLICATION

The pilot project will be completed in two parts, a Finnish project that will utilize Finnish tools and methods, which is also incorporated in the Finnish nuclear safety research program (SAFIR), and a Swedish project, which will utilize Swedish tools and methods. The Finnish project began during 2013, while the Swedish portion of the project will begin during 2014. This section details the progress of the Finnish project during this past year. Some of the overall goals with the pilot application are; to clarify the insights given by Level 3 PSA; demonstrate required resources; get clearer understanding of key uncertainties; provide more knowledge about how current Level 2 release categories structure fits into off-site consequence modeling needs; gain insights in the use of proposed risk metrics; and to support the guidance document and provide practical background to the guidance.

5.1. Specific goals of the Finnish pilot study

The goals of the Finnish pilot study are:
- to gain experience in the application of the IDPSA (Integrated Deterministic and Probabilistic Safety Assessment, sometimes referred to as "Dynamic PSA") methodology (originally developed for Level 2 PSA) to Level 3 PSA studies, and to evaluate its usefulness on Level 3 PSA,
- to apply and evaluate risk metrics identified in Task 1,
- to develop methods for taking into account multiple source terms at different times and from different sources (as was the case in Fukushima),
- to gain experience in conducting Level 3 analysis for the development of a new Level 3 code, and
- to study how uncertainties proliferate through Level 3 analysis

The pilot allows also other uses. For example, comparisons between the IDPSA approach and the current Swedish approach might be made. The pilot will also give perspective on what input should be expected from PSA level 2 analyses. Such uses may be implemented in later years. The goal of the first year in Task 4 was to create a plan for the Finnish pilot study.

5.2. Finnish pilot study in brief

Within the plan for the Finnish pilot study it has been decided that it will be applied to Fukushima Daiichi NPP disaster utilizing IDPSA methodology. There are several issues concerning Fukushima. The first is that there were several source terms at different times from different sources (reactors and used fuel storage). Significant sources of uncertainty include source terms, and the amount of population in the affected area (much of the area was depopulated after the tsunami). All of these issues have to be addressed computationally in the pilot.

In IDPSA, deterministic methods and tools are used to address computationally heavy parts of the system (such as plant response on Level 2 PSA), and probabilistic methods are used to handle uncertainty. Normally the deterministic and probabilistic parts are integrated in the way that the needs of the probabilistic part determine what kind of computations are done in the deterministic part, and some central results of the deterministic part (such as timing information) are fed to the probabilistic part. In the pilot, atmospheric dispersion computations and dose calculations are handled deterministically, and source terms, meteorological conditions, countermeasures and population behavior probabilistically.

The scope of the Finnish pilot study is to estimate population doses and related health effects caused by atmospheric dispersion of the radioactive release in the selected case. Emphasis will be on short-term health effects. Another metric that will be studied is the averted dose, that is, the dose averted by the population due to countermeasure(s). Also the number of persons whose received dose exceeds a certain limit will be examined as a metric. Other consequences, such as land contamination through radioactive fall-out, may be considered.

6. CONCLUSIONS

The first phase of this scoping study of Level 3 PSA has been completed. Focus during this phase has been to get input from different stakeholders on the needs and challenges with Level 3 PSA, to study possible risk metrics that can be used and to participate in international guidance development in the area of Level 3 PSA. The work is planned to continue for two more years and the focus will now be to gain experience by performing pilot applications and to develop a Nordic practicable guidance document.

Acknowledgements

The working group in this project would like to acknowledge the funding organizations that stand behind this project. Funders are found in several organizations such as the Nordic PSA Group represented by the Swedish utilities Forsmark (FKA), Ringhals (RAB) and Oskarshamn (OKG) and the Swedish Radiation Safety Authority (SSM), funding is also provided by the Nordic Nuclear Safety Research group (NKS) and the Finnish Research Programme on Nuclear Power Plant Safety (SAFIR2014). NKS conveys its gratitude to all organizations and persons who by means of financial support or contributions in kind have made the work presented in this paper possible.

References

[1]   Holmberg, J.-E., Knochenhauer, M. *Probabilistic Safety Goals. Phase 1 – Status and Experiences in Sweden and Finland*, SKI Report 2007:06
[2]   A. Caldwell, *Addressing Off-site Consequence Criteria Using Level 3 Probabilistic Safety Assessment*. 2012, ISSN 0280-316X
[3]   *Probabilistic Risk Criteria and Safety Goals*, NEA/CSNI/R(2009)16, OECD/NEA, Paris, 2009.
[4]   *Use and development of probabilistic safety assessment*, NEA/CSNI/R(2007)12, OECD/NEA, Paris, 2007.
[5]   *Utsläppsbegränsande åtgärder vid svåra härdhaverier*, SKI ref 7.1.24 1082/85, SKI/SSI, 1985
[6]   *Massive radiological releases profoundly differ from controlled releases*, Eurosafe, IRSN.
[7]   Keeney, R.L., von Winterfeldt, D. *Managing Nuclear Waste from Power Plants*, Risk Analysis, Volume 14, Issue 1, pages 107–130, February 1994.

A Unified Approach to PSA Accident Sequence Model Quantification

Donald J. Wakefield and James C. Lin
ABSG Consulting Inc. (ABS Consulting), Irvine, CA, USA

Abstract: Existing, fault tree linking models and large, event tree linking models for nuclear power plants are so large that they challenge computer memory limits and/or require excessive run times to fully quantify at the frequency cut-offs required for convergence. In some software quantification tools, the amount of frequency cut-off is not known, and for others, the sheer size of the models becomes unwieldy. The conceptual approach described here is to make use of Monte Carlo simulation. The simulation is one which treats a series of initiating event challenges to the logic model as constants and each challenge assesses whether the logic model end states are true or not. The logic model may be a single fault tree, a single event tree with branch probabilities, or a combination of fault trees and event trees. The outcomes of each challenge are tallied at the end of the simulation to obtain conditional end state probabilities and then combined with the initiating event frequencies to obtain accident sequence frequencies. Quantification cut-offs are not used for this approach and there are no restrictions on the use of NOT gates. Convergence of the Monte Carlo simulation would be the main issue.

Keywords: PSA, Monte Carlo, Simulation, Sequences, Fault Trees.

1. BACKGROUND

Currently, linked fault tree models and linked event tree models have become so large that the time to quantify a large, nuclear plant probabilistic safety assessment (PSA) model has become excessive and/or challenges the available computer memory. Both of these issues limit the frequency truncation cut-offs that can be practically applied for accident sequence frequency quantification. Currently, the cut-offs that can be practically used are not sufficient to mathematically assure adequate convergence of the model end state frequencies.

The use of NOT logic in fault tree models is generally restricted by the features within the available quantification tools; e.g., a popular fault tree quantification engine FTREX [1] limits the level of NOT gates permitted to just 1. Large fault trees are used in both fault tree linking and event tree linking approaches. The advent of the Binary Decision Diagram approach [2] makes it possible to have many levels of NOT logic for modest size fault trees, but not for large fault tree linking models.

For maximum speed, it is desired to create a fully separable approach to sequence quantification to which multi-processors can be applied keeping each processor fully employed right up to the end of the quantification and without much penalty for assigning tasks to each processor. Assigning processors to separate initiating events is one way to accomplish this, but this approach is limited when only one or a small number of initiators account for most of the processing time.

Further, it is desired that both approaches to sequence modeling, be quantified using the same evaluation tool. The proposed simulation approach could be developed in such a way to accomplish this final objective.

2. OVERVIEW OF THE APPROACH

The proposed approach is to make use of Monte Carlo simulation. A simulation consists of a series of challenges to the logic model, where for each challenge we randomly sample the status of each probabilistic input to the model and assesses whether the logic model end states are true or not for that challenge. The outcomes of many challenges are tallied at the end of the simulation to obtain the probabilities of each end state of the logic model given the type of challenge posed. Typically one type of a challenge would be posed for a given simulation. These challenge types would correspond to initiating event types such as a small loss of coolant accident, steam generator tube rupture, earthquake of a selected size, or a simple reactor trip. As many simulations would be run as there are initiating events posed. Since the rate of a specific challenge type occurs at a given frequency, this is like asking the outcome of millions of years of plant operation, where we assume that plant performance is represented by our logic models. The simulation envisioned here is not one that simulates many years of time directly, but rather simply many challenges from a constant frequency initiating event. The simulation approach is easily suited for parallel processing since the challenges can be evaluated independently and the results combined at the end of the simulation.

The evaluation of minimal cutsets and multiplication of basic events which make up each cutset (for fault tree linking models), or multiplication of split fraction probabilities to obtain a sequence probability conditional on the initiating event challenge (for large event tree linking models), is not what we mean here by assessing the logic model. Rather, we mean that given the true or false state of each logic model, probabilistic input for a given challenge, to assess whether the logic model top event or end states are true for that set of input states. Whether the logic model consists of a fault tree, an event tree, or some combination of the two, the logic model end state or end states are determined as true or false for each challenge.

The idea is that each logic model evaluation of a single challenge, as described above, should be very fast. The current logic model quantification approaches that use Boolean reduction and full event tree and fault tree walks with frequency truncation would not be necessary. However, in the proposed approach, many such challenges would have to be evaluated to complete a full simulation. The proposed approach does not rely on truncation, but does rely on Monte Carlo simulation and so is subject to convergence issues depending on the nature of the problem; i.e., high conditional failure probability outcomes should converge more quickly and with fewer challenges than those whose end state conditional probabilities are exceptionally rare events.

This approach has already been used in software developed by others for fault tree analysis; e.g., References [3] and [4]. Reference [3] is such a software tool that was developed in 2005. Tests of that code on a modern computer, and without crediting parallel processing, reveal that 300 million samples of an admittedly simple fault tree involving just 54 minimal cutsets can be performed in less than 5 seconds; and that quantification time includes the time to identify contributors and compute basic event importance measures as well as the probability of the fault tree top event probability. The development of Monte Carlo simulation approaches for fault trees in the past has been to handle more complex, often time-dependent gates that are not employed in most complex accident sequence models; e.g., for nuclear power plants.

3. ADVANTAGES OF A SIMULATION APPROACH

Before describing the proposed approach in more detail, we first list the advantages expected from the simulation approach.

1. For fault tree-linking models:
   a. There is no frequency truncation used to perform the simulations.
   b. There are no approximations required for cutset totaling.
   c. NOT logic gates can be accommodated easily and for many levels. This would permit exclusive event logic to be incorporated directly in the linked fault tree and also allow the incorporation of exclusive maintenance combinations and human action dependencies directly into the models.
   d. For fault tree linking models represented by a single top event, not only can the top event probability be reported but also the fractional importance of any gates the user chooses. Therefore, for models linking many individual sequences under a large OR

2

gate, the contribution of each sequence can also be evaluated, effectively allowing fault trees to compute many end states in a single pass.

- e. Sequence groups can be defined in terms of intermediate gate states and importance measures computed for those groups as well.

2. For event tree linking models:
   - a. It is possible to eliminate or at least minimize the split fraction development, associated frequency quantifications, and to avoid the need for complex split fraction assignment logic rules.
   - b. This approach allows separate top events in the sequences to be explicitly dependent on the same basic events, while still preserving the time sequencing of the event tree tops.
   - c. For large, event tree linked model evaluations, the split fraction representation of sequences and the basic event cutset contributors are saved, and importance measures for both split fractions and basic events may be computed directly.
3. Simulations can be developed in such a way to evaluate both fault tree linking and event tree linking models, allowing the best of both approaches to be obtained.
4. Parallel processing can be easily employed since each processor can be assigned to its own simulation batch of challenges and its own random number seed, and the simulation results then directly added at the end.
5. Partial correlation between probabilistic input states can be modeled efficiently by simulation techniques.
6. Additional gate types may be developed for complex accident sequence models that could prove useful; e.g., recovery gates which consider the timing of failures within the assumed mission time.

4. SIMULATION OF PROBABILISTIC INPUT STATES

The fault tree or event tree probabilistic inputs are either basic events, or split fractions, or both. We first need to sample whether these probabilistic inputs are true on a particular logic model challenge. The most brute force approach is to evaluate the status of every basic event or split fraction each challenge using a random number to decide which of the inputs are true for that challenge. This is one of the approaches adopted in Reference [3] for fault tree simulations. However, other approaches can greatly reduce the number of input state evaluations on any given challenge.

A second approach adopted in [3] first orders the probabilistic inputs and then uses their probabilities of being true to compute the first of the ordered inputs to be true for that challenge. This can be generalized to select the second event in the ordered list to be true and so on, minimizing the number of random numbers required to assess the entire list of ordered input states; i.e., those not evaluated to be true are then all false.

Another way to approach the evaluation of probabilistic inputs is to initially evaluate each probabilistic input as if it were to be repeatedly challenged and use a random number to determine which challenge number it is first to be assumed true. By applying this to all inputs, an ordered array by challenge number could list the inputs that are true for each challenge. All other inputs would be false. As the ordered list of challenges are evaluated in turn, new input state evaluations would only be performed for those input states that were true on the current challenge, the rest being known to be true only in later challenges. Note that in all these approaches while the input state probabilities are used in the calculation of which are true for a given challenge, they are not used in the logic model evaluation.

An obvious feature of this approach is that if there are challenges, in which there are no probabilistic states that are true, then the logic model solution for all states being false can be cited with no need to assess the status of the logic model end states again.

3

Additional heuristics are envisioned that would make this process more efficient and to limit the number of unique evaluations of the logic models.

5. SIMULATION APPLIED TO FAULT TREES

The determination of the status of the probabilistic events (i.e., basic events for fault trees) for each challenge was discussed in the previous section. Here we discuss the evaluation of the fault tree logic when it is known which basic events are true or false. It appears most efficient if the assessments used for one basic event change save intermediate information that could be used for the next basic event change in state. Of course if no basic events fail then there is no need to walk the tree logic because we already know the state of the fault tree assuming all basic events successful. Figure 1 provides an illustration.

Figure 1: Example Evaluation of a Fault Tree

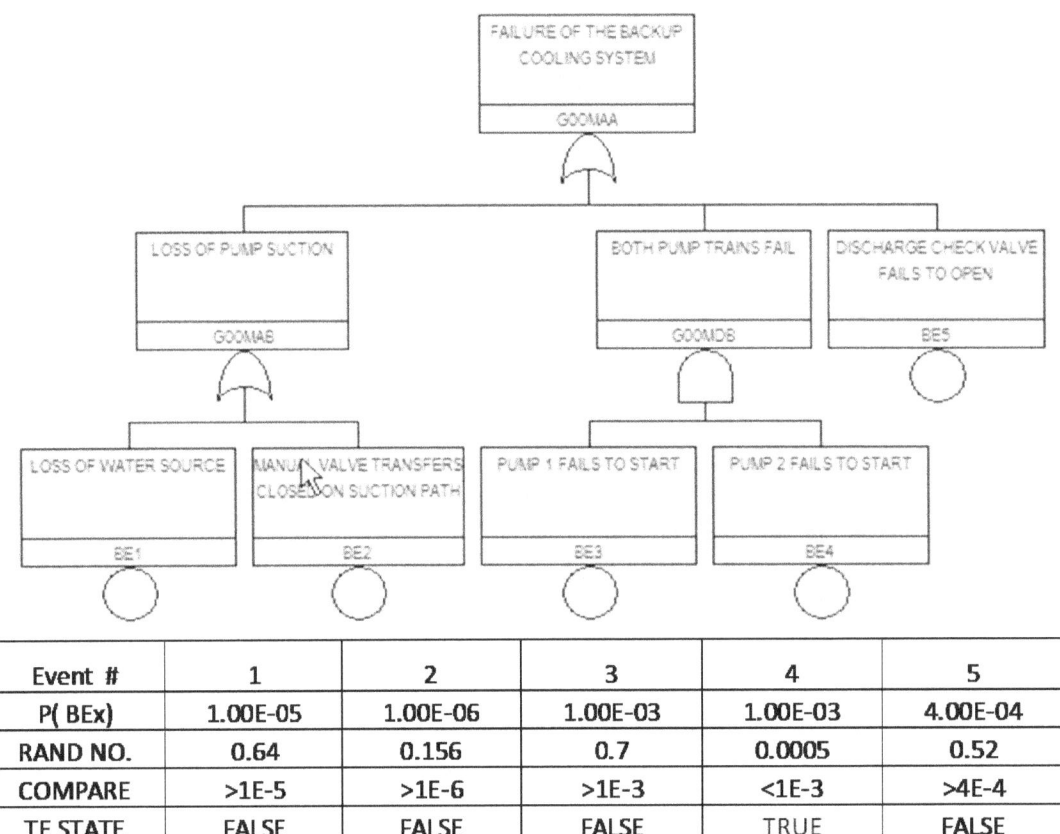

Event #	1	2	3	4	5
P(BEx)	1.00E-05	1.00E-06	1.00E-03	1.00E-03	4.00E-04
RAND NO.	0.64	0.156	0.7	0.0005	0.52
COMPARE	>1E-5	>1E-6	>1E-3	<1E-3	>4E-4
TE STATE	FALSE	FALSE	FALSE	TRUE	FALSE

In this illustration, the brute force approach to determining the basic events' status for one challenge is used. The status of the fault tree with all inputs set to false is also false. Of the five events considered, only one, Basic Event 4, was found to be true for the challenge represented. Basic Event 4 only feeds the AND Gate G00MDB. Since the input from Basic Event 3 to the same gate is false, the AND gate is also false. There is no need to evaluate the higher levels of the fault tree because no further changes in gate states are possible. This is an important point, it will often be the case that the entire fault tree need not be evaluated because of the limited impacts of the basic events that are set to true each challenge; i.e., the higher level gates remain at their initial state. If no events are set to true, then no evaluation at all is required. Some fault tree pre-processing of the OR and AND gates to subsume unnecessary levels and to remove basic events whose values are 1.0 or 0, will also speed the fault tree evaluation process.

For the bottom-up fault tree walk, it would be useful to have a database of which gates each basic event enters since each occurrence of a failed basic event changes that gate's input states. Similarly, a

4

database of gates each gate enters and which gates and basic events enter each gate could be prepared prior to the start of the simulation.

It would appear straightforward to walk the fault tree logic one basic event state change to a given gate at a time. The cumulative effects of such gate state changes could then be determined as all basic events that change state and all the gates they feed are evaluated. Reference [3] implemented a clever bit-map approach for fault trees with just OR and AND gates. This would have to be generalized to k/n gates and to NOT logic for our purposes.

It is worth mentioning that some dependencies between basic events may best be considered using NOT logic directly in the fault trees. Other basic event dependencies, such as for mutually exclusive system alignments, may instead be accounted for by the manner in which the basic event states are sampled.

Contributors to the top event results can be saved if desired. Such failure combinations may not be minimal. We could determine if they are minimal by removing each basic event one at time from the cutset to see if the top event is still failed. If still failed then the counter for that basic event would not be incremented for that challenge after all. In the simulation approach, the computation of fractional basic event importance does not depend on whether the cutsets are minimal or not. In Reference [3], the cutsets identified by simulation are minimized at the end of the computation.

The cutset probabilities can be computed to determine which are to be saved. If the cutset probability is above a saved cutset cut-off specified by the user, then it would be saved for review. Since the highest probability cutsets may appear many times among all the challenges in the simulation, a process to remove duplicates is necessary.

For basic event importance, each basic event failed in a cutset that causes the top event to fail, could be tracked and a counter for each basic event incremented for each challenge that it fails and the top event is true. Unlike for the classical approach to fault tree solutions, where information about the status of gates is lost, for a simulation the approach to importance can be generalized to gates as well.

6. SIMULATION APPLIED TO EVENT TREES

For event tree logic models challenged by constant initiating event frequencies, the approach is similar to that for fault trees, but not exactly the same. The branch point probabilities, or split fractions, become the probabilistic input states and the challenge is to split fractions whose success or failure state is to be determined. However, for evaluation of event trees, just a single path through the event tree is evaluated; i.e., the plant has only one response to the challenging initiating event and its outcome is the end state assigned to that sequence. The logic evaluation would walk the event tree first deciding which split fraction applies at each event tree branch node, and then determine whether to choose the success or failure path from that node based on a random number selected for that top event. For split fractions that are always 1.0 or 0, the end state for that top event is always the same and no random number need be selected. A key difference from the fault tree evaluation approach discussed in Section 4, is that here, only the split fractions which are all pre-calculated and are actually used in the single sequence path, are then evaluated for that challenge; i.e., at most one random number for each of the top events in the linked event tree. Sequence frequency truncation would not be used; i.e., all sequences would be computed end-to-end regardless of their frequency. This simulation approach is then performed separately for each initiating event.

5

Figure 2: Simulation of Linked Event Trees

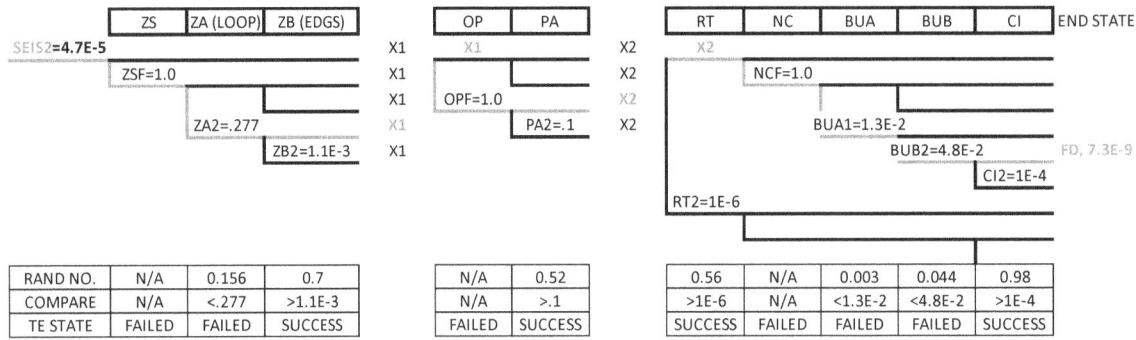

RAND NO.	N/A	0.156	0.7
COMPARE	N/A	<.277	>1.1E-3
TE STATE	FAILED	FAILED	SUCCESS

N/A	0.52
N/A	>.1
FAILED	SUCCESS

0.56	N/A	0.003	0.044	0.98
>1E-6	N/A	<1.3E-2	<4.8E-2	>1E-4
SUCCESS	FAILED	FAILED	FAILED	SUCCESS

Figure 2 illustrates the evaluation of linked event trees by this approach. Random numbers are rolled for each split fraction whose values are neither 1.0 nor 0 along the sampled path. These random numbers are compared with the split fraction values to determine the branch outcome; i.e., success or failure. In the single challenge example of Figure 2, three of the top events are guaranteed failed and three others fail probabilistically; i.e., a particularly unlucky day. The red highlighted path shows the sample sequence.

A very good feature of this approach is that it can be implemented on existing large event tree linking models developed in RISKMAN™ [5], without any modeling changes.

Similar to fault tree simulations, the sequence groups and split fraction importance information can be tracked for each of the split fractions along the single evaluated sequence, exactly as is done now within RISKMAN; i.e., as a table of split fraction names and a count of their status for each challenge evaluated. Here the counts stored would be for the total number of sampled sequences involving the success or failure of the split fractions to each sequence group, instead of the frequency of the sequences involving the split fractions to each bin or group.

A key issue is how to save the sequence representations without saving too many. Recall that since many simulations are performed, saving every sequence outcome would be excessive, even if just for a single initiator. In general, for RISKMAN sequences we only need to save the top "n" or so sequences to each end state; i.e., just those that are to be reviewed by the user. There are several things we can do to address this but the key is how to look for simulation outcomes that result in the same sequence, and to remove the second and other occurrences of that same sequence. One approach is to compute the sequence frequency for each sequence simulated (we already know the split fractions and their values at every node of the linked event tree sequence) assigned to a given end state. These can be compared against the sequences saved previously for that end state and only the highest ranked sequences would be retained. For sequences that have exactly the same frequency as those computed previously, these could be compared top event state by top event state to see if they are duplicate sequences, and the duplicates removed if a match of sequences is found. Since only one initiator is computed at a time, the comparison of sequences would not have to examine sequences starting with different initiators.

For multi-state top events, the approach is slightly different. For most problems the sum of the multi-state split fraction probabilities must sum to 1.0. Therefore we could sample a random number every challenge to determine the single path from among the multi-state paths of that top event to pursue. The split fraction values for the multi-state top event would determine the relative split to be sampled.

Note that no frequency truncation is used for this approach. Instead, it is the convergence of each end state frequency that is of interest. For nuclear power plant applications, the end states may be just success or core damage, or may include an extensive list of end states such as are required to define release categories in a Level 2 assessment.

6

The event tree logic model is evaluated conditional on the occurrence of the constant initiating event. To convert the challenge results to end state frequencies, one just multiplies the conditional end state probabilities evaluated in the simulation by the initiating event frequency to obtain the end state or more generally any sequence group frequencies for that initiating event.

Estimates of the error in the sample mean of each end state or sequence group can be made and are useful to demonstrate that sufficient simulations are performed. For the variation in an end state mean, one can assume the samples are distributed normally and use the estimates for the 95% confidence interval; i.e., between the 2.5% and 97.5%. The 95% confidence interval is then:

$$=2*1.96*SQRT\ [\mu*(1-\mu)/M]$$

where M is the number of trials and μ is the proportion of challenges with the end state true divided by the number of challenges M.

From the above, we see that for smaller values of μ larger values of M are required to achieve the same absolute value of the 95% confidence interval. This implies that the simulation approach works best when the conditional probabilities of the logic model end state are relatively high; so that fewer challenges are required for convergence. However, experience shows that the longest running RISKMAN sequences using classical techniques are precisely those which have high conditional probabilities of the end states.

7. SIMULATION APPLIED TO COMBINED EVENT TREE-FAULT TREE MODELS

Consider the simple example in Figure 3 which has two fault trees to represent the failure event combinations for System A and for System B. The simulation approach for logic models that use both event trees and fault tree models starts out the same as that used for just those with fault trees. The basic event states, used in any of the fault trees, are evaluated for each challenge using any of the approaches identified above in Section 4. The key here is that the basic event states that are determined for a given challenge are applicable to all the fault trees used in the event tree for that challenge.

We postulate an initiating event challenge and evaluate the fault tree logic under the first Top Event A in the event tree sequence. For the single challenge example in Figure 3, Event X is true and Events Y and Z are false on that challenge. From the fault tree for System A, Event X feeds an OR gate which must then be true; i.e., System A fails the first challenge. We are therefore on the down branch of the event tree under System A. System B is governed by an AND gate that requires both X and Z to be true to cause System B to fail. Event X is true but not Z. Therefore, System B does not fail and the three is walked on the upward, success, branch under System B to Sequence ID 3; i.e., the up branch under System B following failure of System A. This single path through the event trees is developed and the associated end state assigned for the first challenge. The simulation repeats this process for many challenges tracking the sequence counts ending in each of the end states and other results of interest. After multiplying by the initiator frequency, we can save the highest frequency sequences and also track the fractional importance of basic events, top event failures, and other gate states or event combinations as desired.

7

Figure 3: Simulation of Combined Event Tree-Fault Tree Models

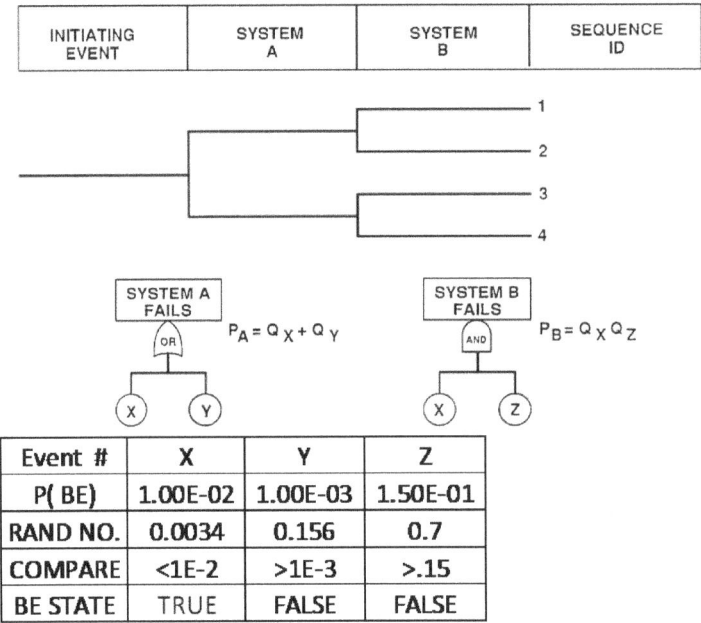

Event #	X	Y	Z
P(BE)	1.00E-02	1.00E-03	1.50E-01
RAND NO.	0.0034	0.156	0.7
COMPARE	<1E-2	>1E-3	>.15
BE STATE	TRUE	FALSE	FALSE

For fault-tree linking models, the problem formulation is largely as presented in Figure 3. The fault tree may be the same for each node under a given top event, or may differ for each node. The fault trees for different systems often share basic events and even high level gates are repeated between the system models; for example to model common support systems. The evaluation of the individual fault trees are to be resolved only along the one sequence path taken for that challenge. These separate fault trees may be evaluated completely separately or their shared gates first evaluated and then the states of these shared gates introduced to the higher level fault trees for each top event. It is not necessary to evaluate these shared gates via parallel processing because the multi-processors are more efficiently assigned to their own set of challenges to integrate into the overall simulation.

For large event tree linking models, such as evaluated using RISKMAN [5], the event tree branch probabilities are evaluated conditionally on the status of earlier branches of the event tree. This requires many split fractions for each top event to be defined in terms of the states of earlier top events and each split fraction is then quantified separately ahead of time. The applicable split fraction values, or one minus their values along success branches, are then multiplied along each sequence to derive the individual sequence frequency.

For simulation of large event tree models, the classical approach to sequence frequency quantification would instead be dramatically changed. The fault tree top events could be used as is in the simulation without having to define split fractions, to define conditional split fractions on other split fractions, nor to quantify them ahead of time. Instead the states of preceding top events in the event tree would be associated with house events in the fault trees used for later top events. The states of all house events in a top event fault tree could then be resolved as part of the fault tree evaluation based on the earlier sequence path in the event tree. The states of such preceding top events can just be added to the list of basic event impacts assumed to be true for that challenge. In this way, the relatively small fault trees of the large linked event tree models are evaluated separately but only for the conditions defined by the specific sequence being walked for the given challenge. Since they are typically small fault trees, they would be expected to be resolved relatively quickly.

The large event trees in the event tree linking approach were designed to achieve finer granularity to the events in the accident sequences; i.e., to preserve event time sequencing and at a level of detail near the level of events seen in plant emergency procedures and hence what the operators are used to discussing. However, to accomplish this greater degree of granularity, additional top events have to be added to the sequence models to ensure that any shared equipment between top events appear

8

earlier as separate top events. The result of a properly constructed event tree linking model is that any basic event appears no more than once in the fault trees for top events describing an entire accident sequence. Further, trains of supporting systems must also be separated into their own top events so as to properly track the dependencies of other systems on these train separated support systems. The point of mentioning this is that with the simulation approach, it will no longer be necessary to add these top events. The status of all basic events in the model for each challenge is known prior to the event tree walk. Basic events can therefore appear in multiple top events without fear of double counting since we are only determining whether a top event is successful or failed and not what its occurrence probability is determined to be. The need for Boolean rules to choose the appropriate split fraction would also be eliminated. We can track the conditional probability of an end state or of individual sequences without having to compute the failure probabilities of individual top events. It is still possible, however, to determine the importance of individual basic events, top event failures, or gate failures to the end states and sequence groups of interest.

We see then that with the simulation approach, the large event tree linking and large fault tree linking methods for sequence quantification are coming closer together. Instead of combining the fault tree linking style models into a single top event, we could retain their relatively smaller event tree sequence models and evaluate the one event tree path per simulation just as for large event tree linking models. Whichever approach is used to build the top event node fault trees (i.e., use house events and evaluate them conditional on preceding top event states, or repeat the supporting logic explicitly within the top event fault tree itself) could be supported using simulation.

In fact, it becomes more a matter of modeling style as to how many top events to include in an event tree set for each initiating event. It is only conjecture at this point as to whether the evaluation of numerous small fault trees along an accident sequence, each depending on relatively few basic events, would be quicker than a smaller number of larger fault trees that are more typical of fault tree linking models.

8. CONCLUSIONS

In this paper, Monte Carlo simulation approaches to sequence frequency quantification have been discussed as may be applied to standard fault tree linking and large event tree linking accident sequence models for nuclear power plants. The flexibility of simulation techniques to more accurately and without frequency truncation evaluate such large logic models, including models that extensively utilize NOT logic; along with the ability to easily separate such problems for parallel processing, offers two powerful incentives to explore such techniques further. Convergence of the Monte Carlo simulation would be the main concern; i.e., can the processing time to evaluate the logic models for each challenge be made acceptable when the conditional probabilities of the end states are quite low?

We see that the proposed approach, lends itself to quantifying both linked fault tree models and linked event tree models. It does so in ways that bring these two sequence logic model constructs closer together, perhaps even allowing an easy model transition from one to the other, depending on the needs of the user.

Further, simulation approaches involving new gate types may possibly be formulated to address the timing of accident failures in ways that permit more affective modeling of recoveries.

Acknowledgements

The author would like to thank the encouragement and comments by the probabilistic risk assessment staff at ABS Consulting, headed by Dr. David H. Johnson, for their contributions to this paper.

9

References

[1] Electric Power Research Institute, "*Fault Tree Reliability Evaluation eXpert (FTREX 1.6)*", (2011).

[2] A. Rauzy, "*Aralia User Manual*", ARBoost Technologies, (2005).

[3] Formal Software Construction Limited, "*Open FTA User Manual Version 1.0*", (2005).

[4] K. Durga Rao, et al., "*Dynamic fault tree analysis using Monte Carlo simulation in probabilistic safety assessment*", Reliability Engineering and System Safety, Vol. 94, pp. 872–883, (2009).

[5] D.J. Wakefield, et al., "*RISKMAN, Celebrating 20+ Years of Excellence!*", ABS Consulting, presented at PSAM10, (2010), Seattle, Washington.

10

Earthquake Risk Perception: The Case of Mexico City

Tatiana Gouzeva[a], Galdino Santos-Reyes, and Jaime Santos-Reyes[a*]

[a] SARACS Research Group, SEPI-ESIME, IPN, Mexico City, Mexico

Abstract: Given the concerns of society in relation to natural hazards, nowadays the analysis of risk perception and communication play an important role in decision making of those in charge, for example, of Civil Protection. The analysis of risk perception and communication may be regarded not only as a presentation of the scientific calculations of risk, but also a forum for discussion on issues on broader ethical and moral concerns. The paper present some preliminary findings of the ongoing research project on earthquake risk perception of the population of Mexico City. It is hoped that the results of the research project may help to understand, to some extent, the degree of knowledge of the study population in terms of earthquake risk perception and preparedness, so that the impact of earthquakes could be mitigated.

Keywords: Earthquake, Mexico City, Risk Perception.

1. INTRODUCTION

1.1. Earthquakes

Earthquakes may regarded as one of the most deadliest natural hazards on earth. Literally in seconds thousands of lives can be (and have been) lost due to its considerable amount of force of destruction. According to the data being registered since 1900, one earthquake of magnitude 8 or greater, 15 earthquakes of magnitude 7-7.9 and 134 earthquakes of magnitude 6-6.9 on the Richter scale are expected each year, worldwide [1]. It is also believed that the number of large earthquakes has remained relatively constant; however, the observed number of smaller earthquakes (of magnitude lesser than 6) has increased each year [2]. According to the International Federation of Red Cross (IFRC), during the first decade of the 21 century, 4,022 natural disasters have been reported, 284 (7%) of which were earthquakes [3]. Although they constitute a small share among the number of disasters, earthquakes are the major cause of death (55.7%) and cause US$232,070 million worth of damage (22%) comparing to other natural phenomena [3].

1.1. Risk Perception and Preparedness

A natural hazard only becomes a disaster when it affects a human community that is exposed and vulnerable. Entities at risk are humans, infrastructure, buildings, utilities, etc. Seismic vulnerability of a community is "the degree of loss to a given element of risk or set of such elements" [4]. Moreover, some authors argued that "an earthquake is an event that can be prepared for in advance" [5]. Governments, local communities, and social organizations all should undertake measures for major earthquakes. Individuals also reduce the impacts of earthquake disasters by learning what to do before, during and after earthquakes and by taking a variety of personal safety measures [6-9].

Overall, risk, risk perception, and risk communication has been deal with from different perspectives; that is, risk from a quantitative perspective [10] and as a threat [11,12]. Slovic [11], for example, has described risk perception from different sources; i.e., geography, sociology, political science, anthropology and psychology [11]. The geographical perspective focuses in trying to understand human behavior for natural hazards; the sociological and anthropological approaches, on the other hand, have shown that perception and acceptance of risk have their roots in social and cultural factors.

---

* E-mail address: jrsantosr@hotmail.com

Finally, the psychological aspect addresses the fear level prior to the event and the confidence in a person's available resources. Slovic [13] argues that risk perception is related to three major factors: dread, familiarity and exposure. Other authors have found that factors affecting risk perception are usually not independent and vary across different hazard types and people [14].

A number of studies have been conducted worldwide and reported in the literature addressing knowledge on seismic risk, earthquake risk perception and willingness to take action to reduce seismic risk [15-21]. However, there is not such a study related to Mexico City. The authors are part of a research team conducting research on risk perception and risk communication related to seismic risk in Mexico City and other states of the country. The paper addresses some results of a pre-test of a survey instrument being designed for a big scale application.

2. THE SEISMICITY OF MEXICO AND THE 1985 EARTHQUAKE

2.1. Seismicity of Mexico

Situated atop three of the large tectonic plates that constitute the earth's surface, Mexico is one of the most seismologically active regions on earth. The motion of these plates causes earthquakes and volcanic activity. Most of the Mexican landmass rests on the westward moving North American plate. The Pacific Ocean floor off southern Mexico, however, is being carried northeast by the underlying motion of the Cocos plate. Ocean floor material is relatively dense; when it strikes the lighter granite of the Mexican landmass, the ocean floor is forced under the landmass, creating the deep Middle American trench that lies off Mexico's southern coast (Fig. 1). The westward moving land atop the North American plate is slowed and crumpled where it meets the Cocos plate, creating the mountain ranges of southern Mexico. The subduction of the Cocos plate accounts for the frequency of earthquakes near Mexico's southern coast. As the rocks constituting the ocean floor are forced down, they melt, and the molten material is forced up through weaknesses in the surface rock, creating the volcanoes in the Cordillera Neovolcánica across central Mexico [22].

Figure 1: Mexico and tectonic plates. [23].

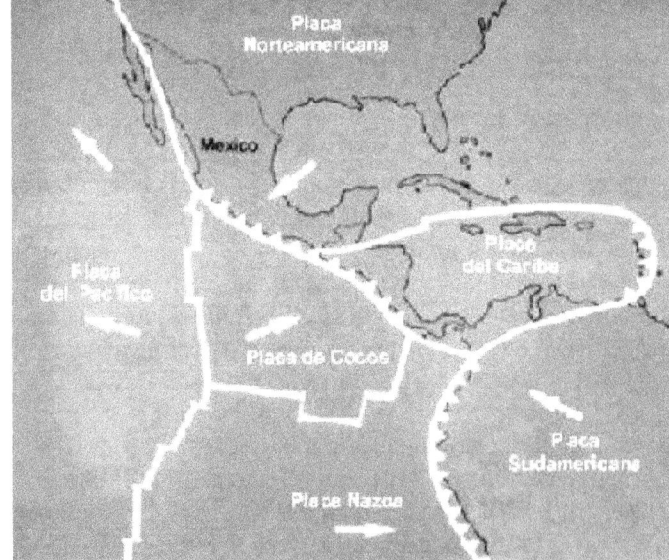

2.2. The 1985 earthquake

On September 19, 1985, at 7:19 local time, an intense earthquake with a magnitude of 8.1 on the Richter scale struck the country. The epicentre was located near the coast of the state of Guerrero, about 400 kilometres southeast of Mexico City, at 17.8 degrees north latitude and 102.3 degrees west longitude. The global area affected by the seismic shock waves was estimated at 800,000 square kilometres making this earthquake one of the most powerful in the history of the country (Fig. 2) [24]. One of the most affected was Mexico City (Fig. 3).

Figure 2: Affected area of the seismic shock waves. [24].

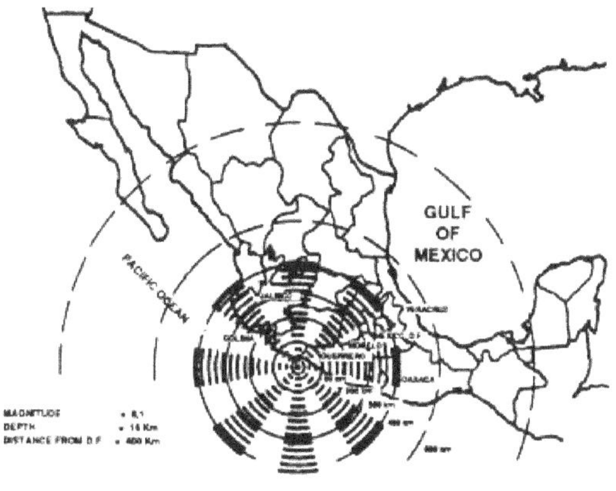

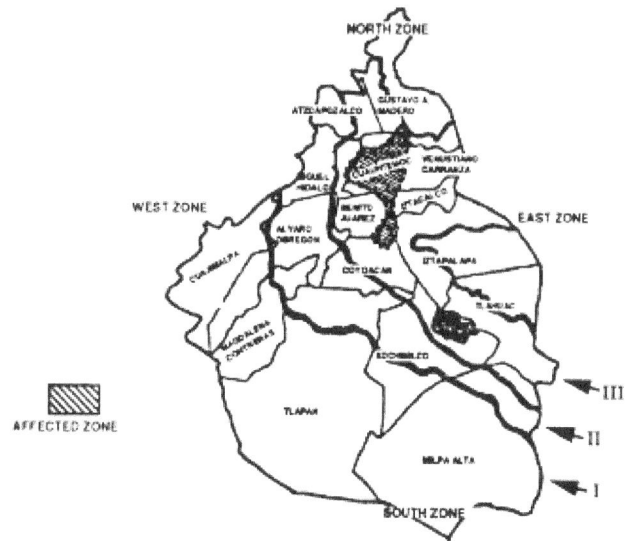

Figure 3: The affected areas of Mexico City. [24].

The following day, at 19:40 p.m. local time, a second seismic movement measuring 7.5 on the Richter scale, with an epicentre at the same place, caused panic in the population despite the fact that the damages were minor when compared to the devastating magnitude of the first. It is believed that during the following 45 days after the earthquakes of September 19 and 20, more than 150 secondary earthquakes were registered, with varying magnitudes between 3.5 and 5 on the Richter scale. [24].

According to the official figures the earthquake caused the death of 6000 people and about 30,000 people were injured, and 150,000 were left homelessness. [23].

3. METHODS

Interviews were conducted by using a version of the questionnaire being designed for a much bigger scale. Essentially the first version of the questionnaire was tested and administered in November 2013 by student interviewers from the Institute. The instrument was divided into three parts: (1) demography (age, gender, occupation, religiosity, economic status, where they live, etc.), (2) perception of seismic risk, and other hazards; the possibility of being affected by it and to suffer losses; the degree of expectation for getting support from authorities, and (3) the level of adaptation to seismic risk, including the education about minimizing risk.

The sample size was randomly composed and the subjects were self-selected as willing to talk about earthquakes. This may not be representative of the City population; however, we are hoping to conduct a probabilistic sample and to make the findings representative for the population under study.

4. THE RESULTS

The results of some of the entries of the survey instrument are presented in the section; the result are descriptive in nature.

4.1. Demographics

Fig. 4 shows the demographic characteristics of the respondents. The sample size was 410 participants. A total of 225 (54.87%) were male and 185 (45.12%) female.

Figure 4: Gender of the participants.

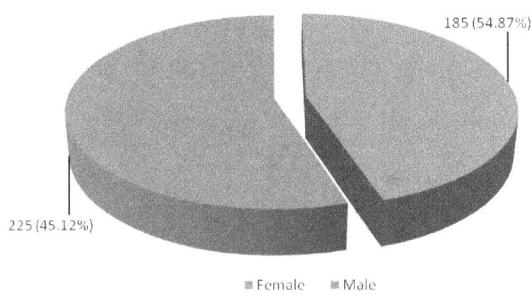

The questionnaire included a question with several categories regarding the age of the respondents. According to the results shown in Fig. 5, it can be seen that most of the respondents were over 43 years old (27.07 %), which could be interesting in the study; this is because these participants have experienced the 1985 earthquake (i.e., a person of 43 years old was about 16 when the earthquake struck the capital City) and their experience is important in our research. There has been some studies that have found that past experience are more proactive, for example, in taking preventive measures. [15,18].

Figure 5: Age of the participants.

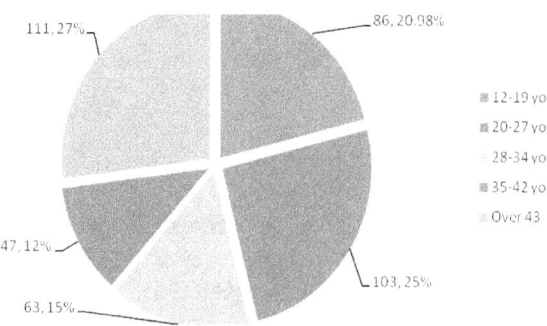

25.12 % of respondents aged between 20 to 27 years old. This percentage of participants did not experienced an earthquake of the magnitude similar to the 1985 earthquake; however, there have been several earthquakes in recent years. A total of 86 (20.98%) teenagers participated in the survey. Their responses are of great interest in our study. According to studies conducted elsewhere show that youngsters with earthquake hazard education, discussed such issues with their parents and this effectively encourage adult participation and preparedness in case of an earthquake [19]. In fact, there is evidence that youngsters that received earthquake education preparedness contributed significantly in sharing their knowledge with their family members during and after the L'Aquila earthquake in 2009 [19].

Fig. 6 shows the level of education of the participants. Education is one of the key variables that needs to be consider when assessing the preparedness of seismic risk. Research has shown that educated people tend to implement proactive measures to seismic risk [15,16]; for example, a study conducted in Turkey [15] showed that well educated people are retrofitting their houses as a proactive measure to withstand earthquakes. In the same study, the results showed that less educated people are not willing to retrofitting their houses.

Figure 6: Educational level of the participants.

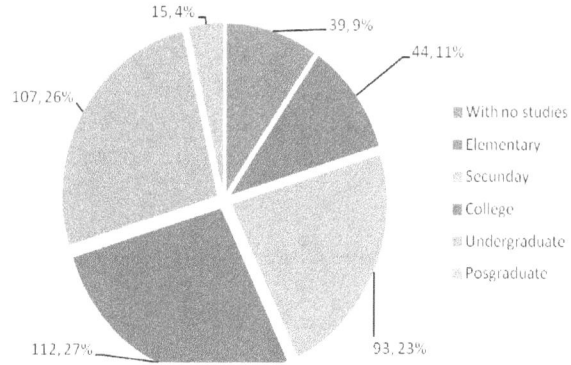

The education level can also contribute to the perception of seismic risk and consequently the required level of awareness of seismic risk. For example, the results of a research conducted on earthquake risk perception in Morocco concluded that less educated people took a fatalist attitude towards seismic risk.; for example, the study found that the less educated "were more likely to deny the significance of scientific assessment and forecasting, and that level of protection from devotion and/or prayer was above all more important and effective [20].

Fig. 6 shows that 79% of the participants have at least the elemental level of education; 20%, on the other hand, do not have any. In our study we are very interested in their seismic risk perception given the fact that they may be more vulnerable to earthquake preparedness.

4.2. Earthquake And Other Hazards Risk Perception

The results of the risk perception to six categories of hazards is shown in Fig. 7. As expected, seismic risk is the highest concern of the participants of the survey with 33% (137). This may be due to the fact that recently there were three earthquakes with magnitudes 6.5, 6.3 & 6.2, on the Richter scale, respectively. Moreover, the news about the devastating consequences of recent earthquakes such as Haiti (2010), China (2008), among others, influenced their seismic risk perceptions.

Figure 7: Risk perception.

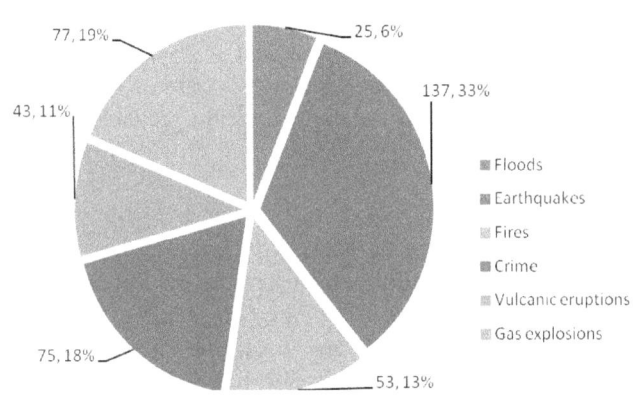

The second concern of the respondents is 'Gas explosions'. This was quite surprising for us given the fact that there are not that many related accidents, at least not in Mexico City. However, one reason for this could be the devastating consequences of two major gas explosion events; the first, an explosion occurred in San Juanico in 1984 where about 500 people lost their lives [25]; The second, an accident related to gas pipeline explosions in Guadalajara in 1992 (212 people were killed, 69 missing, 1470 were injured) [26]. Crime came third in the list; this also was very surprising, given the fact that the crime committed by the drug cartels is very much in the news every day. Fire risk and volcanic eruptions came fourth and fifth, respectively. Finally, 'flood' occupy the last in the list with 6.2 %.

4.3. Knowledge On Seismic Risk And Preparedness

In the following entry: "Many small earthquakes avoid a large one", the participants were asked to respond according to the following options: "False", "True", and "I don´t know". This was included in the questionnaire survey because in Mexico City exists the myth that "it is better to have many small earthquakes instead of a large one"; Fig. 8 shows the results.

46% (189) of the respondents considered the statement as "False"; this could be considered as a correct answer. That is, in order to understand the Richter scale, it may be helpful if it is compared to the energy released by an atomic explosion. It is estimated that an atomic bomb of 13 kilotons (13,000 tons of TNT) releases energy equivalent to a magnitude 5 tremor. While 32 atomic bombs equivalent to the energy released during an earthquake of magnitude 6, so a Grade 8 earthquake energy is equivalent to 32,000 tremors grade 5 so we would have to withstand 32,000 tremors to release the energy of a major event. [23].

Figure 8: Risk perception on the frequency of earthquakes.

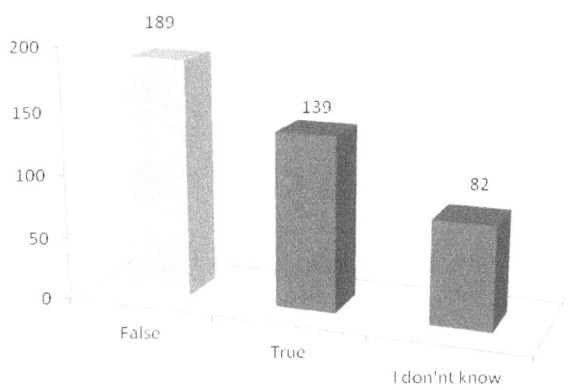

There has also been evidence that supports that this statement is incorrect. For example, in 2009, a 'Risk Committee', composed of scientific experts on earthquakes, met in L' Aquila (Italy) on March 31, 2009 given the fact that there were continuous tremors that shook central Italy. After 45 minutes of discussion, they concluded that there was no real danger. However, on the night of 6th April, an earthquake struck killing 308 people, injured 1,500, 65,000 were left homeless and 20,000 buildings collapsed. [19].

Since then, it has been a great debate in that country about the irresponsibility of the Civil Protection authorities. Recently, it has been reported that [27]:

"The six scientists and a former government official were all members of the Major Risks Committee which met in the central Italian city on March 31, 2009, after several small tremors had been recorded in the region. At the time, they ruled that it was impossible to determine whether the tremors would be followed by a large quake, in a judgment which reassured residents. One of the group famously advised them to relax with a glass of wine. Just six days later, a 6.3 magnitude quake devastated L'Aquila."

The questionnaire included the following statement: "During an earthquake, it is best to leave the building immediately", the participants were asked to respond according to the following options: "False", "True", and "I don´t know". The results are shown in Fig. 9.

Figure 9: What to do during an earthquake.

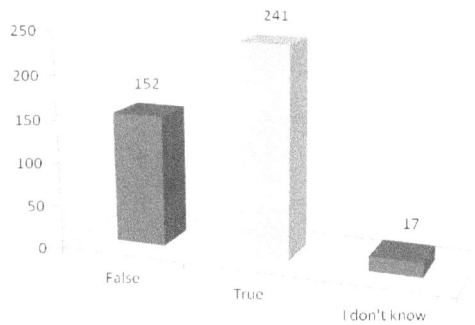

Two of the recommendations issued by the Mexico City's Civil Protection organization about what to do during an earthquake are listed in Table 1.

Table 1: Two examples of what to do during an earthquake.

No	Description
1	Keep calm and get in security zones during an earthquake and try to protect yourself as best as possible. Most of those injured in an earthquake occurred when people tried to enter or leave the house or buildings.
2	If you are in a building , stay where you are , do not try to use the elevators or stairs during the quake.

58.8 %, the majority of respondents, answered as "true" to the statement. Only 37 % responded as 'false' (but this not necessarily means that they knew the answer) .

Figure 10: People who experienced the 1985 earthquake (left) vs people who did not (right).

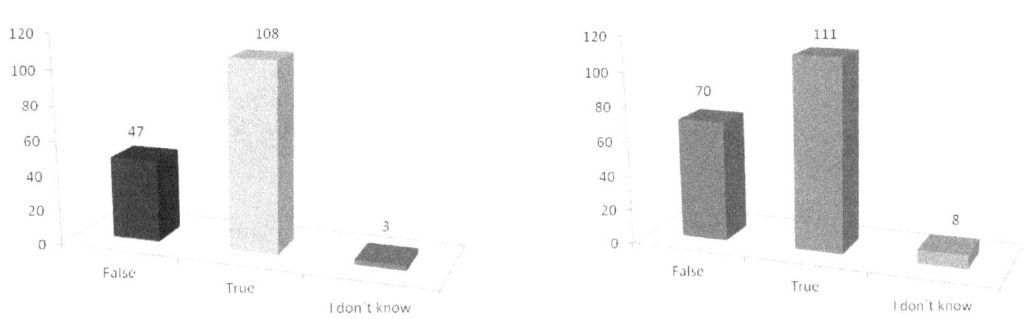

Fig. 10 shows the results for the same statement for those respondents that experienced the 1985 earthquake (Fig. 10- Left) and the respondents that did not (Fig. 10-Right). The results show that the majority of both categories of the respondents have not learned the lesson (108 and 111, respectively). It can be argued that such a recommendation if followed, could save many lives.

Figure 11: Falling objects during an earthquake.

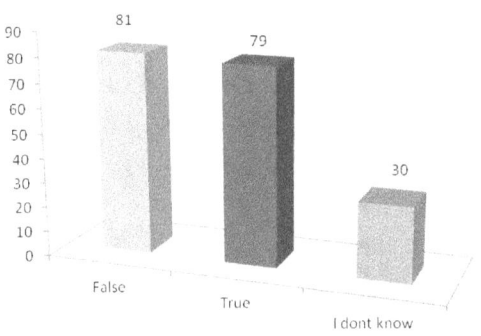

According to all the photographic evidence that have been revised as well as the relevant literature in the present research project on earthquakes worldwide; a key factor appears apparent and it is associated with 'falling objects'. Falling objects may be considered as one of the factors that cause deaths during earthquakes. Given this, the following statement was included in the questionnaire: "Falling objects are the most dangerous during earthquakes". Similarly, the participants were asked to respond according to the following options: "False", "True", and "I don´t know"; the results are shown in Fig. 11. The results show that the majority of the participants are well aware of this fact.

5. CONCLUSION

As mentioned in the Introduction section, the results presented here are not conclusive; however, they represent a starting point for a much bigger scale study. In fact, we tested the instrument and found several limitations in relation to, for example, the wording of the statements were confusing for many of the participants. However, from the results presented here we can say that:

{a} The preliminary results show that the survey participants do not have the culture of prevention in relation to seismic risk.

{b} The participants also showed insufficient knowledge about the right actions to take during an earthquake.

{c} The participants (those who experienced the 1985 earthquake as those who did not) have not learned the basic lesson (as recommended by Civil Protection) not to immediately leave the building during the quake.

{d} Finally, the above sums up well the preliminary conclusion that the sector of the population surveyed are not prepared for these events.

Acknowledgements

The research project was funded by SIP-IPN-No. 20141500.

References

[1] USGS (United States Geological Survey). *"Earthquake Facts and Statistics."* Available at:http://earthquake.usgs.gov/earthquakes/eqarchives/year/eqstats.php. (2014). USA.

[2] USGS. (United States Geological Survey). *"Are Earthquakes Really on the Increase?"* Available at:http://earthquake.usgs.gov/learn/topics/increase_in_earthquakes.php. (2014). USA.

[3] International Federation of Red Cross. *"World Disaster Report."* IFRC, pp. 112-123, (2011), Switzerland.

[4] K. Granger, T. Jones, M. Leiba and G. Scott. *"Community Risk in Cairns: A Provisional Multi Hazard Risk Assessment"*. AGSO Cities Project Report No. 1.Australian Geological Survey Organisation, (1999). Canberra, Australia.

[5] R.H. Turner. *"Earthquake prediction and public policy: Disillusions from a National Academy of Sciences report (1)"*. Mass Emergencies, 1, 179–202. (1976).

[6] D.R. Lehman and S.E. Taylor. *"Date with an earthquake: Coping with a probable, unpredictable disaster"*. Personality and Social Psychology Bulletin, 13, 546–555. (1987).

[7] National Research Council (Ed.). *"Practical lessons from the Loma Prieta earthquake: Report from a symposium sponsored by the geotechnical board and the board on natural disasters of the national research council."* Washington, DC: National Academy Press. (1994), USA.

[8] K.J. Tierney, M.K. Lindell and R.W. Perry. (Eds.). *"Facing the unexpected: disaster preparedness and response in the United States."* Washington, DC: Joseph Henry Press. (2001), USA.

[9] R.H. Turner, J.M. Nigg and D.H. Paz. *"Waiting for disaster: Earthquake watch in California".* Berkeley: University of California Press. (1996), USA.

[10] A.N. Beard. *"Risk assessment assumptions."* Civil Engineering and Environmental Systems, 21(1), pp. 19-31. (2004).

[11] P. Slovic. *"Perception of risk."* Science, vol. 236, 280-285. (1987).

[12] P. Slovic and E. Weber. *"Perception of risk posed by extreme events"*. In: Proc. of Conference on Risk management Strategies in an Uncertain World. New York. (2002), USA.

[13] P. Slovik, B. Fishhoff and S. Lichtenstein. *"Perceived risk: psychological factors and social implications."* In: Proc. of the Royal Society of London, Vol. A376, 17-34. (1981).

[14] M.K. Lindell. *"Perceived characteristics of environmental hazards."* International journal of mass emergencies and disasters, vol. 12, No. 3, 303-326. (1994).

[15] S. Tekeli-Yeşil, N. Dedeoğlu, M. Tanner, C. Braun-Fahrlaender and B. Obrist. *"Individual preparedness and mitigation actions for a predicted earthquake in Istanbul"*. Disasters, 2010, 34(4): 910−930. (2010).

[16] P.B. Kanti and R. Hossain Bhuiyan. *"Urban earthquake hazard: perceived seismic risk and preparedness in Dhaka City, Bangladesh."* Disasters, 2010, 34(2): 337−359. (2010).

[17] E.B. Isabelle, A. Poyaud, P.A. Davoine, S. Chardonnel and C. Lutoff. *"Risk perception and social vulnerability to earthquakes in Grenoble (French Alps)"*, Journal of Risk Research, 15:10, pp. 1245-1260. (2012).

[18] A.O. Taghizadeh, A, Mostafa Hosseini, I. Navidi, A. Mahaki, A., Hassan Ammari and A. Ardalan. *"Knowledge, Attitude and Practice of Tehran's Inhabitants for an Earthquake and Related Determinants."* PLoS Curr. 2012 August 6; 4. (2012).

[19] F. Marincioni, F. Appiotti, M. Ferretti, C. Antinori, P. Melonaro, A. Pusceddu and R. Oreficini-Rosib. *"Perception and Communication of Seismic Risk: The 6 April 2009 L'Aquila Earthquake Case Study."* Earthquake Spectra, Volume 28, No. 1, pp. 159–183. (2012).

[20] R.T. Paradise. *"Perception of seismic risk in a Muslim city,"* The Journal of North African Studies, 11:3, pp. 243-262, USA.

[21] R. Palm. *"Urban earthquake hazards The impacts of culture on perceived risk and response in the USA and Japan."* Applied Geography, Vol. 18. No. I, pp. 3546, (1998).

[22] USGS (United States Geological Survey). *"Earthquake Facts and Statistics."* Available at: http://earthquake.usgs.gov/earthquakes/world/index.php?region=Mexico (2014).

[23] C. Gutierrez, M. Santoyo, R. Quaas, M. Ordaz, y S.K. Singh. *"Fascículo Sismos"*, CENAPRED, 5a. Edición, (2008). México.

[24] OPS. *"Terremoto Ciudad de México 1985."* Organización Panamericana de la Salud (OPS). (1985). Mexico.

[25] IChemE (Institution of Chemical Engineers). *"Explosions in the Process Industries,"* Major Hazards Monograph IchemE. (1994). London, UK.

[26] J. Dugal. *"Guadalajara gas explosion disaster."* Disaster recovery journal. Accessed at: http://www.drj.com/drworld/content/w2_028.htm. (1999), USA.

[27] J. McKenna and N. Collins. *"L'Aquila earthquake scientists sentenced to six years in jail-Six Italian scientists and a government official have been found guilty in a watershed trial of multiple manslaughter for underestimating the risks of the L'Aquila 2009 earthquake."* Daily Telegraph. Accessed at: http://www.telegraph.co.uk/news/worldnews/europe/italy/9626075/LAquila-earthquake-scientists-sentenced-to-six-years-in-jail.html. (2012), UK.

Reliability Analysis of Core Protection Calculator System using Petri Net

Hyejin Kim[*a], Jonghyun Kim[b]

[a] KEPCO Nuclear Fuel, Daejeon-si, Korea
[b] KEPCO International Nuclear Graduate School, Ulsan-si, Korea

Abstract: As digital systems are introduced to nuclear power plants, issues related with reliability analyses of these digital systems are being raised. One of these issues is that static Fault Tree (FT) and Event Tree (ET) approach cannot properly account for dynamic interactions in the digital systems, such as multiple top events, logic loops and time delay. This study proposes an approach to analyzing the reliability of Core Protection Calculator System (CPCS) using Petri Net (PN) modeling. The PN, one of the dynamic methodologies, allows modeling event dependencies and interaction to represent the time sequence and delay time for dynamic events. This study applies the approach to the reliability analysis of CPCS. In order to analyze the digital system modeling, further studies are required with the dynamic modeling methods and the software in the digital system. Modeling of digital systems should be realistic to account for the system characteristics and be able to predict system behavior.

Keywords: Dynamic PSA, Petri Net, CPCS.

1. INTRODUCTION

Digital technology is replacing the analog Instrumentation and Control (I&C) systems in both new and upgraded nuclear power plants. As digital systems are introduced to nuclear power plants, issues related with reliability analyses of these digital systems are being raised. One of these issues is that static FT and ET approach cannot properly account for dynamic interactions in the digital systems, such as multiple top events, logic loops and time delay [1].

Given the limitations of static modeling methods when applied to dynamic systems, several studies on dynamic modeling methods have been conducted. Most research has involved existing modeling methods in order to integrate easily with static methods, which have already been widely implemented in the system modeling field. Some of dynamic methods have been upgraded making it possible to analyze the dynamic interactions between components of dynamic systems [2].

The presence of the interaction among complex hardware, software and physical processes in digital I&C systems may necessitate the use of dynamic methodologies for dependable results. Dynamic methodologies are defined as those that can account for the coupling between the triggered or stochastic logical events in system reliability modeling, through explicit consideration of the time element in system evolution. Many methods have been proposed to solve the problems, but there is no single method that is universally accepted for the application to the current generation Probabilistic Safety Analysis (PSA)[3].

On the other hand, some assumptions are used to analyze the reliability of Reactor Protection System (RPS), which is one of the digital I&C system, there is no consideration except for ex-core signal as an input for Core Protection Calculator System (CPCS) to avoid the complexity for modeling the RPS system in Shin-kori 3&4 PSA Report [4]. CPCS is a digital computer system which continuously calculates Departure from Nucleate Boiling Ratio (DNBR) and Linear Power Density (LPD) to initiate a reactor trip when needed during certain transients to prevent violation of the DNB and LPD safety limits.

In this study, PN is extended to model system failures. PN method for qualitative and quantitative

---

[*] *Corresponding author : heyjin@knfc.co.kr*

analysis of CPCS in Advanced Power Reactor 1400 (APR1400) is presented.

2. PN MODELING METHOD

The PN, one of the dynamic methodologies, allows modelling the event dependencies and interaction, to represent the time sequence and delay time, and to model assumptions for dynamic events [5].

The PN is a directed graph consisting of two types of nodes, called places and transitions. Systems are modeled as a set of conditions and events. Places represent conditions in the process, and transitions represent events. Transitions can be immediate, deterministically time-delayed, or time-delayed based on a probability distribution defined by the user [5]. Also, the PN model allows explicit representation of the time elements of system with the use of a dynamic system model and subsequently is capable of simulation of concurrent and dynamic activities and time-delays [6].

Events such as 'transmitter fails' are represented by nodes. Arcs connect either transitions to nodes or nodes to transitions. It uses tokens that can move when the PN is executed for the representation information flow through the net. A token moves from a node or place and is consumed by a transition. When a transition fires, it produces tokens in places that it connects to and consumes one token in each of the places that connect to it. In order for a transition to fire it must have at least one token on each of its input places. Transition delays and timed transitions can be represented [6]. The state of a net is modeled by the presence or absence of a token in the places. An event occurs only when the preconditions are met and is represented by an enabled transition. The firing of a transition changes the marking of its input and output places, modeling a change in its precondition and post-conditions.

PN consists of four basic elements [4] as stated in Table 1:

Table 1: Basic Elements of PN

Figure	Name of element	Drawn as	denotation
⬭	Place	Circle	Event
▭	Transition	Cube or Bar	Event transfer
→	Arc	Arrow	Transfer between places and transitions
•	Token	Dot	Data

Places indicate failures, transition expresses the event transfer and delay time, and token indicates the condition of failure in the PN modeling in this study.

3. CASE STUDIES: RELIABILITY ANALYSIS OF CPCS

3.1 CPCS Design Basis [7]

The CPC design basis requires that the system calculate conservative, but relatively accurate, values of Departure from Nucleate Boiling Ratio (DNBR) and Local Power Density (LPD). In order to achieve a system time response sufficient to accommodate the limiting design basis events, additional dynamic calculations of DNBR and LPD are required. The dynamic calculations must provide conservative estimates of DNBR and peak linear heat rate based on changes in the process variables between successive detailed calculations of DNBR and LPD.

The resultant protection software consists of six interdependent programs, five of which are resident in the CPC processor, and the sixth in each Control Element Assembly Calculator (CEAC) processor:

- Coolant Mass Flow Program (FLOW)
- DNBR and Power Density Update Program (UPDATE)
- Power Distribution Program (POWER)
- Static DNBR and Power Density Program (STATIC)
- Trip Sequence Program (TRIPSEQ)
- CEAC Penalty Factor (PF) Program (CEAC).

In the CPCS, the TRIPSEQ Program shall compare the DNBR and LPD to their respective pretrip and trip setpoints. Whenever a setpoint is violated, the appropriate contact output is actuated.

Figure 1 shows the CPCS Algorithm Diagram among the six programs.

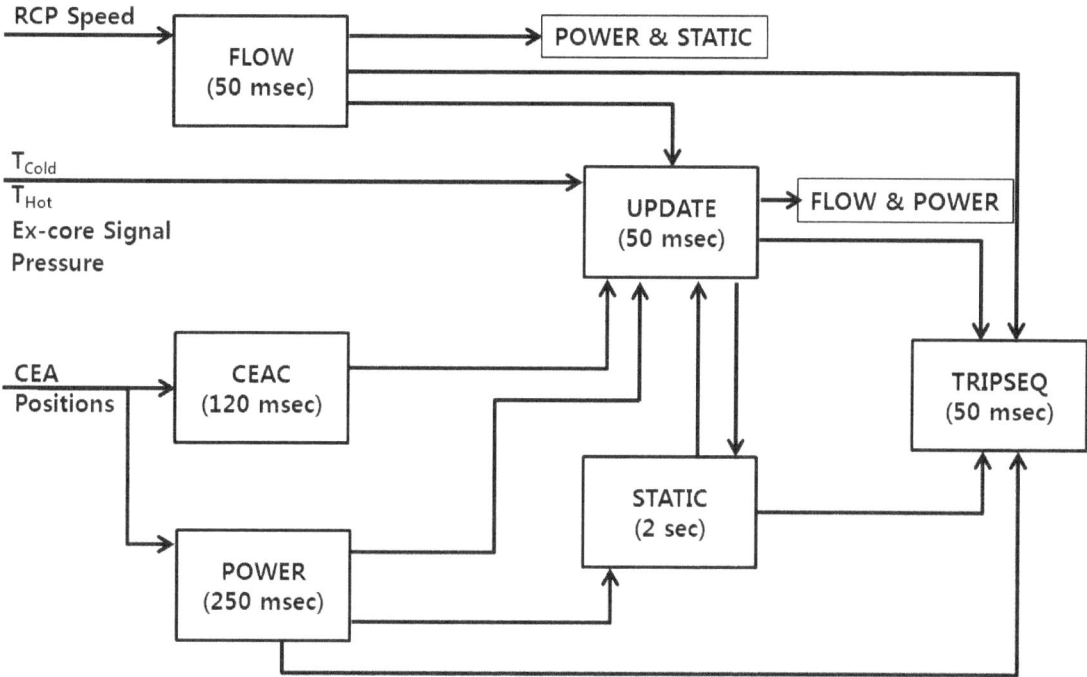

Figure 1: CPCS Algorithm Diagram

3.2 PN Modeling

This paper applies the proposed approach to analyzing the reliability of CPCS. More specifically, the scope of modeling is processing the pressurizer pressure signal for generating the DNBR trip signal to Plant Protection System (PPS). PN modeling includes the transmitter, converter, analog input card module, processing module, and contact to PPS. Continuously repeated execution is expressed using the place 'E' and arc 'e'. Some of the delay times are expressed as 'DT1' and 'DT2' for convenience in Figure 2.

Figure 2 shows the PN to represent the failure of trip signals to PPS from pressurizer pressure transmitter for pressurizer pressure signals. The transmitter senses the pressure inputs and transforms it to the current signal. The I/E converter transforms the signal from current to voltage. The AI685 analog input card module continuously scans, stores, and transforms the analog input to digital values.

The UPDATE program reads the digital values from the AI685 and calculates the DNBR, and the TRIPSEQ program compares the DNBR to its setpoint values. If the DNBR is lower than the setpoint, a trip signal is generated.

Figure 2 include: 1) the time sequence from pressurizer pressure transmitter to contact to PPS, 2) the interaction between hardware and software, 3) the time-delay to process the input signals, e.g. T5 and T7, 4) the continuous scanning and memory update, e.g. P8, and 5) the execution of processing modules, e.g. P7.

This modeling was done by using the Colored PN Ver. 4.0. [8].

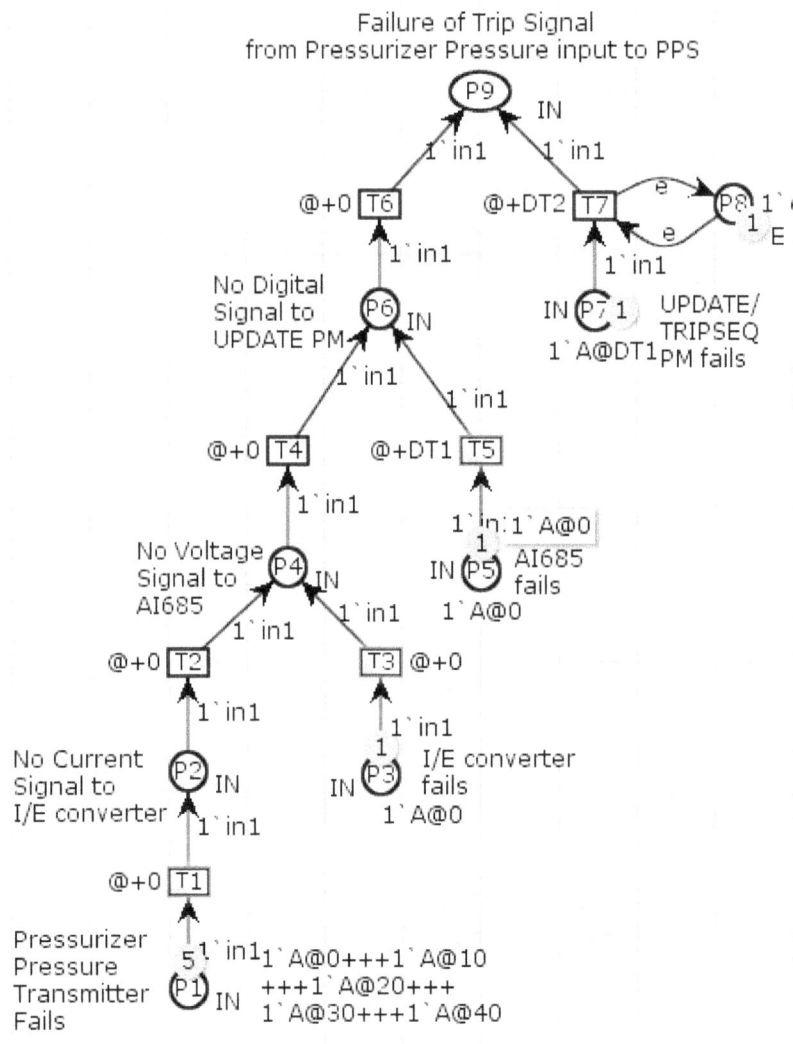

Figure 2: PN modeling for Pressurizer Pressure signal

3.3 Interaction

The presence of the interaction among complex hardware, software and physical processes in digital I&C systems may necessitate the use of dynamic methodologies for dependable results. The PN allows modeling event dependencies and interaction to represent the time sequence and delay time for dynamic events.

Figure 3 represents the interaction among Hardware/Software/Processing modules in CPCS. The interaction 1) between hardware, e.g. P1, P2, P3, P4, P5 and P6, and software, e.g. P16, and 2) among the processing modules, e.g. P11, P13, P15, P16, P19, and P20, are presented.

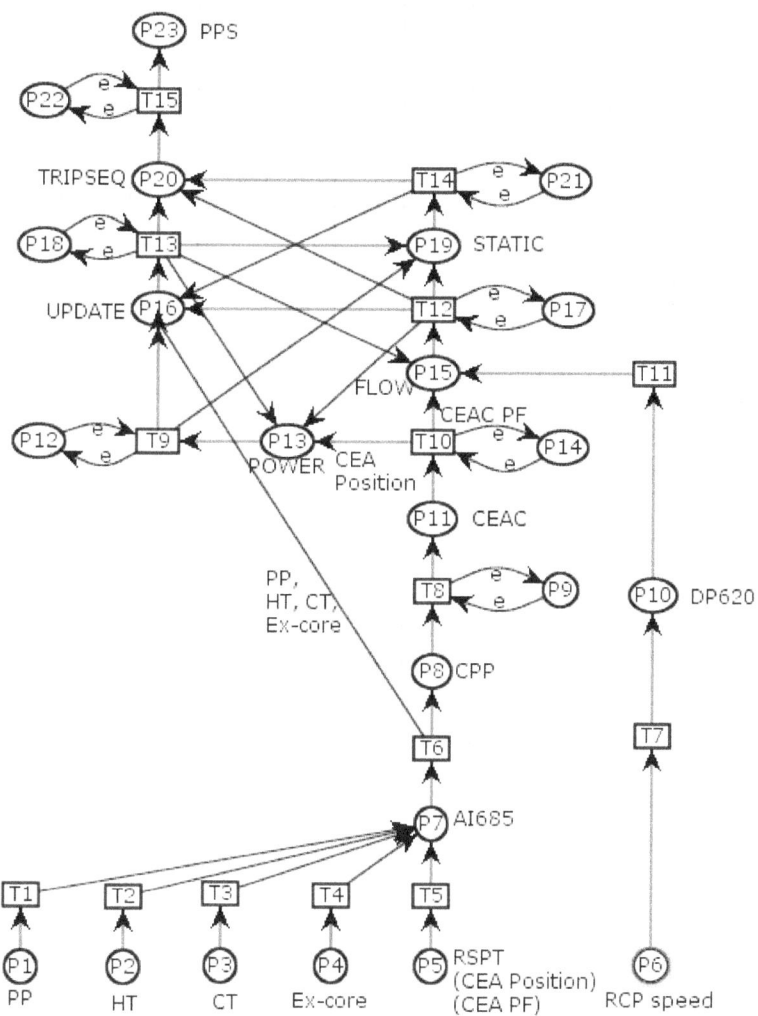

Figure 3: Interaction among Hardware/Software/Processing module in CPCS

3.4 PN Quantification

3.4.1 Failure Probability Calculation

As long as the failure probability of basic events in a fault tree is given, the reliability or failure probability of the top event can he calculated. It is known that calculating failure probability for a fault tree depends on gates. However, for PN it depends on symbols of transitions, places and arcs [9].

The failure probability P_f in case of a transition with multi-input places can be written as:

$$P_f = \prod_{i=1}^{n} P_{fi} \qquad (1)$$

where, P_{fi} denotes the failure probability of the i_{th} event.

By contrast, the failure probability for a place with multi-input transitions is written as:

$$P_f = 1 - \prod_{i=1}^{n}(1 - P_{fi}) \qquad (2)$$

It is not necessary to calculate for the transition or place with single input.

3.3.2 Minimal Cut Set – Matrix Method

Minimal cut sets can be found at the same time using the present matrix method to analyze the PN from a top place to basic places. This method proceeds as follows [9]:

1. Write down the numbers of places by making a horizontal arrangement if the output place is connected by multi-arcs to transitions.
2. Write down the numbers of places by making a vertical arrangement if the output place is connected by an arc to a common transition.
3. When all places are replaced by basic places, a matrix is established. If there is common entry located between rows or columns, it is the entry shared for each row or column, the column vectors of the matrix represent cut sets.
4. Remove the supersets to obtain the minimal cut sets.

From the PN modeling, cutsets and failure probability are calculated using the matrix methods [9].

Figure 4 gives minimal cutsets for the PN depicted in Figure 2. Consequently, minimal cut sets are [P1], [P3], [P5], and [P7] and they are obtained by using matrix method.

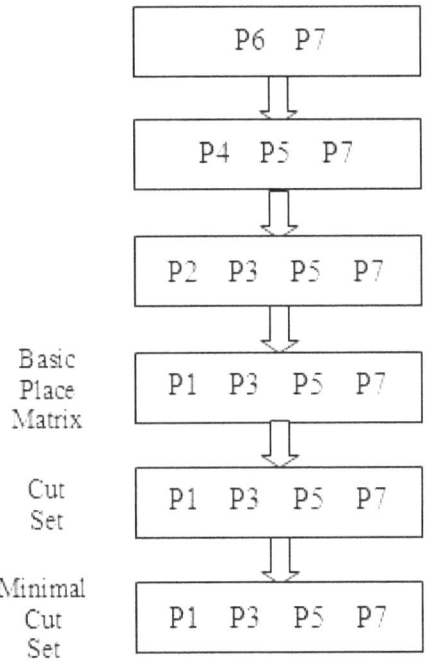

Figure 4: Matrix method for Pressurizer Pressure Input

The failure probability for a place with multi-input transitions is written as:

$$P_f = 1 - \prod_{i=1}^{n}(1 - P_{fi}) = 1 - (1 - P1)(1 - P3)(1 - P5)(1 - P7) \qquad (3)$$

Using the methods mentioned above, failure probability for CPCS is calculated. Some failure probability data are used as it is in Shin-kori 3&4 PSA Report [4] and some are assumed. Software failure probability is assumed that the contribution from software failure to total failure probability is 10% of the hardware failure probabilities [10].

Following the failure probability data and equations, failure probability for the "Failure of Trip Signal (DNBR) from CPCS input signals to PPS is calculated as 3.73605E-3.

4. CONCLUSION

Until a recent date, static modelling, e.g. FT and ET, are extensively used in PSA to model and evaluate the probability and consequence of failures of nuclear power plants. Given the limitations of static modeling methods when applied to dynamic systems, several studies on dynamic modeling methods have been conducted.

To overcome the limitations of static modeling methodologies for the digital system, this study proposes an approach to analyzing the reliability of CPCS using PN method. Both qualitative, e.g. PN modelling, and quantitative, e.g. failure probability calculation, analyses for CPCS in APR1400 are presented. Failure probability and minimal cutset are also presented from the PN modeling by using the equation and matrix method.

In order to analyze the digital system modeling, further studies are required with the dynamic modeling methods. In addition, the study on the software in the digital system are also recommended to consider the precise failure probability, common cause failure, human error, software HAZOP (Hazard and Operability) results, and Failure Mode Effect Analysis (FMEA) for the reliability analysis. Modeling of digital systems should be realistic to account for the system characteristics and be able to predict system behavior.

References

[1]   US NRC, "Dynamic Reliability Modeling of Digital Instrumentation and Control Systems for Nuclear Reactor Probabilistic Risk Assessments", NUREG/CR-6942, (2000).
[2]   Shin, S.K. and Seong, P.H., "Review of various dynamic modeling methods and development of an intuitive modeling method for dynamic systems", Nuclear Engineering and Technology, Vol. 40, No.5, (2008).
[3]   Lu, L, "An Overview of Digital I&C System Reliability Analysis in Nuclear Power Plants", NPIC&HMIT, Albuquerque, (2006).
[4]   KHNP, "Shin Kori 3&4 Probabilistic Safety Assessment (PSA) report", (2011).
[5]   Lee, A and Lu, L, "Petri Net Modeling for Probabilistic Safety Assessment and its application in the air lock system of a CANDU nuclear power plant", Procedia Engineering, Vol. 45, pp11 – 20, (2012).
[6]   US NRC, "Dynamic Reliability Modeling of Digital Instrumentation and Control Systems for Nuclear Reactor Probabilistic Risk Assessments", NUREG/CR-6901, (2006).
[7]   KEPCO Nuclear Fuel, "Functional Design Requirements for a Core Protection Calculator System for Shinkori Nuclear Power Plant Units 3&4", (2010).
[8]   Coloured Petri Net, Version 4.0. http://cpntools.org/.
[9]   Liu, T.S. and Chiou, S.B, "The Application of Petri Nets to Failure Analysis", Reliability Engineering and System Safety, Vol. 57, pp. 129-142, (1997).

[10] Authen, S. and Holmberg, J. E, "Reliability Analysis of Digital Systems in a Probabilistic Risk Analysis for Nuclear Power Plants", Nuclear Engineering and Technology, Vol. 44, No.5, (2012).

Degradation Modeling and Algorithm for On-line System Health Management using Dynamic Hybrid Bayesian Network

Chonlagarn Iamsumang, Ali Mosleh, Mohammad Modarres

The Center for Risk and Reliability

University of Maryland College Park, Maryland, USA

Abstract: This paper presents a new modeling method and computational algorithm for reliability inference with dynamic hybrid Bayesian network. It features a component-based algorithm and structure to represent complex engineering systems characterized by discrete functional states (including degraded states), and models of underlying physics of failure, with continuous variables. The methodology is designed to be flexible and intuitive, and scalable from small localized functionality to large complex dynamic systems. In System Health Management applications, this method introduces a well-defined interface between continuous system component status and discrete system functionality within the network model. Markov Chain Monte Carlo (MCMC) inference is optimized using pre-computation and dynamic programming for real-time monitoring of system health. The scope of this research includes new modeling approach, computation algorithm, and an example application for on-line System Health Management.

Keywords: On-line System Health Management, Dynamic Hybrid Bayesian Network

1. INTRODUCTION

With increasing complexity of today's engineering systems that contain various component dependencies and degradation behaviors, there has been increasing interest in real-time System Health Management (SHM) capability to continuously monitor sensors, software, and hardware components for detection and diagnostic of safety-critical systems. The modeling framework should be flexible to accommodate the complexity of component dependencies and failure behaviors, such as sequence-dependent failures, functional dependencies, etc.

Bayesian Networks (BN) [1][2] and their extension for time-series modeling known as Dynamic Bayesian Network (DBN) [3][4] have been shown by recent studies to be capable of providing a unified framework for system health diagnosis and prognosis [5][6][7]. Bayesian Network has many modeling features, such as multi-state variables, noisy gates, dependent failures, and general posterior analysis [8][9][10]. It also allows a compact representation of the temporal and functional dependencies among system components [11][12].

The main advantage of using BN in system reliability is its simplicity to represent systems and the efficiency for obtaining component associations. Another important benefit of BNs is that they enable us to integrate information from different sources, including experimental data, historical data, and prior expert opinion. This feature is particularly useful for the reliability assessment of fault tolerant systems, where failure data from tests and field operations are sparse and obtained from diverse source of information. Bayesian networks are particularly well suited to modeling systems that we need to monitor, diagnose, and make predictions about, all under the presence of uncertainty.

However, one of the barriers to applying BN to real-world problems is to be able to adequately handle the "hybrid models", which contain both discrete and continuous variables with general static and time-dependent failure distributions. Despite the advances in BN researches, the previous applications of BNs as mainstream technology for SHM problems remain modest. To date, the BN framework has only partially addressed these limitations [13][14][15][16]. The vast majority of BNs used in real world applications are either purely discrete or purely continuous.

For hybrid BNs containing mixtures of discrete and continuous nodes with non-Gaussian distributions, exact inference becomes computationally intractable [17]. The common approach to handling (non-

Gaussian) continuous nodes is to discretize them using some pre-defined range and intervals [18]. This is cumbersome, error prone and usually inaccurate.

Even though a universal framework for hybrid BN is currently impracticable, a special case algorithm can be effective in SHM where a relatively small subset of possible values covers a large proportion of all possible values typically encountered. This paper presents a hybrid BN-based methodology for component degradation model and efficient algorithms to apply them in on-line health monitoring of complex systems.

The focus of this research is to enable probabilistic diagnosis and prognosis of system in real-time by optimizing SHM modeling and Markov Chain Monte Carlo inference with pre-computation and dynamic programming to reduce the computation time and number of inferences required. Efficient computation allows on-line system monitoring and provides on-demand system health inquiry for operators to make maintenance decision and to prioritize which part of the system to investigate to avoid an accident.

2. PROPOSED METHODOLOGY AND ALGORITHM

2.1. Hybrid Bayesian Network

For SHM modeling, it is advantageous and intuitive to consider a hybrid system, typically with the continuous variables being modeled as continuous and the system's functionality probability being discrete.

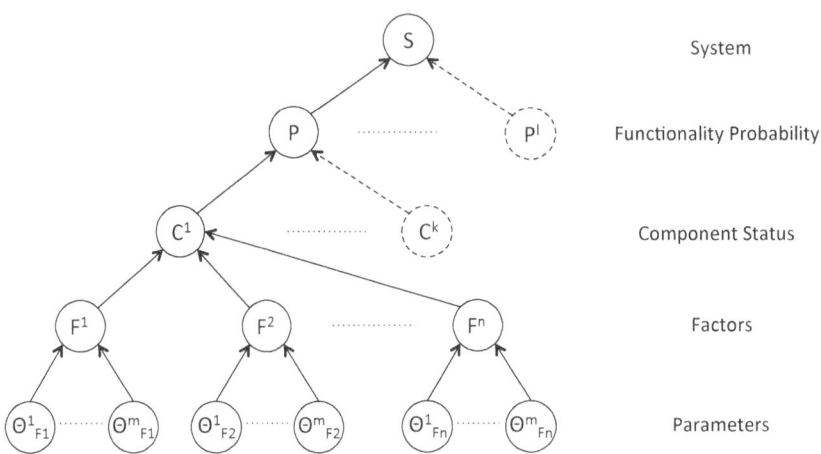

Figure 1: Overview of different levels in SHM Bayesian Network

The proposed complex system hybrid BN can be separated into 5 levels as shown in Figure 1, according to the typical characteristics of the nodes. The BN combines high-level functionality nodes with low-level physical of failure nodes. Here are the descriptions of each level:

1. System node: this is the highest level of the BN nodes (it has no children), it represents the state of the whole system and usually indicates whether or not the system is working as intended.
2. Functionality probability nodes: these nodes are designed to be abstract discrete nodes that represent various functionalities, which are required for the system to operate. The node can be requirement for operation of a single component, or multiple components.
3. Component status nodes: these are continuous nodes representing states of physical components susceptible to specific failure mechanisms in the system. These values should be measurable directly or indirectly, and they are expected to degrade over the lives of the components.

4. Factor nodes: these nodes contribute to the degradation of the components. They can be component internal factors related to material properties or physical characters, or they can be external factors such as environmental stresses or temperature.
5. Parameter nodes: these nodes are hyper-parameters that describe probability distributions of the factors.

It is to be noted that each level of the HBN could by itself be represented as a complex BN model. It can contain a combination of different layers of nodes that have the same type.

Reliability concerns arise when some critically important materials or devices degrade with time. Let C represent a critically important material/device parameter. This parameter degrades over the life of the component. The value itself can either increase (threshold voltage of a semiconductor device, increase in leakage of a capacitor, increase in resistance of a conductor) or decrease (decrease of pressure in a vessel, decrease of spacing between mechanical components, decrease in lubricating properties of a fluid). Figure 2 presents the SHM BN at a specific time, t. The shaded areas show continuous nodes that are related to each component.

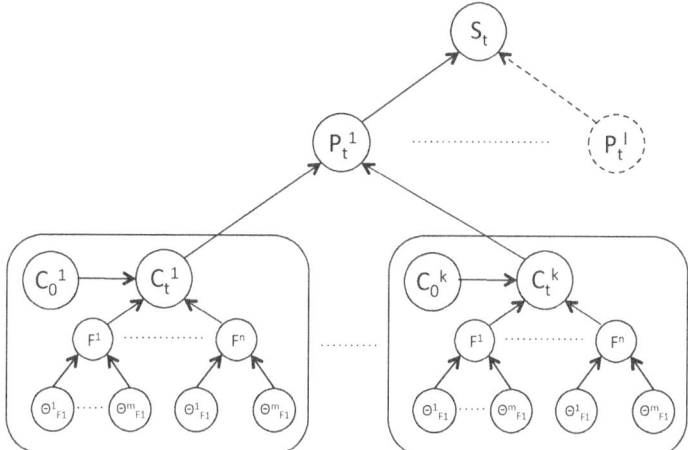

Figure 2: SHM Bayesian network at specific time t.

A Taylor expansion about t=0 produces the Maclaurin Series, assuming that C changes monotonically and relatively slowly over the lifetime of the material/device:

$$C(t) = C_{t=0} + \left(\frac{\partial C}{\partial t}\right)_{t=0} t + \frac{1}{2}\left(\frac{\partial^2 C}{\partial t^2}\right)_{t=0} t^2 + \cdots \tag{1}$$

By assuming that the higher order terms in the expansion can be approximated by simply modeling degradation of component/device parameter C with a power-law equation:

$$C = C_0[1 \pm A_0(F_1, \dots, F_n)t^m] \tag{2}$$

where, C_0 is the value of C at $t = 0$. Summation (+) is used when the parameter C increases with time, while subtraction (-) is used when the parameter C decreases with time. Parameter A_0 is generally material/microstructure dependent. It is not only a function of material variations, but also a function of other factors, such electrical, thermal, mechanical and chemical environments to which the device is exposed. The parameter m and other parameters are considered to be constant for the component/device. Considering a BN at a time slice of a given system, t indicates the current life of the component/device.

For a component/device to fail, the amount of degradation must reach a critical value, C_{crit}. Therefore, the time to failure, $T_{failure}$, is then:

$$T_{failure} = \left[\frac{1}{\pm A_0(F_1, \dots, F_n)}\left(\frac{C_{crit} - C_0}{C_0}\right)\right]^{1/m} \tag{3}$$

Since the component status and their parents are continuous nodes, and the functionality probability nodes are discrete, the interface between these different types of nodes becomes critical. In general hybrid BN when continuous nodes have discrete parents, there are simple conditional inference techniques such as in conditional linear Gaussian (CLG) model. Difficulty arises when discrete nodes have continuous parents, which is the case for our SHM network. However in this case, even though discrete functionality probability nodes have continuous component status nodes, they are related by degradation thresholds.

Discrete functionality nodes can contain more than 2 states with thresholds between the transitions of one state to the other. Let the threshold value between functionality state i and j be $C_{th,i/j}$. The most common case would be state i denotes the component function, and state j denotes the component does not function. Let P_i be the probability of functionality being in state i. The probability P_i is then the probability that the component status C is lower than the threshold value $C_{th,i/j}$. Figure 3 shows a typical component exponential degradation function and the overlap of probability distributions of C and $C_{th,i/j}$.

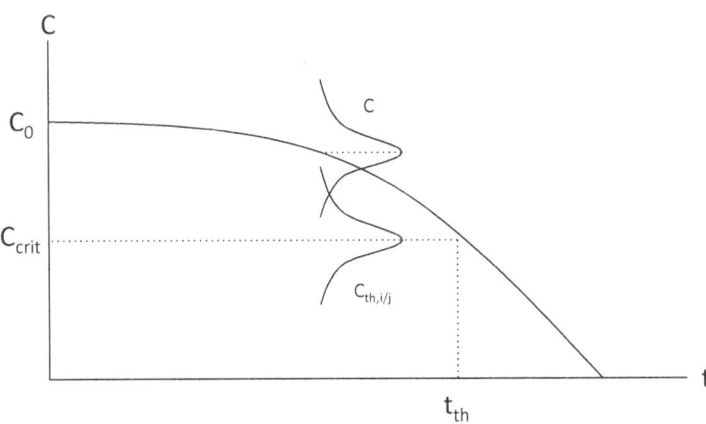

Figure 3: Overlap of probability distribution of component status and its threshold.

Let a functionality node has n states, the probabilities of being in the states are $P_1, \dots, P_n$. Assume the state of the functionality node changes monotonically according to the component degradation status:

$$C_{th,i-1/i} < C_{th,i/i+1} \text{ for } i = 2, \dots, n-1 \tag{4}$$

Therefore,

$$P_i = prob\left(C_{th,i-1/i} < C < C_{th,i/i+1}\right) \tag{5}$$

Analytically, P_i can be calculated from the following convolution equation:

$$P_i = \int_{-\infty}^{C_{th,i/i+1}} \int_{C_{th,i-1/i}}^{\infty} \int_{C_{th,i-1/i}}^{C_{th,i/i+1}} p\left(C_{th,i-1/i}\right) \cdot p(C) \cdot p\left(C_{th,i/i+1}\right) dC \, dC_{th,i/i+1} \, dC_{th,i-1/i} \tag{6}$$

If there are k component status parameters contribute to this functionality then the state of the functionality node conditionally depends on comparison between the status of each component and its threshold values.

$$P_{i^1...i^n} = prob\left(C^1_{th,i^1-1/i^1} < C^1 < C^1_{th,i^1/i^1+1}, ..., C^n_{th,i^n-1/i^n} < C^n < C^n_{th,i^n/i^n+1}\right) \qquad (7)$$

2.2. Dynamic Bayesian Network

Dynamic Bayesian Network (DBN) is a Bayesian network that includes a temporal dimension. This new dimension is managed by time-indexed random value t to indicate time stage of the nodes. A set of nodes at certain stage contains random variables relative to time slice t. An arc that links two variables belonging to different time slices represents a temporal probabilistic dependence between these variables. Variables can be modeled to have impact on the future distribution of the other variables. These impacts are defined as transition probabilities between the stats of variables at time step t and $t + \Delta t$.

A DBN describes the joint distribution of a set of variables $\boldsymbol{\theta}$. This is a complex distribution, but may be simplified by using the Markov assumption. The Markov assumption requires only the present state of the variables $\boldsymbol{\theta}_t$ to estimate $\boldsymbol{\theta}_{t+1}$, i.e. $p(\boldsymbol{\theta}_{t+1}|\boldsymbol{\theta}_0,...,\boldsymbol{\theta}_t) = p(\boldsymbol{\theta}_{t+1}|\boldsymbol{\theta}_t)$ where p indicates a probability density function and bold letters indicate a vector quantity. Additionally, the process is assumed to be stationary, meaning that $p(\boldsymbol{\theta}_{t+1}|\boldsymbol{\theta}_t)$ is independent of t.

For SHM Bayesian network, the main variables that change between time slices are component parameters. Components degrades over time, therefore, the status of components at a certain time slice depend on their status at the previous time slice and the factors affecting the degradation processes during that transition.

$$p(C_t) = p(C|C_{t-\Delta t}, \{F_t^1, ..., F_t^n\}) \qquad (8)$$

Given that F_t^i is the average value of factor i between time slice $t - \Delta t$ and t. Δt should be set according to the system under interest and how often the parameters can be observed, such as frequency of sensor signals. The benefit of continuous monitoring and inferences of variable that cannot be observed directly is to detect anomaly in the system before failure actually occur. Figure 4 shows a two-time-slice representation of a dynamic SHM Bayesian network.

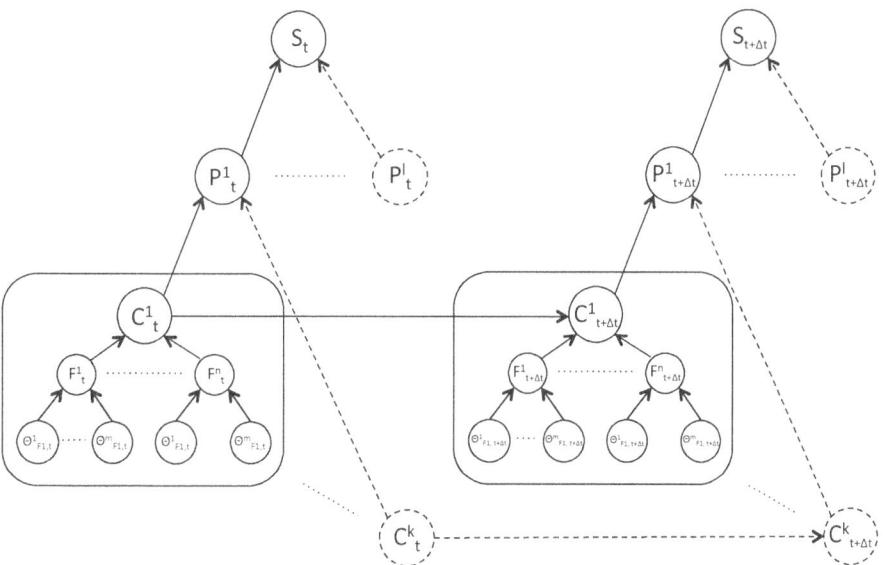

Figure 4: Two-time-slice representation of a dynamic SHM Bayesian network

At any point in time during system operation, any value of variables in the system can be derived by probabilistic inference to compare with its expected value to see if the probability is still in the acceptable range and the system as a whole is working as intended. With continuous monitoring, the trajectory of the degradation processes can be estimated form our knowledge of the health of the system. We can then use this information to estimate remaining useful life (RUL) of components and plan maintenance accordingly.

2.3. Inference

Bayesian network is a complete model for the variables and their relationships. Therefore, it can be used to answer probabilistic queries about them. The main application is to use BN to realize updated knowledge of the state of a subset of variables when other variables (the evidence variables) are observed.

Bayes' rule with continuous variables:

$$p(\theta|D) = \frac{p(D|\theta)p(\theta)}{\int d\theta\, p(D|\theta)p(\theta)} \tag{9}$$

Let θ be a parameter value and D is data value of the evidence, $p(\theta|D)$ is then the posterior probability of getting parameter value θ when data value D is presented.

In real world SHM applications, there are various types of parameter distributions, which make it difficult to calculate full marginal distributions analytically. Therefore, sampling techniques can be used to approximate the distributions instead. Expected values of a distribution can be estimated as follow:

$$E[p(\theta|D)] \approx \frac{1}{N}\sum_{n=1}^{N} p\left(\theta^{(n)}|D\right) \tag{10}$$

Where $\theta^{(1)}, \dots, \theta^{(n)}$ are the sample values of parameter θ.

There are many ways to sample these values, the key idea is to let θ values be points in state space and find a way to walk around so that the likelihood of visiting any point θ is proportional to $p(\theta)$. Therefore, the sampler will spend more time sampling from the distribution where the probability is high, and spending less time sampling from where the probability is low. This can be achieved by using Markov chain Monte Carlo (MCMC) algorithm [19][20].

The procedure for updating the belief about the system state as new information becomes available is called Bayesian recursive filtering.

$$p(\theta_t|D_{1:t}) = \frac{p(D_t|\theta_t)p(\theta_t|D_{1:t-1})}{\int d\theta\, p(D_t|\theta_t)p(\theta_t|D_{1:t-1})} \tag{11}$$

Under certain assumptions, such as when the system is linear Gaussian, the belief state will be of a known parametric form and computationally efficient solutions to the filtering problem (e.g. Kalman filter, extended Kalman filter, unscented Kalman filter) are available. Outside such assumptions, a computationally feasible method for inference in the DBN is particle filtering, a form of sequential Monte Carlo based on Bayesian recursive filtering. Common particle filtering methods are based on sequential importance sampling (SIS) [21].

2.4. Computational Algorithm for On-line SHM

In highly complex systems, MCMC algorithm requires large amount of computational time for inference in hybrid DBN. The computation time grows exponentially with each additional layer of network and becomes infeasible with large number nodes. The computation time makes it impossible for on-line health monitoring of complex systems. To solve this problem, special case algorithm for SHM is introduced to reduce the number of computations and the amount of time required for each computation.

One of the main characteristics of SHM in contrast of other applications is that during a normal operation, the environmental factors that affect component degradation process are expected to be roughly the same and predictable. Therefore, instead of performing Bayesian updating at a specific time interval, it only needs to be done when a factor value changes outside of expected range.

$$|f_t - f_{t-1}| > \epsilon_f \tag{12}$$

Where ϵ_f depends on the sensitivity of component status due to the change in value of that factor. Please note that this is possible because component status is a function of time. Therefore, the degradation of a component between time period t_i to t_j where the change in factor value is less than ϵ_f will take a normal distribution $\mathcal{N}(\mu_f, \sigma_f)$ for $\Delta t = t_j - t_i$.

Since the values are predicted to be in certain ranges, it is possible to perform pre-computation for all combinations of possible values in the ranges before the system is in operation. The results are then stored in a database, such that they can be pulled quickly to approximate the inferences in real-time. More computation should be conducted and more results should be added to the database as the health of the system is being monitored such that the database will cover all the possible computations that may be needed in the future.

With continuous range of parameter values, it is impossible to pre-compute every possible outcome. The goal of pre-computation is to cover enough values of observable parameters, so that the values of unobservable parameters can be accurately interpolated from the results. There are two factors in considering the selection of possible values.

First is the range of observable parameters after a time period Δt. The selections should cover full range of possible values. There should be at least one selected value at lower bound and one selected value at upper bound. The common range is from 5th percentile to 95th percentile, or more accurately 0.5th percentile to 99.5th percentile.

Second is the number of selections within the bound: the higher the number of selections, the more accurate results from interpolation will be. The density of selections should be proportional to the probabilistic density of the observable parameters. For example, if 19 values should be selected, then they should be the values at 5th, 10th, ... , 90th, 95th percentiles. Therefore, for a given measurement interval Δt, we can estimate the set of possible values and use those values to pre-computed possible outcomes.

There are two different types of observable parameters. The first one is the parameters that change over time. This is usually the case for component status parameters. For pre-computation to be feasible, the changes must be predictable. For a component status parameter, the change in value can be computed from its degradation equation for a given Δt. Figure 5 shows example expected value, 5th percentile, and 95th percentile values.

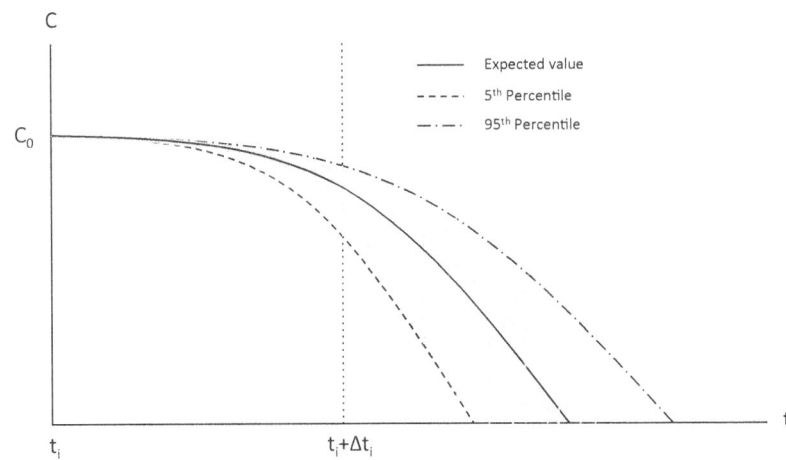

Figure 5: Example component status degradation with 5th percentile, and 95th percentile values.

For this case, the range of possible values grows over time. Therefore, the number of selection should increase proportionally with the range to keep the interval between selected values the same, thus, keep the accuracy of interpolation constant. For example, if there is N number of selections per variable, the selections are:

$$C_S = \left\{ C^{p_{low}th}, C^{p_{low}+\delta th}, C^{p_{low}+2\delta th}, \ldots C^{p_{high}th} \right\} \tag{13}$$

$$\delta = \frac{p_{high} - p_{low}}{N_{selections} - 1} \tag{14}$$

The other type of observable parameters is constant parameters. These parameters are usually Gaussian distributed. For this case, the range always stay constant, therefore, the selections remain the same throughout the life of the component.

If the observed values are always in the predicted range, the accuracy of the results depends upon the number of selections for pre-computation. The number of selections is the number of selections at each time-slice multiplies be the number of measurement intervals. The number of pre-computations is then the number selections for each observable times the number of observables parameters.

$$N_{pre-computation} = \sum_{j=0}^{T_c/\Delta t} \left(\prod_{i=1}^{n} N_{selections,i,t+(j\Delta t)} \right) \tag{15}$$

Where $N_{selections,i,t}$ is the number of selections of observable parameter i at time t. n is the number of observable parameters. T_c is the component life. The total computation time then can be estimated.

$$T_{pre-computation} = N_{pre-computation} \cdot T_{average-per-computation} \tag{16}$$

For MCMC computation, the average computation time is proportional to the number iterations. The higher the number of iterations, the higher accuracy of the result will be. Therefore, there is a trade-off between computation time and accuracy. For pre-computation, the decision between higher number of value selections or higher number of iteration per computation must be made.

One advantage of the isolation among component sub-tree is that time intervals do not have to be uniform for all components. Measurement/inspection intervals can be based on the rate of component degradation and possible change to component parameters. They can also be dynamically changed during the life a component depending on its status. For example there can be less frequency of measurements during the early life of a component due to less probability of failure. Then increase the frequency when the component approaches the end of life. The time interval between measurements, Δt, should then be inverse proportional to the amount of change of the parameter C, $\Delta t \propto 1/\Delta C$. Thus, the sampling rate around a certain evidence value will be proportional to the probability that the evidence value could happen and how much different in values to the possible values around it at certain period of time.

Dynamic programming is a method for solving complex problems by breaking them down into simpler subproblems. It is applicable to problems exhibiting the properties of overlapping subproblems and optimal substructure. When applicable, the method takes far less time than naive methods that don't take advantage of the subproblem overlap. In general, to solve a given problem, we need to solve different parts of the problem (subproblems), then combine the solutions of the subproblems to reach an overall solution. Often when using a more naive method, many of the subproblems are generated and solved many times. The dynamic programming approach seeks to solve each subproblem only once, thus reducing the number of computations: once the solution to a given subproblem has been computed, it is stored the next time the same solution is needed, it is simply looked up. This approach is especially useful when the number of repeating subproblems grows exponentially as a function of the size of the input.

Using dynamic programming can reduce the precomputation time for Bayesian Network inference drastically. Instead of computing full inferences for each set of evidence values, dynamic programming algorithm retain marginal results that can be reused with similar set of evidence values.. There are three steps for the algorithm:

1. Use logic-sampling algorithm and degradation model to generate all possible evidence values according to its probability of occurring. Not all evidence nodes have to be instantiated for each case, only the evidence nodes that are required for observing nodes are instantiated.
2. Check and construct a cache by comparing each generated case to those already in the cache. If the case is found to be new, this algorithm determines, the joint probability of the case's evidence using the algorithm in the third step.
3. The marginal posterior-probability distributions over the diagnosis nods are determine, then the values of the evidence nodes, the joint probability of the evidence set, and the marginal posterior-probability distributions for the diagnosis node are stored in the cache.

Figure 6 shows two example cases where dynamic programming can reduce the number of computation. The first case is when nodes have the same set of parent nodes, thus the same sets of possible marginal probability distributions for discrete nodes. The second case is when continuous parameters have several trajectories that can reach the same values after some period of time. In addition, if more computations are needed during an operation in the event where evidence values reaches the bound of expected values, dynamic programming provide a set of marginal results that can be used for possible faster inference of values outside the pre-computed cache.

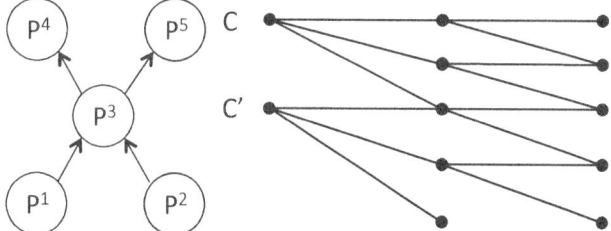

Figure 6: Example cases where dynamic programming reduces number of computations

Since both deterministic and approximate inference were found to be NP-hard [22][23], the computation complexity for both discrete functionality and continuous component degradation model are exponential in the network's treewidth. Figure 7 shows a plot presenting differences between pre-computation time with and without dynamic programming.

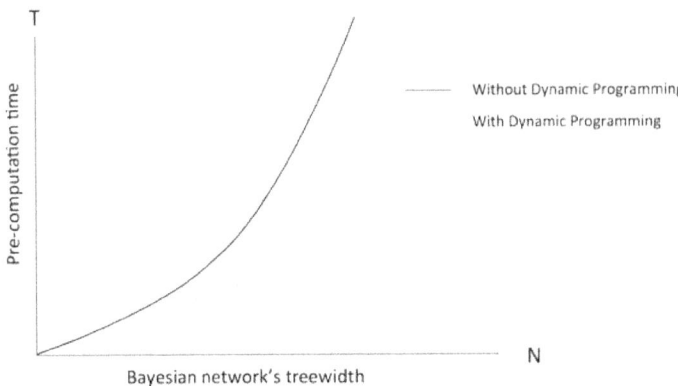

Figure 7: Inference pre-computation time with and without dynamic programming.

3. EXAMPLE APPLICATION

Electromigration (EM) in advanced integrated circuits (ICs) was considered to demonstrate the proposed methodology. For both Al-based and Cu-based metallization, EM has historically been a significant reliability concern.

The model generally used to describe EM time-to-failure takes the form:

$$TF = B_0 \left(J^{(e)} - J^{(e)}_{crit} \right)^{-n} \exp\left(\frac{Q}{K_B T}\right), \tag{17}$$

where: TF is the component time to failure. B_0 is a process/material-dependent coefficient. $J^{(e)}$ is the electron current density. $J^{(e)}_{crit}$ is a critical (threshold) current density which must be exceeded before significant EM is expected. n is the current density exponent. Q is the activation energy.

Using degradation model of component/device parameter C with the power-law equation:

$$C = C_0 \left[1 - A_0 \left(J^{(e)}, T \right) t^m \right] \tag{18}$$

We can derive at the following relationship:

$$C = C_0 \left[1 - A_0 \cdot \left(J^{(e)} - J^{(e)}_{crit} \right)^r \cdot \exp\left(\frac{-Q}{K_B T}\right) t^m \right] \tag{19}$$

Since both current density $J^{(e)}$ and temperature T are expected to be normally distributed between time t-1 to t,

$$\begin{aligned} J^{(e)} &= \mathcal{N}\left(\mu_J, \sigma_J\right) \\ T &= \mathcal{N}\left(\mu_T, \sigma_T\right) \end{aligned} \tag{20}$$

In the context of simple health monitoring in this example, $A_0, Q, r,$ and m are considered to be constant parameters representing material/device internal factors. These parameters can also be modeled with probabilistic distributions. The BN model of a component affected by EM is shown in Figure 8.

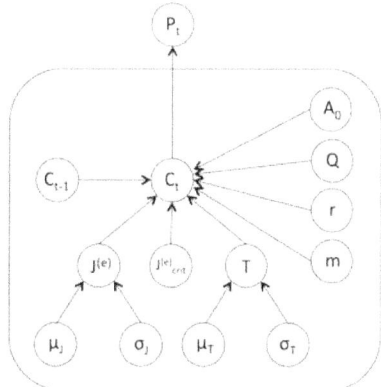

Figure 8: BN of a component affected by EM.

Consider an Al-alloy under high temperature operation, with current density $J = 2\times10^6$ A/cm^2 and at a metal temperature T = 200 °C. Assuming an activation energy of Q = 0.8 eV and the current density exponent of n = 2. Using conservative design approach, assume $J_{crit} = 0$. Consider the data set shown in Figure 9 of $J^{(e)}$ and T during an operation.

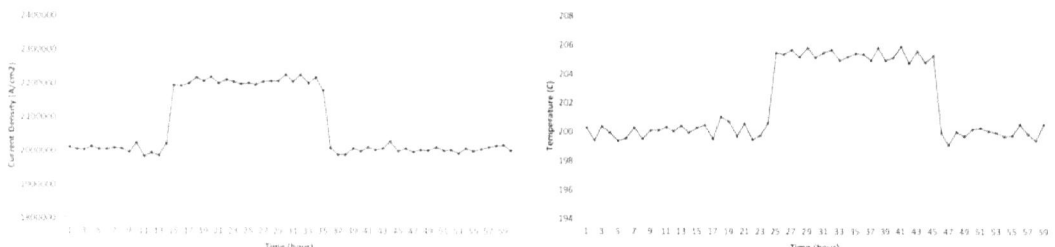

Figure 9: Current density and temperature data set

The data is retrieved once per minute during one hour of operation. In traditional Bayesian updating method, a calculation is required at each time step, which means 60 inferences have to be performed. With the proposed algorithm, only 4 inferences are needed when the values of current density and temperature changes out of ϵ_f range. Approximate inference of component status is available almost instantly with pre-computation of C_t at t = 1,...,60, with the range of J between 1.8×10^6 A/cm^2 to 2.2×10^6 A/cm^2, and T between 90°C to 120°C. Figure 10 shows an example plot of component status degradation under electromigration vs. time at different current density and temperature, including from the data set.

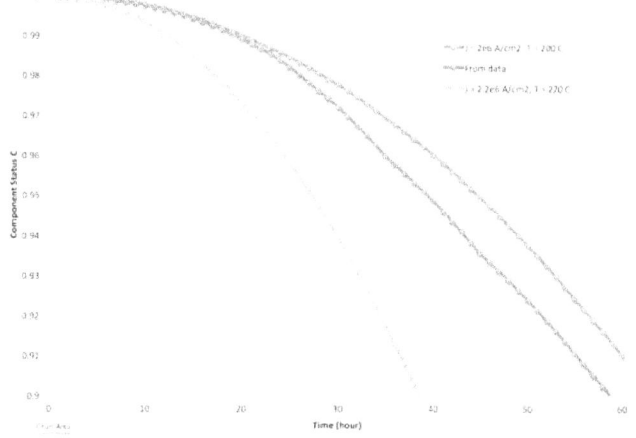

Figure 10: Plot of component status under electromigration vs. time at different current density and temperature, including from the data set

4. CONCLUSION

This research presents new modeling approach, computational algorithms, and an example application for on-line System Health Management. New method of using dynamic hybrid Bayesian Network were introduced with component-based algorithm and structure to represent complex engineering systems in a way that it allows accurate representation of underlying physics of failure by using empirical degradation model with continuous variables.

With dynamic hybrid Bayesian Network model requiring Markov Chain Monte Carlo for probabilistic inference, this paper develops computational algorithms that enables monitoring and diagnosing complex systems in real-time. The algorithms use the characteristics of System Health Management applications to allow reduction of number of inference required and reduce the calculation time by the means of pre-computation and dynamic programming.

References

[1] Pearl, J. (1986). Fusion, Propagation and Structuring in Belief Networks. *Artificial Intelligent , 29*, 241-288.
[2] Jensen, F. (2001). *Bayesian Networks and Decision Graph* . Springer.
[3] Friedman, N. (1998). The Bayesian structural EM algorithm. In G. F. Cooper (Ed.), *Proceedings of the Fourteenth Conference on Uncertainty in Artificial Intelligence (UAI-98)* (pp. 129-138). Morgan Kaufmann.
[4] Murphy, K. (2002). *Dynamic Bayesian Networks: Representation, Inference and Learning.* PhD thesis, UC Berkeley , Dept. Computer Science .
[5] Ferreiro, S., Arnaiz, A., Sierra, B., & Irigoien, I. (2011). A Bayesian network model integrated in a prognostics and health management system for aircraft line maintenance. *Proceedings of the Institution of Mechanical Engineers, Part G: Journal of Aerospace Engineering , 225* (8), 886-901.
[6] Tobon-Mejia, D. A., Medjaher, K., Zerhouni, N., & Tripot, G. (2012). A Data-Driven Failure Prognostics Method Based on Mixture of Gaussians Hidden Markov Models. *Reliability, IEEE Transactions on , 61* (2), 491-503.
[7] Schumann, J., Rozier, K. Y., Reinbacher, T., Mengshoel, O. J., Mbaya, T., & Ippolito, C. (2013). Towards Real-time, On-board, Hardware-supported Sensor and Software Health Management for Unmanned Aerial Systems. *Annual Conference of the Prognostics and Health Management Society.*
[8] Wilson, A. G., & Huzurbazar, A. V. (2007). Bayesian networks for multilevel system reliability. *Reliability Engineering & System Safety , 92* (10), 1413-1420.
[9] Langseth, H., & Portinale, L. (2007). Bayesian networks in reliability. *Reliability Engineering & System Safety , 92* (1), 92-108.
[10] Doguc, O., & Ramirez-Marquez, J. E. (2009). A generic method for estimating system reliability using Bayesian networks. *Reliability Engineering & System Safety , 94* (2), 542-550.
[11] Boudali, H., & Dugan, J. B. (2006). A continuous-time Bayesian network reliability modeling, and analysis framework. *Reliability, IEEE Transactions on , 55* (1), 86-97.
[12] Weber, P., & Jouffe, L. (2006). Complex system reliability modelling with dynamic object oriented bayesian networks (DOOBN). *Reliability Engineering & System Safety , 91* (2), 149-162.
[13] Lauritzen, S. L., & Jensen, F. (2001). Stable Local Computation with Conditional Gaussian Distributions. *Stat. & Comp. , 11*, 191-203.
[14] Moral, S., Rumi, R., & Salmeron, A. (2001). Mixtures of Truncated Exponentials in Hybrid Bayesian Networks. In *ECSQARU 2001. LNCS (LNAI)* (Vol. 2143, pp. 156-167). Springer, Heidelberg.
[15] Lerner, U. N. (2002). *Hybrid Bayesian Networks for Reasoning About Complex Systems.* Standford University, Dep. of Comp. Sci. Stanford.
[16] Shenoy, P. P. (2006). Inference in Hybrid Bayesian Networks Using Mixtures of Gaussians. In *Uncertainty in Artificial Intelligence* (pp. 428-436). AUAI Press, Corvallis.
[17] Boyen, X., & Koller, D. (1998). Tractable inference for complex stochastic processes. *Proceedings of the Fourteenth conference on Uncertainty in artificial intelligence* .
[18] Neil, M., Tailor, M., Marquez, D., Fenton, N., & Hear. (2007). Inference in Bayesian Networks using dynamic discretisation. *Statistics and Computing , 17* (3), 219-233.
[19] Cousins, S. B., Chena, W., & Frisse, M. E. (1993). A tutorial introduction to stochastic simulation algorithms for belief networks. *Artificial Intelligence in Medicine , 5* (4), 315-340.
[20] Dagum, P., & Horvitz, E. (1993). A Bayesian analysis of simulation algorithms for inference in belief networks. Networks. *23* (5), 499-516.
[21] Chen, Z. (2003). Bayesian filtering: From Kalman filters to particle filters, and beyond. *Statistics* , 1-69.
[22] Cooper, G. F. (1990). The computational complexity of probabilistic inference using Bayesian belief networks. *Artificial Intelligence , 42* (2-3), 393-405.
[23] Dagum, P., & Luby, M. (1993). Approximating probabilistic inference in Bayesian belief networks is NP-hard . *Artificial Intelligence , 60* (1), 141-154.

Survivability Evaluation of Disaster Tolerant Cloud Computing Systems

Bruno Silva[a]*, Paulo Romero Martins Maciel[a], Armin Zimmermann[b] and Jonathan Brilhante[a]

[a]Federal University of Pernambuco, Recife, Brasil
[b]Ilmenau University of Technology, Ilmenau, Germany

Abstract: A prominent type of cloud service is the Infrastructure-as-a-Service (IaaS), which delivers, on-demand, computing resources in the form of virtual machines (VMs) satisfying user needs. In such systems, penalties may be applied if the defined quality level of service level agreement (SLA) is not satisfied. Therefore, high availability is a critical requirement of these systems. A strategy to protect such systems from natural or manmade disasters corresponds to the utilization of multiple data centers located into different geographical locations to provide the service. Considering such systems, redundancy mechanisms can be adopted to receive copies of VM images during data center operation. Hence, whenever a disaster makes one data center unavailable, the VMs can be re-instantiated in other operational data center. Modeling techniques, with a strong mathematical foundation, such as Stochastic Petri Nets (SPN) can be adopted to evaluate survivability in these complex infrastructures. This work presents SPN models to evaluate survivability metrics in IaaS systems deployed into geographically distributed data centers taking into account disaster occurrences. Using the proposed models, IaaS providers can evaluate the impact of VM transmission time and the VM backup period on survivability metrics. A case study is provided to illustrate the effectiveness of the proposed work.

Keywords: survivability, IaaS systems, disaster recovery plan, stochastic Petri nets.

1. INTRODUCTION

Cloud computing has driven the new wave of Internet-based applications by providing computing as a service [1]. Nowadays, common business applications (e.g., spreadsheets, text editors) are provided as cloud computing services, in the sense that they are often accessed using a web browser, and their respective software/data reside on remote servers. This approach has affected all fields of the computational research, from users to hardware manufacturers [2]. Such paradigm is attractive for a number of reasons: (i) it frees users from installing, configuring and updating the software applications; (ii) it offers advantages in terms of mobility as well as collaboration; and (iii) updates and bug fixes can be deployed in minutes, simultaneously affecting all users around the globe [3]. An important type of cloud service is the Infrastructure-as-a-Service (IaaS), such as Amazon EC2 [4] and IBM Smart Business Cloud [5]. IaaS delivers, on-demand, computing resources in the form of virtual machines (VMs) running on the cloud provider's data center, satisfying user needs. User requests are provisioned depending on the data center capacity in terms of physical machines.

For prominent IaaS providers, the quality level is regulated by adopting a Service Level Agreement (SLA), which specifies, for instance, the maximum downtime per year. Penalties may be applied if the defined quality level is not satisfied. Thus, to meet SLA requirements, IaaS providers need to evaluate their environment, considering, also, the possibility of disasters. Therefore, a disaster recovery plan requires the utilization of different data centers located far enough apart to mitigate the effects of unforeseen disasters (e.g., earthquakes) [6]. Considering such systems, redundant data centers can be adopted to receive copies of VM images during data center operation. Hence, whenever a disaster

---

* Corresponding author, bs@cin.ufpe.br

makes one data center unavailable, the VMs can be re-instantiated in other operational data center. Unfortunately, some data between the last VM backup and the disaster may be lost and it is necessary some time to restart the operation after a failure. However, this may be traded-off changing the time between backups and the distance between data centers. Two metrics can be utilized to evaluate system survivability: (i) Recovery Point Objective (RPO), which corresponds to the maximum age of the most recent backup prior to disaster and (ii) Recovery Time Objective (RTO) that specifies the maximum time to repair the service after a disaster occurrence. Modeling techniques, with a strong mathematical foundation, such as Stochastic Petri Nets (SPN) [7] can be adopted to evaluate survivability in complex infrastructures.

This work presents an approach to evaluate survivability in IaaS systems deployed into geographically distributed data centers as well as taking into account disaster occurrence. The proposed approach contemplates state-based models (SPN - Stochastic Petri Nets) to determinate the probability of IaaS systems meet their survivability objectives. Using the proposed approach, IaaS providers can evaluate the system distributed in different data centers and the impact of VM backup time on these metrics. The paper is organized as follows. Section 2 highlights the related works. Section 3 describes the cloud computing system considered. Then, basic concepts about SPN models are introduced in Section 4. Section 5 presents the survivability parameters adopted in this work. In Section 6, the SPN models adopted to evaluate IaaS survivability is presented. Finally, Section 8 shows the adopted case study and Section 8 concludes this paper.

2. RELATED WORK

Over the last years, some authors have been devoting efforts to study dependability issues on cloud computing systems. Longo et al. [8] proposed an approach for availability analysis of cloud computing systems based on Petri nets and Markov chains. The authors also developed closed-form equations and demonstrated that their approach can scale for large systems. In [9], a performability analysis for cloud systems is presented. The authors quantify the effects of variations in workload, failure rate and system capacity on service quality. In [10], the authors investigate the aging effects on the Eucalyptus framework [11], and they also propose a strategy to mitigate such issues during system execution.

[12] describes a system design approach for supporting transparent migration of virtual machines that adopt local storage for their persistent state. The approach is transparent to the migrated VM, and it does not interrupt open network connections during VM migration. In [13], the authors present a case study that quantifies the effect of VM live migrations in the performance of an Internet application. Such study helps data center designers to plan environments in which metrics, such as service availability and responsiveness, are driven by Service Level Agreements. Dantas et al. [14] present a study of warm-standby mechanisms in Eucalyptus framework. Their results demonstrate that replacing machines by more reliable counterparts would not produce improvements in system availability, whereas some techniques of fault-tolerance can indeed increase dependability levels.

In [15], the authors adopted model checking algorithms to decide if a given system is survivable. The logic CSL was adopted to represent and estimate survivability metrics in GOOD (given-occurrence-of-disaster) models. Unlike previous works, this paper proposes performability models for evaluating cloud computing systems deployed into geographically distributed data centers, considering VM transfer data and disasters occurrence.

3. SYSTEM ARCHITECTURE OF RELIABLE DISTRIBUTED DATA CENTERS

This section presents an overview of the cloud computing system considered in this work, which consists of a set of distributed data centers (Figure 1). The system is composed of D data centers. A Backup Server (BS) is assumed to provide backup of VM data, which periodically receives a copy of each VM image during data center operation. Hence, whenever a disaster makes one data center unavailable, BS sends VM copies to operational data centers. In this work, the number of running VMs (w) is compared with a threshold (k) to evaluate the availability of cloud computing system. Hence, if $w \geq k$ the system is assumed operational.

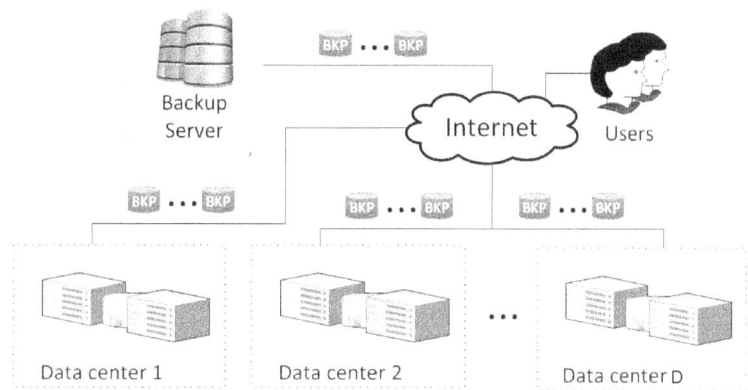

Figure 1: Distributed Cloud System Example

4. SPN MODELS

Petri nets (PN) [16] are a family of formalisms very well suited for modeling several system types, since concurrency, synchronization, communication mechanisms as well as deterministic and probabilistic delays are naturally represented. This work adopts a particular extension, namely, Stochastic Petri Nets (SPN) [17], which allows the association of stochastic delays to timed transitions using the exponential distribution, and the respective state space can be converted into continuous-time Markov chains (CTMC) [18]. Figure 2 depicts an example of a SPN model. Places are represented by circles, whereas transitions are depicted as filled rectangles (immediate transitions) or hollow rectangles (timed transitions).

Arcs (directed edges) connect places to transitions and vice-versa. Tokens (small filled circles) may reside in places, which denote the state (i.e., marking) of a SPN. An inhibitor arc is a special arc type that depicts a small white circle at one edge, instead of an arrow, and they usually are used to disable transitions if there are tokens present in a place. The behaviour of a SPN is defined in terms of a token flow, in the sense that tokens are created and destroyed according to the transition firings [19]. Immediate transitions represent instantaneous activities, and they have higher firing priority than timed transitions. Besides, such transitions may contain a guard condition, and a user may specificy a different firing priority among other immediate transitions. SPNs also allow the adoption of simulation techniques for obtaining dependability metrics, as an alternative to the generation of a CTMC. Regarding SPN formal definitions and semantic, the reader is referred to [17].

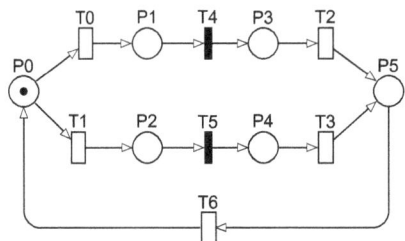

Figure 2: SPN model example

4.1. Distribution Moment Matching

A well-established method that considers *expolynomial distribution* random variables is based on distribution moment-matching . The moment matching process presented in [20] and considers that Hypoexponential and Erlangian distributions have the average delay (μ) greater than the standard-deviation (σ) -$\mu > \sigma$-, and Hyperexponential distributions have $\mu < \sigma$, in order to represent an activity with a generally distributed delay as an Erlangian or a Hyperexponential subnet referred to as s-transition. One should note that in cases where these distributions have $\mu = \sigma$, they are, indeed, equivalent to an exponential distribution with parameter equal to $\frac{1}{\mu}$. Therefore, according to the coefficient of variation associated with an activity's delay, an appropriate s-transition implementation model could be chosen. For each s-transition implementation model (see Figure 3), a set of parameters should be configured for matching their first and second moments. In other words, an associated delay distribution (it might have been obtained by a measuring process) of the original activity is matched with the first and second moments of s-transition (*expolynomial distribution*). According to the aforementioned method, one activity with $\mu < \sigma$ is approximated by a two-phase Hyperexponential distribution with parameters

$$r_1 = \frac{2\mu^2}{(\mu^2 + \sigma^2)},\tag{1}$$

$$r_2 = 1 - r_1\tag{2}$$

and

$$\lambda = \frac{2\mu}{(\mu^2 + \sigma^2)}.\tag{3}$$

where λ is the rate associated to phase 1, r_1 is the probability of related to this phase, and r_2 is the probability assigned to phase 2. In this particular model, the rate assigned to phase 2 is assumed to be infinity, that is, the related average delay is zero.

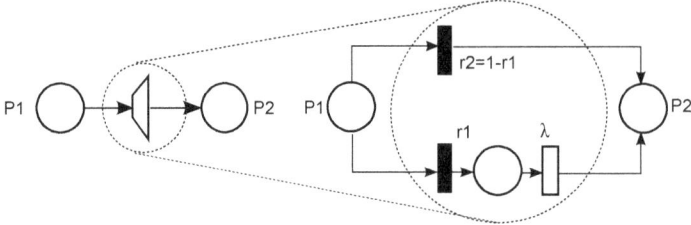

Figure 3: Hyperexponential Model

Activities with coefficients of variation less than one might be mapped either to Hypoexponential or Erlangian s-transitions. If $\frac{\mu}{\sigma} \notin \mathbb{N}, \frac{\mu}{\sigma} \neq 1, (\mu, \sigma \neq 0)$, the respective activity is represented by a Hypoexponential distribution with parameters λ_1, λ_2 (exponential rates); and γ, the integer representing the number of phases with rate equal to λ_2, whereas the number of phases with rate equal to λ_1 is one. In other words, the s-transition is represented by a subnet composed of two exponential and one immediate transitions. The average delay assigned to the exponential transition t_1 is equal to μ_1 ($\lambda_1 = 1/\mu_1$), and the respective average delay assigned to the exponential transition t_2 is $\mu_2(\lambda_2 = 1/\mu_2)$. γ is the integer value considered as the weight assigned to the output arc of transition t_1 as well as the input arc weight value of the immediate transition t_3 (see Figure 4). These parameters are calculated by the following expressions:

$$(\frac{\mu}{\sigma})^2 - 1 \leq \gamma < (\frac{\mu}{\sigma})^2, \tag{4}$$

$$\lambda_1 = \frac{1}{\mu_1} \ and \ _2 = \frac{1}{\mu_2}, \tag{5}$$

where

$$\mu_1 = \frac{\mu \pm \sqrt{\gamma(\gamma+1)\sigma^2 - \gamma\mu^2}}{\gamma+1}, \tag{6}$$

$$\mu_2 = \frac{\gamma\mu \mp \sqrt{\gamma(\gamma+1)\sigma^2 - \gamma\mu^2}}{\gamma+1} \tag{7}$$

If $\frac{\mu}{\sigma} \in \mathbb{N}, \frac{\mu}{\sigma} \neq 1, (\mu, \sigma \neq 0)$, an Erlangian s-transition with two parameters, $\gamma = (\frac{\mu}{\sigma})^2$ is an integer representing the number of phases of this distribution; and $\mu_1 = \mu/\gamma$, where $\mu_1(1/\lambda_1)$ is the average delay value of each phase. The Erlangian model is a particular case of a Hypoexponential model, in which each individual phase rate has the same value. The reader should refer to [20] for details regarding the representation of expolinomial distributions using SPN.

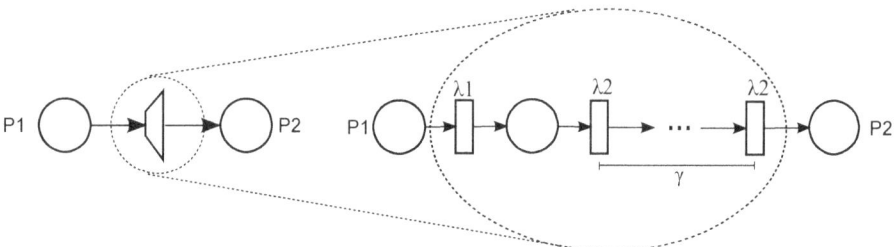

Figure 4: Hypoexponential Model

5. BUSINESS CONTINUITY OBJECTIVES

Survivability can be defined as the ability of a system to recover a predefined service level in a timely manner after the occurrence of disasters [15]. In this context, companies are adopting Business Continuity Management (BCM) [21] to support the ability to operate in spite of unforeseen events and recover in a short time frame. The main BCM's outcome is the Business Continuity Plan (BCP) or Disaster Recovery Plan (DRP), which is a document that describes the business continuity process in order to reduces or minimizes impact of events that disrupt critical services and their supporting resources [21]. Two indexes are utilized to define survivability objectives: (i) Recovery Point Objective (RPO), which corresponds to the time limit of the most recent backup prior to disaster and (ii) Recovery Time Objective (RTO) that specifies the maximum time to repair the service after a disaster occurs.

These objectives are based on business decisions that contemplate costs of inactivity periods and data loss. Additionally, technological factors (e.g., system performance) must be considered to stablish these parameters [22, 23].

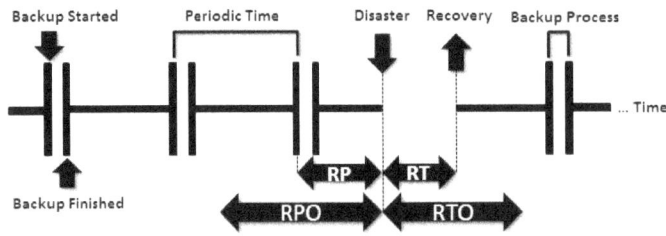

Figure 5: RPO and RTO requirements.

Figure 5 illustrates the backup operation along the time for a general system. During the system operation, a backup is periodically performed and whenever a disaster happens, the last backup should be recovered. If the age of the last backup (Recovery Point - RP) is higher than the RPO or the time to recover the system (Recovery Time - RT) is higher than the RTO, then the survivability requirements are not satisfied. In this case (Figure 5), the system meet the requirements. It is important to state that the amount of data that should be restored or backed up is not fixed for some applications. Consequently, the time to perform backup and restore operations is stochastic and depends on the amount of data involved and the technology utilized [23]. Additionally, for some applications (e.g., collaboration websites) the RPO and RTO should not be higher than a few minutes. On the other hand, for other applications (e.g., static websites) these requirements are not so critical.

5.1. Recovery Point Evaluation

In this study, we are interested in evaluating the recovery point considering the worst-case scenario (Figure 6). According to [24], this situation happens when the disaster occurs during the backup process. Observe that, in this case, the Recovery Point (R_p) is equal to the sum of the Backup Period (B_p) and the actual Backup Time (B_t). Consequently, the probability of meeting the RPO (P_{RPO}) in the worst-case scenario is given by:

$$P_{RPO} = P\{R_p \leq RPO\} = P\{B_t \leq RPO - B_p\}.$$

As B_p and the RPO are project decision parameters, the cloud designer must evaluate the behavior of the backup process to check the P_{RPO} metric.

5.2. Recovery Time Evaluation

The process of checking the survivability in terms of RTO is similar to the recovery point evaluation. While the last considers the worst-case scenario, the recovery time evaluation is calculated directly. The probability of the recovery time meet the objective (P_{RTO}) is calculated as follows:

$$P_{RTO} = P\{R_t \leq RTO\}.$$

Where R_t denotes the recovery time.

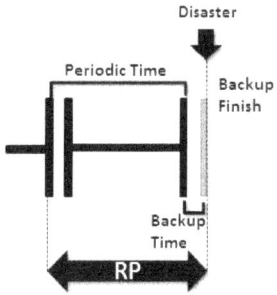

Figure 6: RPO worst case scenario.

6. MODELING APPROACH AND BASIC MODELS

This section presents the adopted hierarchical modeling to evaluate system dependability. The proposed approach (Figure 7) adopts SPN models for estimating survivability parameters in IaaS clouds. Although this work is focused on cloud computing systems, the approach is generic enough to be applied in other disaster recovery systems.

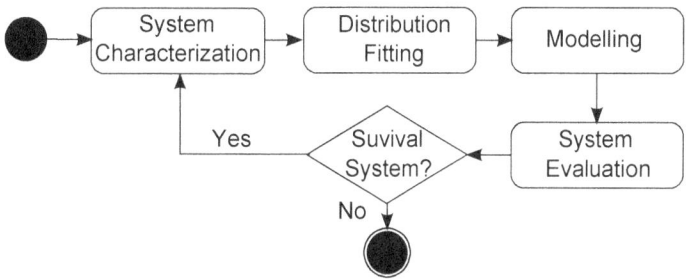

Figure 7: Methodology

The methodology's first step concerns the system characterization. In this phase, the designer must specify the VM characteristics and the number of VMs must be periodically backed up. In this activity, the probable data center locations should be selected and the transfer connection characteristics between the data centers and BS must be collected. Additionally, the backup period should be selected. In the second step, the designer must estimate what is the most suitable expolinomial distribution (Section 4.1) to represent the time to transfer a single VM data. Considering the modeling activity, the model is created with the parameters collected in the previous activities. Finally, the system is evaluated using transient evaluation, which is adopted to observe the behavior of the disaster recovery mechanism along the time [7]. If the evaluated system does not meet the constraints, the disaster recovery mechanisms must be reconfigured (e.g., backup period, data center locations). Henceforth, the following operators are adopted for assessing survivability metrics: $P\{exp\}$ estimates the probability of the inner expression (exp); and $\#p$ denotes the number of tokens in place p.

6.1. RPO Evaluation Model

The RPO evaluation model is presented in Figure 9. It represents the backup process in which a data center transmit VM images to BS. This model is composed of three main sub-models: (i) Data Center Backup, (ii) Transmission and (iii) Store. The first sub-model represents the start of backup operation in a given data center. A token in *BSTR* means that the backup process should start and the firing of *BK_ST* leads to the creation of new tokens (VMs) representing new VM images to be backed up. The new tokens are stored in *DC_TMT*.

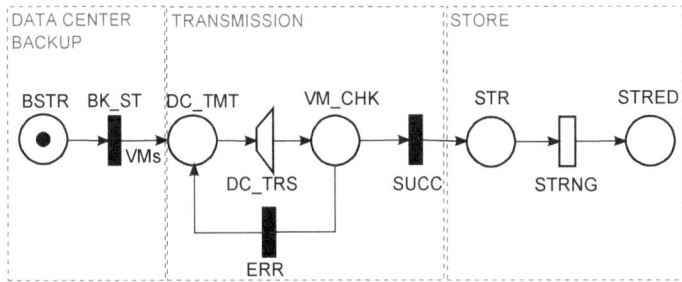

Figure 8: RPO Model

In Transmission sub-model, *DC_TRS* represents the transmission of VMs from the data center to BS. It is important to stress that the moment matching approach (Section 4.1) is adopted to represent expolinomial distributions for this transition. Once the VMs are transmitted, the data integrity of each VM is checked (*VM_CHK*). If the process present errors, the process is restarted (*ERR*). In case of correct transmission, the VMs should be stored/replicated (Store sub-model). Finally, the Store sub-model represents the storing/replication process of the transmitted VMs. It is composed of two places, one that represents the VM images that are about to be stored (*STR*) and another one to model the saved images (*STRED*). The transition *STRNG* models the store/replication process. If the BS has not replication mechanisms, *SUCC* may be connected directly to *STRED*. In this case, STR and STRNG may be discarded. In order to evaluate the system survivability in terms of RPO, a transient evaluation must be performed adopting the metric $P\{\#STRED = VMs\}$ in the time $RPO - B_p$. In other words, we are interested in evaluate the probability of finish the backup process in a specific time ($RPO - B_p$). If the assessed probability is less than the user defined level, the system is not survival.

6.2. RTO Evaluation Model

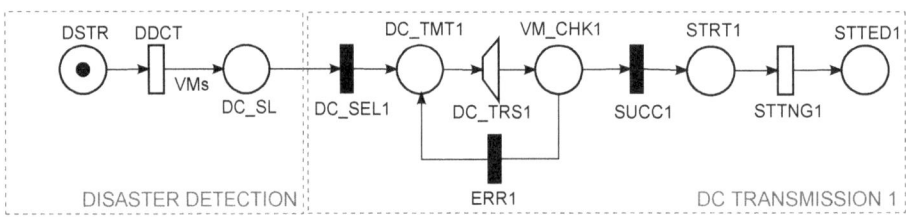

Figure 9: RTO Model

Figure 9 shows the RTO evaluation model which represents the recovery process immediately after the disaster occurrence. The model is composed of two basic sub-models, one representing the disaster detection and other modeling the transmission of VM images to operational data centers. The first sub-model is composed of a place that represents the start of recovery process (*DSTR*), a transition which models the disaster detection (*DDCT*) and a place that denotes the data center selection (*DC_SL*). The number of VM images that will be recovered is represented by VMs (arc multiplicity from *DDCT* to *DC_SL*). The other components of DC Transmission block are analogous to the components of RPO model (Figure 9) and will not be explained in details. The difference is that the three last components (STRT1, STTNG1 and STTED1) represents virtual machine instantiation instead of data store. Similarly to RPO evaluation model, a transient evaluation should be conducted to assess the probability to recover the system respecting the RTO limit. The observed metric is $P\{\#STTED1 = VMs\}$ in time RTO. If the evaluated probability is less than the requirement, the system is not survival.

7. CASE STUDY

In order to demonstrate the feasibility of the proposed work, this work presents a case study which evaluates survivability parameters in a IaaS environment. The environment is composed of five data centers and a BS. The data centers are located in the following cities: (i) New York (USA), (ii) Rio de Janeiro (Brazil), (iii) Zurich (Switzerland), (iv) Vienna (Austria) and Sydney (Australia). The backup server is located in Ilmenau (Germany). This experiment evaluates the recovery and backup process by using the modelling approach presented in Section 6. In this case study, we assume that 512 MB should be transmitted to BS to synchronize the VM data (backup process) and each VM image has 4 GB (recovery process). To estimate the time to transmit the VM data between the data centers and BS, a measurement process has been conducted to characterize the transfer rate between the backup server and the data centers. Mercury-ASTRO [25] and TimeNET [26] tools have been adopted to perform the evaluation. The transfer rates between BS and each data center is presented in Table 1.

Table 1: Transfer rates between BS and each data center.

Data Center Location	Rate (MB/s)	Standard Deviation (MB/s)
Zurich	2.0701	0.4019
Vienna	1.7412	0.3270
New York	1.1253	0.2126
Rio de Janeiro	0.6859	0.1351
Sydney	0.5659	0.1088

To evaluate the recovery process, the behavior of each data center to receive and instantiate five VM images (4 GB each) is considered. The mean time to detect the disaster is 30 minutes and the mean instantiation time is five minutes. The transmission success probability considered is 99.9%. For this particular experiment, the transition *DC_TRS* was converted to Hypoexponential subnets for all data centers (Section 4.1). The evaluation results for each data center are presented in Figure 10 and some important points are summarized in Table 2. For instance, considering that the RTO is two hours, the data center located in Austria can be a good option to recover the service if we assume that probability to recovery should be higher than 93%. On the other hand, if the RTO is four hours and the minimum probability to recovery is 99%, all data centers can be adopted to restart the affected VMs.

Table 2: Probability to recover the service for different data centers

Time(h)	Zurich	Vienna	New York	Rio de Janeiro	Sydney
1.0	0.5539	0.4370	0.2028	0.0027	0.0003
2.0	0.9369	0.9187	0.9183	0.6636	0.5424
3.0	0.9946	0.9909	0.9837	0.9501	0.9326
4.0	0.9998	0.9989	0.9973	0.9913	0.9903

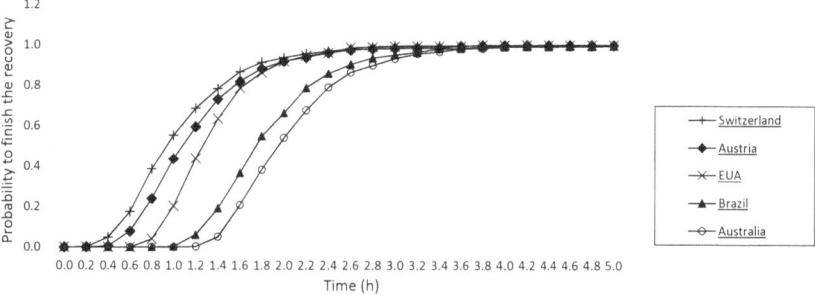

Figure 10: Recovery probability along the time

A similar evaluation was performed considering the backup process. In this case, each data center

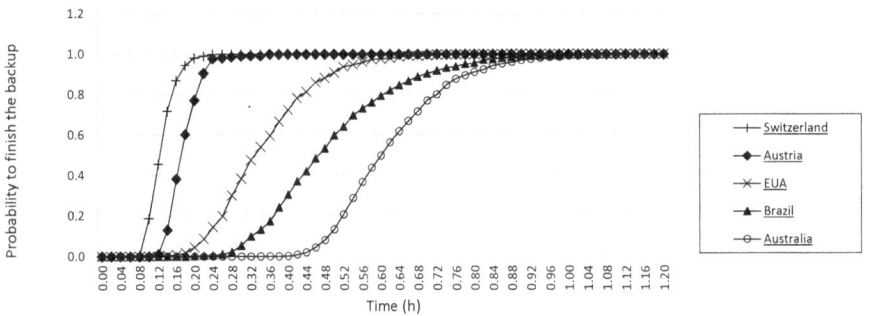

Figure 11: Backup probability along the time

synchronizes the data of five VMs (512 MB each) to BS and the replication process takes one minute. Figure 11 presents the evaluation results. Additionally, Table 3 shows important points considered in this evaluation. For the worst case scenario, if the difference between the RPO and the backup period is 0.2 hours and minimum probability to backup the VMs is equal to 0.98, only the data center of Zurich can be adopted. On the other hand, if the difference between the RPO and the backup period is one hour and minimum probability is equal to 0.99, all data centers respect the requirement.

Table 3: Probability to backup the service for different data centers

Time(h)	Zurich	Vienna	New York	Rio de Janeiro	Sydney
0.2	0.9801	0.7715	0.0449	0.0019	0.0001
0.4	0.9999	0.9998	0.7235	0.3062	0.0019
0.6	~1.000	~1.000	0.9771	0.7928	0.4980
0.8	~1.000	~1.000	0.9988	0.9545	0.9100
1.0	~1.000	~1.000	~1.000	~1.000	0.9911

8. CONCLUSION

This work presented models for survivability evaluation of cloud computing systems deployed into geographically distributed data centers. The proposed technique allows the survivability assessment taking into account the distance between data centers, RPO and RTO requirements. Additionally, a case study is provided considering a set data centers located in different places around the world. The results demonstrated the influence of distance, backup time and disaster recovery requirements on system survivability. As future research, we intend to create a tool to conduct the methodology steps automatically.

REFERENCES

[1] M. Armbrust, A. Fox, R. Griffith, A. D. Joseph, R. Katz, A. Konwinski, G. Lee, D. Patterson, A. Rabkin, I. Stoica, and M. Zaharia, "A view of cloud computing," *Commun. ACM*, vol. 53, no. 4, pp. 50–58, Apr. 2010. [Online]. Available: http://doi.acm.org/10.1145/1721654.1721672

[2] D. A. Menasc and P. Ngo, "Understanding cloud computing: Experimentation and capacity planning," 2009.

[3] Q. Zhang, L. Cheng, and R. Boutaba, "Cloud computing: state-of-the-art and research challenges," *Journal of Internet Services and Applications*, vol. 1, pp. 7–18, 2010. [Online]. Available: http://dx.doi.org/10.1007/s13174-010-0007-6

[4] Amazon ec2. [Online]. Available: http://aws.amazon.com/ec2

[5] IBM smart business cloud. [Online]. Available: http://www-935.ibm.com/services/us/igs/cloud-development/

[6] *Hyper-V Live Migration over Distance.* [Online]. Available: http://goo.gl/GzlkNk

[7] P. Maciel, K. S. Trivedi, R. Matias, and D. S. Kim, *Performance and Dependability in Service Computing: Concepts, Techniques and Research Directions*, ser. Premier Reference Source. Igi Global, 2011, ch. Dependability Modeling.

[8] F. Longo, R. Ghosh, V. Naik, and K. Trivedi, "A scalable availability model for infrastructure-as-a-service cloud," in *Dependable Systems Networks (DSN), 2011 IEEE/IFIP 41st International Conference on*, june 2011, pp. 335 –346.

[9] R. Ghosh, K. S. Trivedi, V. K. Naik, and D. S. Kim, "End-to-end performability analysis for infrastructure-as-a-service cloud: An interacting stochastic models approach," in *Proceedings of the 2010 IEEE 16th Pacific Rim International Symposium on Dependable Computing*, ser. PRDC '10. Washington, DC, USA: IEEE Computer Society, 2010, pp. 125–132. [Online]. Available: http://dx.doi.org/10.1109/PRDC.2010.30

[10] J. Araujo, R. Matos, P. Maciel, R. Matias, and I. Beicker, "Experimental evaluation of software aging effects on the Eucalyptus cloud computing infrastructure," in *Proceedings of the Middleware 2011 Industry Track Workshop*, ser. Middleware '11. New York, NY, USA: ACM, 2011, pp. 4:1–4:7. [Online]. Available: http://doi.acm.org/10.1145/2090181.2090185

[11] "Open source private and hybrid clouds from Eucalyptus," http://www.eucalyptus.com.

[12] R. Bradford, E. Kotsovinos, A. Feldmann, and H. Schiöberg, "Live wide-area migration of virtual machines including local persistent state," in *Proceedings of the 3rd international conference on Virtual execution environments*, ser. VEE '07. New York, NY, USA: ACM, 2007, pp. 169–179. [Online]. Available: http://doi.acm.org/10.1145/1254810.1254834

[13] W. Voorsluys, J. Broberg, S. Venugopal, and R. Buyya, "Cost of virtual machine live migration in clouds: A performance evaluation," in *Proceedings of the 1st International Conference on Cloud Computing*, ser. CloudCom '09. Berlin, Heidelberg: Springer-Verlag, 2009, pp. 254–265.

[14] J. Dantas, R. Matos, J. Araujo, and P. Maciel, "An availability model for eucalyptus platform: An analysis of warm-standy replication mechanism," in *Systems, Man, and Cybernetics (SMC), 2012 IEEE International Conference on*, oct. 2012, pp. 1664 –1669.

[15] L. Cloth and B. R. Haverkort, "Model checking for survivability!" in *Quantitative Evaluation of Systems, 2005. Second International Conference on the*. IEEE, 2005, pp. 145–154.

[16] T. Murata, "Petri nets: Properties, analysis and applications," *Proc. IEEE*, vol. 77, no. 4, pp. 541–580, April 1989.

[17] A. Marsan, *Modelling with generalized stochastic Petri nets*, ser. Wiley series in parallel computing. Wiley, 1995.

[18] K. Trivedi, *Probability and Statistics with Reliability, Queueing, and Computer Science Applications*, 2nd ed. Wiley Interscience Publication, 2002.

[19] R. German, *Performance Analysis of Communication Systems with Non-Markovian Stochastic*

Petri Nets. New York, NY, USA: John Wiley & Sons, Inc., 2000.

[20] A. A. Desrochers and R. Y. Al-Jaar, *Applications of Petri nets in manufacturing systems: modeling, control, and performance analysis.* IEEE press Piscataway^ eNJ NJ, 1995, vol. 70.

[21] "Business continuity management: Bs25999-1," British Stansdard, pp. 2–3, 2006. [Online]. Available: https://www.bsigroup.com/

[22] T. Wood, E. Cecchet, K. Ramakrishnan, P. Shenoy, J. Van der Merwe, and A. Venkataramani, "Disaster recovery as a cloud service: Economic benefits & deployment challenges," in *2nd USENIX Workshop on Hot Topics in Cloud Computing*, 2010.

[23] K. Keeton, C. A. Santos, D. Beyer, J. S. Chase, and J. Wilkes, "Designing for disasters." in *FAST*, vol. 4, 2004, pp. 59–62.

[24] Q. Yang, W. Xiao, and J. Ren, "Trap-array: A disk array architecture providing timely recovery to any point-in-time," in *ACM SIGARCH Computer Architecture News*, vol. 34, no. 2. IEEE Computer Society, 2006, pp. 289–301.

[25] B. Silva, G. Callou, E. Tavares, P. Maciel, J. Figueiredo, E. Sousa, C. Araujo, F. Magnani, and F. Neves, "Astro: An integrated environment for dependability and sustainability evaluation," *Sustainable Computing: Informatics and Systems*, 2012.

[26] R. German, C. Kelling, A. Zimmermann, and G. Hommel, "TimeNET: a toolkit for evaluating non-markovian stochastic petri nets," *Performance Evaluation*, vol. 24, no. 1-2, pp. 69 – 87.

Probability of adventitious fuel pin failures in fast breeder reactors and event tree analysis on damage propagation up to severe accident in Monju

Yoshitaka Fukano[a*], Kenichi Naruto[b], Kenichi Kurisaka[a], and Masahiro Nishimura[a]

[a]Japan Atomic Energy Agency, Tsuruga, Japan
[b]NESI Inc., O-arai, Japan

Abstract: Experimental studies, deterministic safety analyses and probabilistic risk assessments (PRAs) on local fault (LF) propagation in sodium cooled fast reactors (SFRs) have been performed in many countries because LFs have been historically considered as one of the possible causes of severe accidents. Adventitious fuel pin failures were considered to be the most dominant initiators of LFs in these PRAs because of high frequency of occurrence during reactor operation and possibility of subsequent pin-to-pin failure propagation. Therefore event tree analysis (ETA) on fuel element failure propagation initiated from adventitious fuel pin failure (FEFPA) in Japanese prototype fast breeder reactor Monju was performed in this study based on state-of-the-art knowledge on experimental and analytical studies on FEFPA and reflecting latest operation procedure at emergency in Monju. Probability of adventitious fuel pin failures in SFRs which is the initiating event of this ETA was also updated in this study. Probability of FEFPA to the peripheral sub-assemblies was quantified to be 1.7×10^{-12} in Monju based on this ETA. It was clarified that FEFPA in Monju was negligible and could be included in core damage fraction of the anticipated transient without scram and protected loss of heat sink in the viewpoint of both probability and consequence.

Keywords: PRA, ETA, SFR, LF

1. INTRODUCTION

Local fault (LF) accidents have been considered as one of the possible causes of core-disruptive accidents or severe accidents in sodium cooled fast reactors (SFRs) for a long time. The fuel element failure propagation (FEFP) was considered to be of greater importance in safety evaluation because fuel elements are generally densely arranged in the subassemblies (SAs) of SFRs and power densities in this reactor type are higher compared with those in light water reactors (LWRs) as shown in Table 1. Therefore probabilistic risk assessments (PRAs) [1-3], deterministic safety analyses and experimental studies on LF accident have been performed in many countries historically.

Table 1   Comparison of power densities among FBR, PWR and BWR

	FBR	PWR	BWR
Power density (kW/l)	350~1000	~100	~50

Table 2 shows frequency of initiating events of LFs for British commercial demonstration fast reactor (CDFR) [1]. Among the different initiators of LFs, adventitious fuel pin failure was most dominant one because of high frequency of occurrence during reactor operation and possibility of pin to pin failure propagation. Therefore event tree analysis (ETA) of FEFP from

Table 2   Frequency of initiating events of LFs for CDFR

Initiating event	Frequency (/ry)
Adventitious fuel pin failure	35
Inlet blockage	10^{-3}
Outlet blockage	10^{-3}
Wrapper split	2×10^{-2}
Overrated sub-assembly loaded	3×10^{-2}
Partially blocked sub-assembly loaded	2×10^{-3}
Oil in sub-assembly	10^{-1}
Non-oil debris in sub-assembly	10^{-1}

adventitious fuel pin failure (FEFPA) is necessary in SFRs. ETA of FEFPA in Japanese prototype fast breeder reactor Monju (Monju) was performed in this study based on latest knowledge of experimental

---

* Contact author: fukano.yoshitaka@jaea.go.jp

and analytical studies on FEFPA and reflecting latest operation procedure at emergency in Monju. Probability of adventitious fuel pin failures in SFRs was also updated based on the state-of-the-art review of open papers concerning fuel pin failure experiences in SFRs, because probabilities of fuel pin failures used in existing PRA [1-3] were based on experiences up to 1985.

2. Updated frequency of initiating event

In order to quantify the frequency of adventitious fuel pin failure, fuel pin failure experiences in SFRs were widely investigated based on open papers [4-13].

Table 3 shows the number of failed fuel pins and related data in SFRs based on this investigation. It should be noted that fuel pin failure experiences in the SFRs of which nominal full power were less than 100 MWth were excluded from this table because these experiences are used in the ETA for Monju of which nominal full power is 714 MWth.

Table 3 The number of failed fuel pins and related data in SFRs

(n) Reactor	JOYO		(3) Phenix	(4) Super Phenix	(5) PFR	(6) FFTF
	(1) Mk-II	(2) Mk-III				
(A) The number of irradiated (driver) fuel pins (-)	43434	16510	166521	98644	98000	47500
(B) The number of failed (driver) fuel pins (-)	0	0	29	0	22	1
(C) Mean residence time (years)	0.7	1.1	1.9	1.8	1.0	1.8
(D) Equivalent full power years (years)	5.0	1.9	10.1	0.9	4.1	6.2
(E) Total number of fuel pins in equilibrium core (-)	8509	10795	22351	98644	23400	15841
(F) Average achieved burnup (GWd/t)	42	68.5	100	60	150	70
(G) Nominal full power (MWth)	100	140	563	2990	650	400

The largest frequency of fuel pin failure in Monju was decided to use conservatively in the ETA because frequency of fuel pin failure could be obtained by the following several methods.

(1) Method 1: The frequency of fuel pin failure in this method (P_1) was calculated by following equation and A, B and C in **Table 3**. The arithmetic average of failed fuel pin and mean residence time were used in this method. It should be noted that the average weighted by irradiated fuel pins was used for mean residence time.

$$P_1 = \frac{\sum B_n}{\sum A_n} \div \frac{\sum A_n C_n}{\sum A_n} \tag{1}$$

The frequency of fuel pin failure in Monju of this method (P_{M1}) can be calculated by the following equation;

$$P_{M1} = P_1 \times N_{SA} \times N_{Pin} \tag{2}$$

N_{SA} : Total number of SAs in the core

N_{Pin} : Total number of fuel pins in one SA

(2) Method 2: The frequency of fuel pin failure in this method (P_2) was calculated by following equation and B, D and E in **Table 3**. The frequency of fuel pin failure for each reactor and the average of irradiated fuel pins weighted by equivalent full power years (EFPYs) were used in this method.

$$P_2 = \frac{\sum(\frac{B_n}{D_n E_n} \times D_n E_n)}{\sum D_n E_n} \tag{3}$$

The frequency of fuel pin failure in Monju of this method (P_{M2}) can be calculated by the following equation assuming that mean load factor of Monju is 71 %.

$$P_{M2} = P_2 \times N_{SA} \times N_{Pin} \times 0.71 \tag{4}$$

(3) Method 3: The frequency of fuel pin failure per burnup (P_3) was calculated by following equation and B, D, E and F in **Table 3**. In addition to method 2, average achieved burnups of irradiated fuel pins were used in this case.

$$P_3 = \frac{\sum(\frac{B_n}{D_n E_n} \times D_n E_n)}{\sum D_n E_n F_n} \tag{5}$$

The frequency of fuel pin failure in Monju of this method (P_{M3}) can be calculated by following equation assuming that the average achieved burnup of Monju is 80 GWd/t.

$$P_{M3} = P_3 \times N_{SA} \times N_{Pin} \times 0.71 \times 80 \tag{6}$$

(4) Method 4: Frequency of fuel pin failure per reactor power (P_4) was calculated by following equation and B, D, E and G in **Table 3**. In addition to method 2, nominal full power of each reactor was used in this case.

$$P_4 = \frac{\sum(\frac{B_n}{D_n E_n} \times D_n E_n)}{\sum D_n E_n G_n} \tag{7}$$

The frequency of fuel pin failure in Monju of this method (P_{M4}) can be calculated by following equation assuming that the nominal full power of Monju is 714 MWth.

$$P_{M4} = P_4 \times N_{SA} \times N_{Pin} \times 0.71 \times 714 \tag{8}$$

Table 4 shows calculated frequency of fuel pin failure for each method based on the above-mentioned equations. The frequency of fuel pin failure in Monju in method 1 was decided to be conservatively used in the following ETA because the frequency was the largest in all cases.

Table 4   Frequency of fuel pin failure by each method

Method	Method 1	Method 2	Method 3	Method 4
Frequency of fuel pin failure	7.2×10^{-5} [/y/pin]	9.1×10^{-5} [/EFPY/pin]	9.9×10^{-7} [/EFPY/pin/(GWd/t)]	1.0×10^{-7} [/EFPY/pin/MWth]
Frequency of fuel pin failure in Monju (/ry)	2.4	2.2	1.9	1.8

3. Event tree analysis for failure propagation from fuel pin failure

The damage propagation from fuel pin failure up to whole core damage and automatic reactor trip by delayed neutron detectors (DNDs) were considered to be main events in the existing PRA [1-3]. Not only automatic rector trip by DNDs but also various reactor shutdown means by several kinds of detectors such as precipitators or NaI detectors in cover gas (CG) method were equipped in Monju as shown in **Figure 1**.

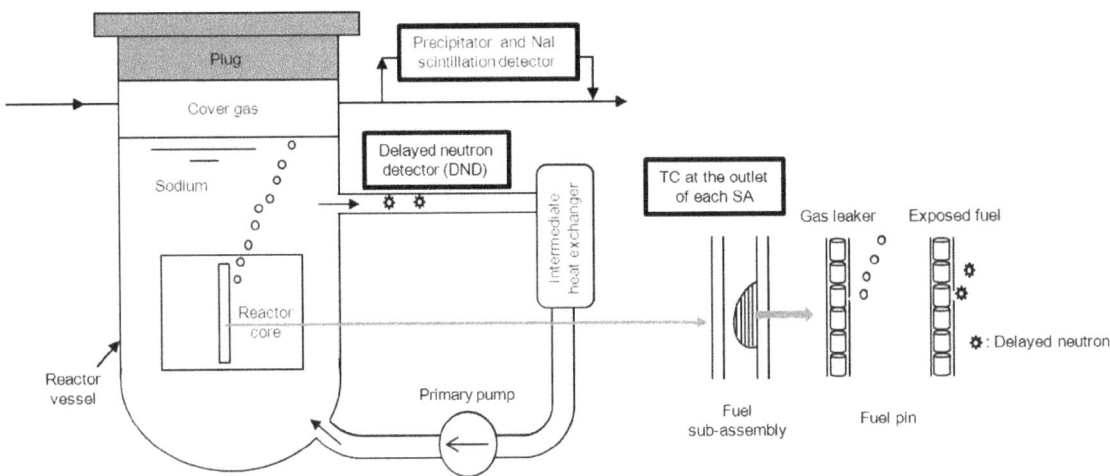

Figure 1   Schematic drawing of failed fuel detectors in Monju

The main flow after fuel pin failure in the operation procedure, alert and reactor trip level of detectors of fuel pin failure are shown in **Figure 2** and **Table 4** respectively.

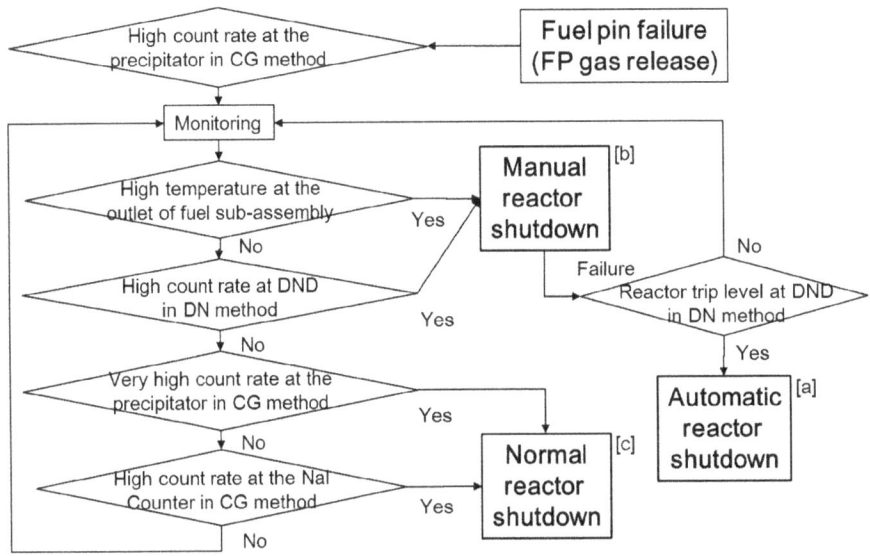

[a] corresponds to reactor trip   [b] corresponds to manual reactor trip
[c] means gradual power decrease until reactor shutdown

Figure 2   Main operation procedure after fuel pin failure in Monju

Table 4   Alert and reactor trip level of detectors for fuel pin failure in Monju

Name of alert or reactor trip	Alert or reactor trip threshold
Alert of high count rate at the precipitators in CG method	FP gas release (0.01% of one fuel pin)
Alert of very high count rate at the precipitators in CG method	Fuel pin failure (0.01% of total fuel pins)
Alert of high count rate at the NaI counter in CG method	Fuel pin failure (0.02% of total fuel pins)
Alert of high count rate at DNDs in DN method	Breached cladding area (over 200mm^2)
Alert from TC at the outlet of sub-assembly	More than 66% of flow blockage within one sub-assembly
Reactor trip at DNDs in DN method	Breached cladding area (over 5,000mm^2)

The ET reflecting the operation procedure after the fuel pin failure and the latest knowledge on experiments and analyses were presented in **Figure 3**. The main characteristics of this ET compared with existing PRA are:

(i) Various measures for failed fuel pin detection and reactor shutdown based on the operation procedure after the fuel pin failure in Monju;

(ii) More detailed development of the ET headings based on the state-of-the-art knowledge on experiments and analyses;

(iii) The possibility that fuel pin failure does not expand to detectable scale even at the end of cycle;

(iv) Removal of damaged SA by refuelling after reactor shutdown owing to detection of fuel pin failure;

(v) The possibility that decay heat is not removed in terms of coolable geometry even after reactor shutdown.

Figure 3   Main ET for FEFPA

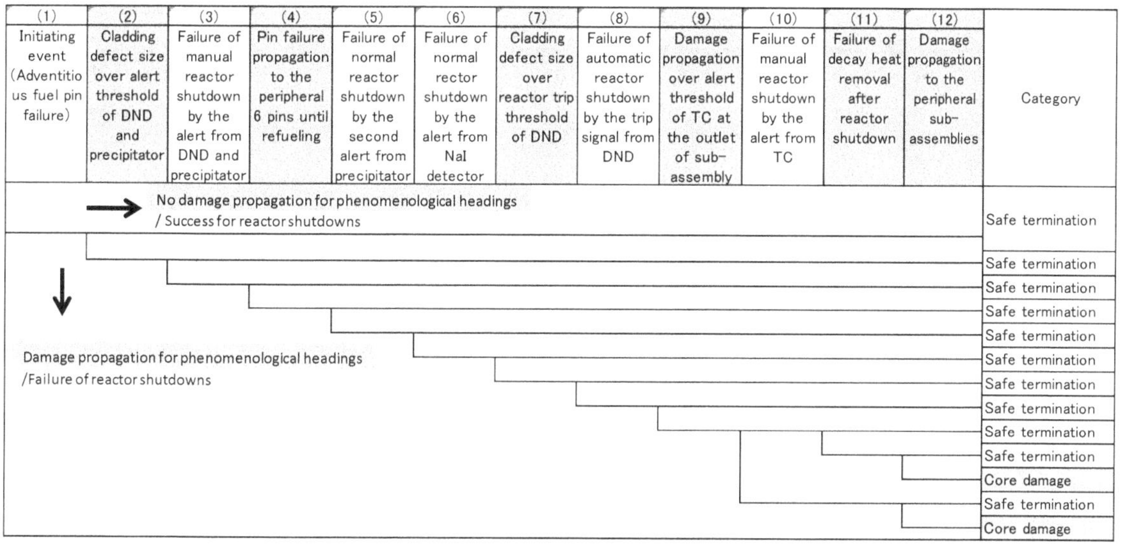

Although the quantification of branch probabilities for phenomenological headings was determined through the engineering judgment based on the knowledge on experiments and analyses, it was standardized using **Table 5** in order to keep consistency in this ETA.

Table 5   Branch probability ranks

Probabilistic rank	Qualitative representation	Representative value	Range of application
1	Indeterminate	0.5	0.7~0.3
2	Unlikely	0.2	0.3~0.1
	Likely	0.8	0.7~0.9
3	Highly Unlikely	0.05	0.1~0.01
	Highly Likely	0.95	0.9~0.99
4	Extremely Unlikely	<0.01	<0.01
	Extremely Likely	>0.99	>0.99
5	Impossible	ε	ε
	Certain	$1-\varepsilon$	$1-\varepsilon$

Each branch probability is described below;

3.1. Cladding defect size over alert threshold or reactor trip threshold in DN method [headings (2) and (7)]

After the fuel pin failure and subsequent FP gas release detection at the precipitators in CG method, manual reactor shutdown will be initiated if cladding defect size exceeds 200 mm^2 which is the alert threshold in DN method. Furthermore automatic reactor shutdown will be initiated if cladding defect size exceeds 5,000 mm^2 which is the reactor trip level in DN method.

Figure 4 shows defect sizes of failed fuel pins in the experiments on run beyond cladding breach (RBCB) [14]. There existed no data which shows cladding defect size over 200 mm^2 even after 200 days RBCB in Mol-7B experiment [15]. Table 6 shows burn-ups of driver fuel pins at the time of the adventitious fuel pin failure [6, 12]. The mean burn-up of the driver fuel pins was approximately 7.2 at.% from this table and was higher compared with that of Mol-7B experiment which was approximately 6.5 at.%.

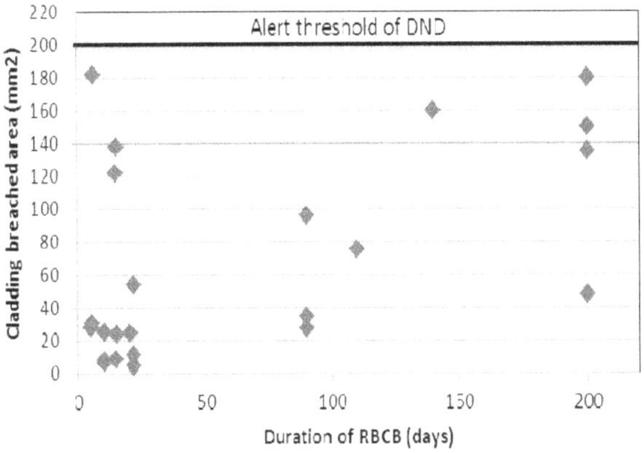

Table 6   Fuel burn-ups at driver fuel pin failures

Figure 4   Cladding breached areas in RBCB experiments

Reactor	No.	Cladding material	Fuel burn-up at fuel pin failure (at.%)
Phenix	1	316CW	8.9
	2	316CW	9.1
	3	15/15Ti	0.6
	4	15/15Ti	5.7
	5	15/15Ti	5.3
	6	15/15Ti	9
PFR	1	316CW	1.2
	2	316CW	10.7
	3	316CW	5.6
	4	316CW	7.6
	5	316CW	10.7
	6	316CW	10.4
	7	316CW	3.14
	8	316CW	9.54
FFTF	1	316SS	10
Average			7.17

Conservatively assuming that the burn-up at the time of the adventitious fuel pin failure in Monju is 6.5 at.% which is same as that in Mol-7B experiment, cladding defect size never exceeds 200 mm^2 at the maximum burn-up of approximately 8 at.% in Monju because cladding defect size was less than 200 mm^2 even at the maximum burn-up of 13 at.% in Mol-7B experiment. Furthermore increase rate of cladding breached area is approximately 10 mm^2/day at most and it will decrease to zero as time goes on [16].Consequently cladding defect size never exceeds 5000 mm^2 until refuelling even assuming initial cladding defect size of 200 mm^2. Therefore the probabilities of cladding defect size over alert threshold and reactor trip threshold by DN method were judged to be highly unlikely (0.05).

3.2. Fuel pin failure propagation to the peripheral 6 pins until refuelling [heading (4)]

Normal reactor shutdown will be initiated after the detection of 0.01 % of total driver fuel pin failures at the precipitators in CG method and after the detection of 0.02 % of total driver fuel pin failures at the NaI detectors in CG method. More than 3 or 6 fuel pin failures are necessary for normal reactor shutdown by precipitators or NaI detectors respectively because the number of driver fuel pin in the core are 33,462 pins in Monju.

Analyses for following two possible causes of FEFPA were performed in this study:
(1) Thermal transient due to FP gas release from adjacent fuel pin;
(2) Flow reduction due to flow blockage.

3.2.1. Thermal transient due to FP gas release from adjacent fuel pin

Figure 5 shows cladding temperature history analysed by FALL code [17-19] of which main analytical conditions are shown in Table 7. Although cladding temperature increased up to approximately 750 degree C during FP gas release in the case that angles with gas blanketing are 360 degree assuming multiple fuel pin failures, duration of FP gas release was estimated at most 104s even under the condition that the reactor operation is kept without removing the initial defect pin from the core. Necessary duration for cladding creep failure of the fuel pin at the end of cycle is approximately 130 hours at 750 degree C from Figure 6 which was used in the licensing document of Monju [17]. Therefore cladding failure due to FP gas release from adjacent fuel pins was judged to be impossible (ε) in Monju.

Table 7　Main analytical conditions for FALL code analysis

Axial position of gas blanketing	Top of fissile column (TFC) / Peak power node (PPN)
Power density of fuel (W/cm^3)	938 at TFC / 1720 at PPN
Angles for gas blanketing (degree)	180 / 360
Released gas temperature (degree C)	660 at TFC / 535 at PPN
Coolant inlet temperature (degree C)	655 at TFC / 525 at PPN
Coolant outlet temperature (degree C)	660 at TFC / 535 at PPN
Number of axial cell	1
Number of radial cell	15
Fuel pellet	10
Fuel-Cladding gap	1
Cladding	3
Coolant	1
Numbers of azimuthal cell	
Angles with gas blanketing	3 for 180degree / 1 for 360degree
Angles without gas blanketing	3 for 180degree / 0 for 360degree

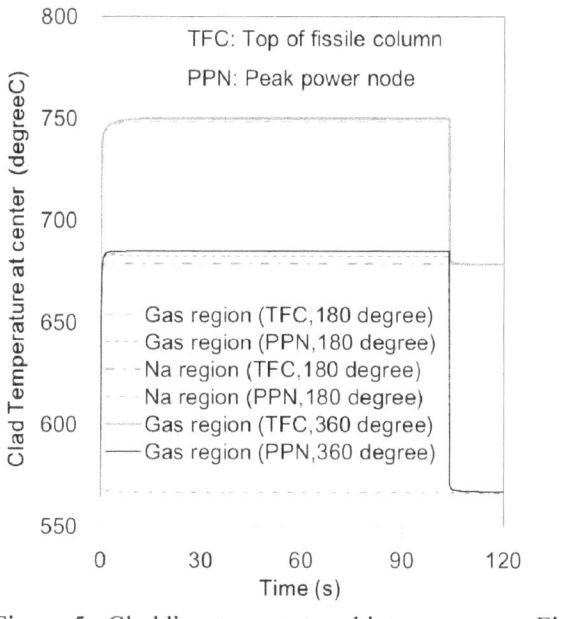

Figure 5　Cladding temperature history

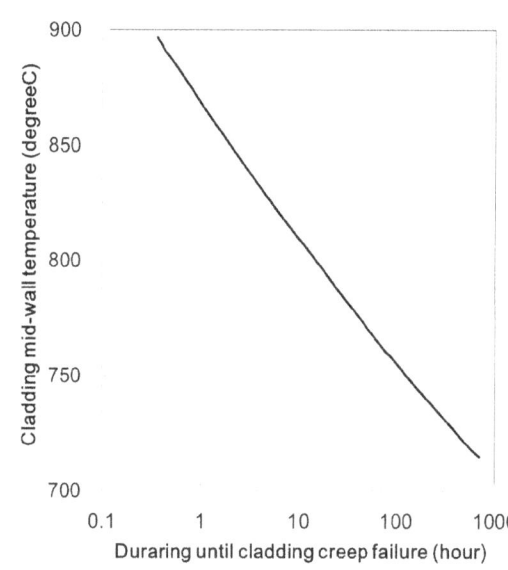

Figure 6　Duration until cladding creep failure

3.2.2. Flow reduction due to flow blockage

Total flow blockage was approximately 38% of free sodium channel at the end of the RBCB in the Mol-7B experiment [15] in the higher power and higher coolant temperature conditions compared with Monju. Seventeen fuel pins of total 18 irradiated fuel pins were failed in this experiment. Therefore flow blockage by each fuel pin was approximately 2.2 % of total flow area. It should be noted that flow blockage rate by each fuel pin in Monju is much smaller because there is 169 fuel pins in one SA. Figure 7 shows cladding and coolant temperatures in case of flow blockage analysed by SEETHE code [17, 18] of which main analytical conditions are shown in Table 8. Up to approximately 30 % of flow blockage, coolant temperatures and cladding temperatures were below boiling point and 830 degree C respectively which were the safety criteria for fuel pin failure in the safety assessment for Monju [17]. Therefore fuel pin failure propagation was judged conservatively to be highly unlikely (0.05) at the 2.2 % blockage of total flow area.

Table 8   Main analytical conditions for SEETHE code analysis

Radial position of blockage	Center
Axial position of blockage	Axial center of the core
Blockage rate (%)	9.5 / 14.3 / 19.0 / 25.4 / 31.7 / 39.7 / 47.6 / 57.1 / 66.6 / 77.7
Number of radial mesh	14
Number of axial mesh	50
Axial height for calculation (m)	1.0
Nominal power condition	
Power of the sub-assembly (kW/m)	39.4
Coolant inlet temperature (degree C)	487
Coolant inlet velocity (m/s)	549
Decay heat power condition	
Power of the sub-assembly (kW/m)	28.0
Coolant inlet temperature (degree C)	468
Coolant inlet velocity (m/s)	0.429

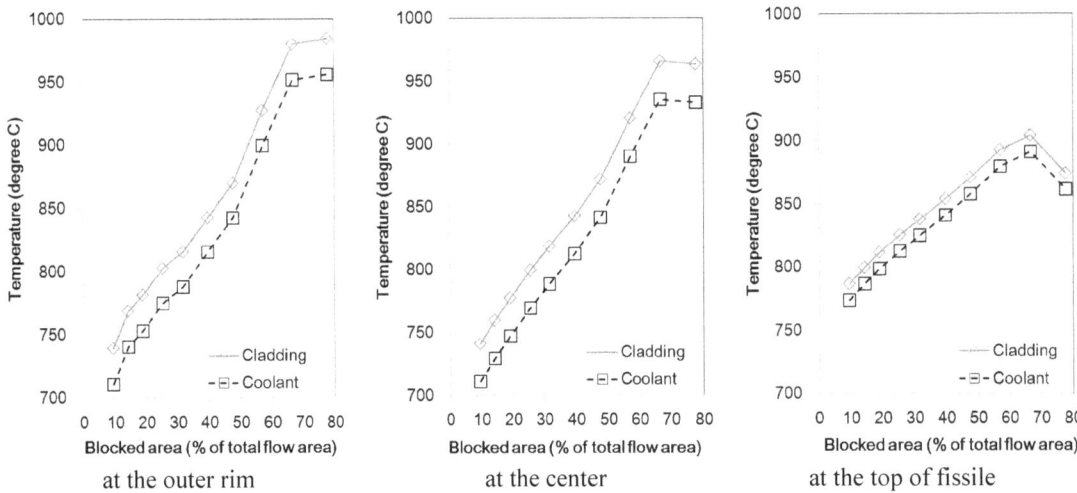

at the outer rim   at the center   at the top of fissile

Figure 7   Cladding and coolant temperature in the downstream of blockage at nominal power

Therefore fuel pin failure propagation to the peripheral 6 pins until refuelling was conservatively judged to be highly unlikely (0.05).

3.3. Damage propagation over alert threshold by TC at the outlet of SA [heading (9)]

Manual reactor shutdown will be initiated if coolant flow blockage exceeds 66% of total flow area in one SA which is the alert threshold of TC at the outlet of SA. Although there is no RBCB experiment with cladding defect size over 5,000 mm^2, probability of damage propagation from cladding defect of which size was over 5,000 mm^2 until flow blockage rate of 67% was quantified based on the analyses and related experimental data.

As described in Sec. 3.2.1, damage propagation is impossible (ε) based on the analysis on FP gas release even assuming multiple pin failures.

The remaining possibility of damage propagation is only in the case of molten fuel ejection into the coolant channel due to flow blockage induced by fuel sodium reaction product. In terms of possibility of molten fuel ejection, there was neither fuel melting nor molten fuel ejection into the coolant channel in the existing RBCB experiments. In addition, Figure 8 shows radial temperature distribution within the fuel calculated conservatively by FALL code in case of 67% of coolant flow blockage Monju. Areal melt fraction was approximately 5% which was far below the molten fuel ejection threshold of at least 20 % [19]. This result shows that there is no molten fuel ejection before heading (9). On one hand, it should be noted that damage propagation will be highly unlikely even in case of small amount of molten fuel ejection [20]. Therefore damage propagation over alert threshold by TC at the outlet of SA was judged to be conservatively highly unlikely (0.05).

3.4. Failure of decay heat removal after reactor scram [heading (11)]

Coolant temperature profile after the reactor scram in case of 67% of coolant flow blockage was calculated by SEETHE code as shown in Figure 9. Maximum coolant temperature was 680 degree C and far below the boiling point of coolant. Fuel and cladding temperature were also calculated by FALL code as shown in Figure 8. Fuel and cladding temperature was far below the fuel melting point and 830 degree C respectively which are the conservative criteria for fuel pin failure. Therefore failure of decay heat removal after reactor scram is judged to be unlikely to occur (0.2).

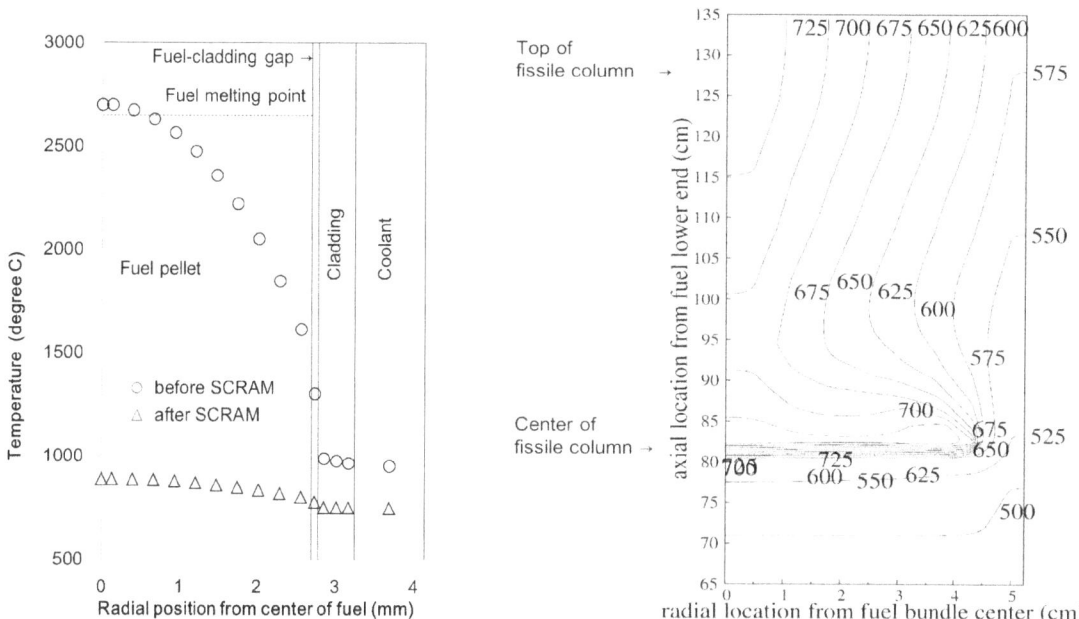

Figure 8 Fuel and cladding temperature profile calculated by FALL code

Figure 9 Coolant temperature profile calculated by SEETHE code

3.5. Damage propagation to the peripheral SAs [heading (12)]

Although damage propagation is unlikely in case of small amount of fuel melting [18], damage propagation is conservatively judged to be likely (0.8) if large amount of fuel melting due to more than 66% of flow blockage. It should be noted that damage propagation might be terminated in the inter SA gap or control rod guiding tube (CRGT) before propagation to the peripheral six SAs.

3.6. Fault tree analysis for other headings [heading (3), (5), (6), (8), and (10)]

There is functional dependency among headings (3), (5), (6), (8), and (10) in Figure 3 because those headings share some support and common systems. It is necessary to consider this functional dependency in the event tree quantification. So, those shared systems were identified as shown in the upper part of Table 9. Since failure of these shared systems causes dependently functional failure in some of those headings, combination of failures of the support and common systems was also developed in support event tree. Then, the functional dependency was considered by combining the main and support event trees with the event tree linking (ETL) method by using RISKMAN® code. In order to obtain the branch probability in those event trees, the fault tree analysis (FTA) was performed in addition to the consideration in the section 3.1 through 3.5. This FTA includes the quantification of human error probability (HEP) on the basis of allowable time estimation for operators. HEP during the allowable time is estimated based on time reliability curves from technique for human error rate prediction (THERP) [21]. Main results in this FTA as shown in Table 9 were applied to calculate the accident sequence probability with the ETL method.

Table 9   Results of FTA

Headings					Probability
Support and common systems for frontline systems	I	Unavailability of support systems (power supply) for manual reactor shutdown			2.30E-06
	II	Unavailability of common system for manual reactor shutdown by the alert from precipitators or NaI detectors (Ar gas sampling line)			7.60E-04
	III	Unavailability of common component for precipitators (3 signal line and calibration error)			4.01E-04
	IV	Unavailability of common component for precipitators (2 signal line)			1.69E-08
	V	Common cause failure and calibration error in DN method for manual and automatic shutdown			4.01E-04
	VI	Unavailability of common system in DN method for manual and automatic shutdown (detectors, control circuits, and power supply for A loop)			2.71E-06
	VII	Unavailability of support system in DN method for manual and automatic shutdown (power supply for B loop)			1.95E-06
	VIII	Unavailability of support system in DN method for manual and automatic shutdown (power supply for C loop)			1.95E-06
	IX	Failures of components necessary only for manual shutdown			2.91E-06
	X	Failures of components for reactor shutdown (reactor trip breakers and control rods)			6.41E-08
Main event tree headings (Frontline systems)	(3)	Failure of manual reactor shutdown by the alert from DNDs and precipitators	(3)-2	Failures of DN detectors	1 out of 2:7.60E-07 2 out of 2:5.77E-13
			(3)-3	Cognitive and decision error against alert	4.26E-04
	(5)	Failure of normal reactor shutdown by the second alert from precipitators	(5)-2	Cognitive and decision error against alert	4.26E-04
	(6)	Failure of normal rector shutdown by the alert from NaI detectors	(6)-1	Failures of NaI detectors	1.16E-06
			(6)-2	Cognitive and decision error against alert	8.27E-04
	(8)	Failure of automatic reactor shutdown by the trip signal from DNDs	(8)-1	Failures of DN detectors	1 out of 2:7.60E-07 2 out of 2:5.77E-13
	(10)	Failure of manual reactor shutdown by the alert from TC	(10)-1	Failures of components related to alert signal lines	9.94E-04
			(10)-2	Cognitive and decision error against alert	4.26E-04

4. RESULTS AND DISCUSSIONS

Table 10 shows the results of ETA for FEFPA compared with those in existing PRA. Probability of damage propagation to the peripheral SAs was estimated to be 1.7×10^{-12} based on the ETA in this study. It should be noted that probability of whole core damage is extremely small because there are following other detection and reactor shutdown systems after damage propagation to the peripheral SAs:

-Manual reactor shutdown owing to the alert from primary argon monitor;

-Automatic reactor shutdown owing to high neutron flux level.

The probability of adventitious fuel pin failure in this study is much smaller than that in existing PRA in the following reasons:

-The probability of fuel pin failure used in this study was derived from pin failure experiences after 1985 in addition to those before 1985 which were used in existing PRA.

-Fuel pin failure experiences in small size reactors of which nominal full power were less than 100 MWth are excluded in this study for Monju of which nominal full power were 280 MWth;

The probability of damage propagation to the peripheral SAs was also much smaller than that in existing PRA because of the following reasons:

-The probability of adventitious fuel pin failure in this study was small.
-Various detection and reactor shutdown systems were taken into account in the ETA of this study;
-Probability of damage propagation within the SA was reduced reflecting latest experimental and analytical knowledge.

Table 10   Result of ETA for FEFPA in Monju compared with those in existing PRA

	Vaughan (CDFR)	Schleisiek (SNR 300)	JNES (Monju)	This study (Monju)
Probability of adventitious fuel pin failure [pin/ry]	35	No estimation (1)	4.4	2.4
Bulk boiling in one SA [ry]	-	6.5×10^{-6}	-	-
Fuel melting in a substantial part of the incident SA [ry]	1.9×10^{-6}	-	-	-
Damage propagation to the peripheral SA [ry]	-	1.0×10^{-7}	-	1.7×10^{-12}
Untripped States (Limited damage) [ry]	2.0×10^{-7}	-	-	-
Damage propagation more than 37 SA [ry]	-	-	5.6×10^{-8}	-
Whole core accident [ry]	1.9×10^{-9}	-	-	-

Table 11 shows the core damage fraction (CDF) of the anticipated transient without scram (ATWS) and protected loss of heat sink (PLOHS) in Monju [22].

Table 11   CDF from FEFPA compared with those from ATWS and PLOHS

	Without scram	With scram (failure of decay heat removal)
CDF from FEFPA (/ry)	$\sim 9.6 \times 10^{-13}$	$\sim 7.3 \times 10^{-13}$
CDF from ATWS and PLOHS (/ry)	$\sim 3 \times 10^{-8}$ in ATWS	$\sim 5 \times 10^{-8}$ in PLOHS

The probability of damage propagation to the peripheral SA without or with scram is smaller than CDFs of ATWS and PLOHS. Furthermore the consequence of whole core accident from adventitious fuel pin failure without or with scram is not greater than that of ATWS or PLOHS because almost all the SA will damaged at ATWS or PLOHS. Therefore FEFPA can be included in CDF of ATWS or PLOHS in the viewpoint of both probability and consequence.

It should be noted that the CDF of FEFPA in the future FBR can become larger than that in Monju because higher fuel burnups may induce more fuel swelling and denser fuel pin arrangement may induce fuel melting.

5. CONCLUSIONS

ETA from adventitious fuel pin failure in Monju was performed in this study based on latest knowledge on experimental and analytical studies for failure propagations and reflecting latest operation manual at emergency in Monju. Probability of damage propagation to the peripheral SAs was quantified to be 1.7×10^{-12} in Monju based on the ETA. Therefore probability of whole core damage is much smaller than this value because there are other detection and reactor shutdown systems after damage propagation to the peripheral SAs. It was clarified in this study that damage propagation from adventitious fuel pin failure in Monju can be included in CDF of ATWS or PLOHS in the viewpoint of both probability and consequence.

Acknowledgements
The authors are grateful to M. Sotsu, Y. Morohashi and S. Suzuki of JAEA for their useful suggestions and comments for FTA. The authors would like to express their gratitude to N. Yoshioka of NESI Inc. who assisted the calculation using FALL and SEETHE codes.

References

[1] K. Schleisiek, "Risk Oriented Analysis of Subassembly Accidents", Proceedings of International Topical Meeting on Fast Reactor Safety, 1985 May 12-16, Guernsey (UK), pp. 141-149.

[2] G. J. Vaughan, "Event Tree Analysis of the Sub-assembly Accident", Proceedings of International Topical Meeting on Fast Reactor Safety, 1985 May 12-16, Guernsey (UK), pp. 457-463.

[3] Japan Nuclear Energy Safety Organization, "Study on analytical method for local subassembly fault in the fast reactor", Japan: Japan Nuclear Energy Safety Organization; JNES/SAE05-108; 2005 [in Japanese].

[4] S. Suzuki, "Summary of the JOYO Mk-III upgrade," Japan: Japan Atomic Energy Agency; JNC TN9200 2003-003; 2003 [in Japanese].

[5] Japan Atomic Energy Commission, "Atomic energy white paper", Japan: Japan Atomic Energy Commission; 2012 [in Japanese]

[6] H. Plitz, G. C. Crittenden, A. Languille, "Experience with failed LMR oxide fuel element performance in European fast reactors", J.Nucl.Mat. 1993;204,238-243.

[7] SFEN, "Power Plant Safety and Fuel performance - The PHENIX Reactor: Assessment of 35 Years' Operation-", France: SFEN; 2009

[8] IAEA, "Status of liquid metal cooled fast reactor technology", International Atomic Energy Agency; IAEA-TECDOC-1083; 1999

[9] IAEA, "Liquid Metal Cooled Reactors: Experience in Design and Operation", International Atomic Energy Agency; IAEA-TECDOC-1569; 2007

[10] IAEA, "Fast reactor database 2006 update", International Atomic Energy Agency; IAEA-TECDOC-1531; 2006

[11] Office for Nuclear Regulation, "MOX FUEL MANUFACTURE AT SELLAFIELD", UK; Office for Nuclear Regulation; 2000 (http://www.hse.gov.uk/nuclear/mox/mox2.htm)

[12] R. B. Baker, F. E. Bard, R. D. Leggett, A. L. Pitner, "Status of fuel, blanket, and absorber testing in the fast flux test facility", J.Nucl.Mat., 1993;204,109-118.

[13] Y. Maeda, Y. Kashimura, T. Suzuki, K. Isozaki, H. Hoshiba, R. Kitamura, T. Nakano, M. Takamatsu, T. Sekine, "Periodic Safety Review of the Experimental Fast Reactor JOYO - Review of the Activity for Safety –", Japan: Japan Atomic Energy Agency; JNC TN9440 2005-001; 2005 [in Japanese].

[14] S. Miyakawa, "Mechanism of fuel release to coolant from breached oxide fuel in liquid metal fast reactor", Transactions of the Atomic Energy Society of Japan, 1994;36,879-888.

[15] P. Weimar, W. Ernst, "Mol-7B - an 18-pin bundle operating 200 days beyond breach", Nuclear Technology. 1982 April; 57:81-89.

[16] K. Haga, "Status of research on local fault and prospect to future work," Japan: Japan Atomic Energy Agency; JNC TN2410 87-002; 1987 [in Japanese].

[17] Japan Atomic Energy Agency. "The licensing document for the construction permit of the prototype FBR Monju", Japan: Japan Atomic Energy Agency; 1980, revised 2006 [in Japanese].

[18] R. Nakai, M. Itoh, K. Terata, Y. Kani, K. Maeda, H. Endo, S. Kondo, K. Aizawa, Y. Ohmori, "Computer codes for safety analysis of LMFBR", Japan: Japan Atomic Energy Agency; PNC TN241 81-28; 1981 [in Japanese].

[19] Y. Fukano, J. Charpenel, "The adventitious-pin-failure study under a slow power ramp", Proc. 12th Int. Conf. on Nuclear Engineering (ICONE12); 2004 April 25–29; Arlington (VA). [CD-ROM]

[20] Y. Fukano, "Comprehensive and consistent interpretation of local fault experiments and application to hypothetical local over-power accident in Monju", J.Nucl.Sci.Technol., 2013;50,950-965.

[21] A. D. Swain, H. E. Guttman, "Handbook of Human Reliability Analysis with Emphasis on Nuclear Power Plant Applications", USA: Nuclear Regulatory Commission NUREG/CR -1278; 1983.

[22] Tsuruga Head Office, Japan Atomic Energy Agency, "Summary of AM report in Monju", 2008 (Available at http://www.jaea.go.jp/04/turuga/jturuga/press/2008/03/p080317.pdf) [in Japanese]

License Application for a Spent Nuclear Fuel Repository in Sweden

Allan Hedin[*]

Swedish Nuclear Fuel and Waste Management Co. (SKB), Stockholm, Sweden

Abstract: The Swedish Nuclear Fuel and Waste Management Co., SKB, has applied for a license to build a final geological repository for spent nuclear fuel at the Forsmark site, situated around 70 miles north of Stockholm, Sweden. A key component in the license application is an assessment of the long-term safety of the repository. Probabilistic radionuclide transport and dose calculations are at the core of the analysis. The license application is currently (Spring 2014) under review by the Swedish Radiation Safety Authority, SSM, and a report from SSM to the Swedish Government is expected in 2015.

This paper *i*) gives an overview of the probabilistic dose calculations of the safety assessment in the license application and *ii*) presents some new results related to the probabilistic calculations obtained after the completion of the assessment.

Keywords: Nuclear Waste, Spent Nuclear Fuel, Probabilistic Safety Assessment, Sweden, SR-Site.

1. INTRODUCTION

The Swedish Nuclear Fuel and Waste Management Co., SKB, jointly owned by the nuclear reactor owners, is responsible for the management of the nuclear waste arising from the nuclear power operations in Sweden. Within SKB's program for the management of spent nuclear fuel, an interim storage facility and a transportation system are today (Spring 2014) in operation. A principal remaining task in the program is to build and operate a final repository for the spent nuclear fuel.

In April 2011, SKB applied for a license to build a final geological repository for spent nuclear fuel at the Forsmark site, situated around 70 miles north of Stockholm, Sweden. A key component in the license application is an assessment of the long-term safety of the repository [1], called the SR-Site assessment. Probabilistic radionuclide transport and dose calculations are at the core of the analysis. The license application is currently under review by the Swedish Radiation Safety Authority, SSM, and a report from SSM to the Swedish Government is expected in 2016.

Several decades of research and development has led SKB to put forward the KBS-3 method for the final stage of spent nuclear fuel management. In this method, copper canisters with a cast iron insert containing spent nuclear fuel are surrounded by bentonite clay and deposited at approximately 500 m depth in groundwater saturated, granitic rock, see Figure 1. The purpose of the KBS-3 repository is to isolate the nuclear waste from man and the environment for very long times. The primary safety function of the repository is to completely contain the nuclear fuel within the canisters. Around 12,000 tonnes of spent nuclear fuel is forecasted to arise from the currently approved Swedish nuclear power program (where the last of the 10 operating reactors is planned to end operation in 2045), corresponding to roughly 6,000 canisters in a KBS-3 repository.

The principal acceptance criterion according to Swedish legislation is a requirement that the annual risk of individuals in the most exposed group of contracting cancer or hereditary effects from repository derived radionuclides must not exceed one in a million. The risk is obtained as a time-dependent mean value of probabilistically calculated doses, multiplied by a dose-to-risk conversion factor of 0.073/Sv, provided the Swedish regulator and in agreement with recommendations from the International Commission on Radiological Protection, ICRP. The assessment is required to cover one million years after repository closure.

[*] E-mail allan.hedin@skb.se

Figure 1: The KBS-3 concept for disposal of spent nuclear fuel.

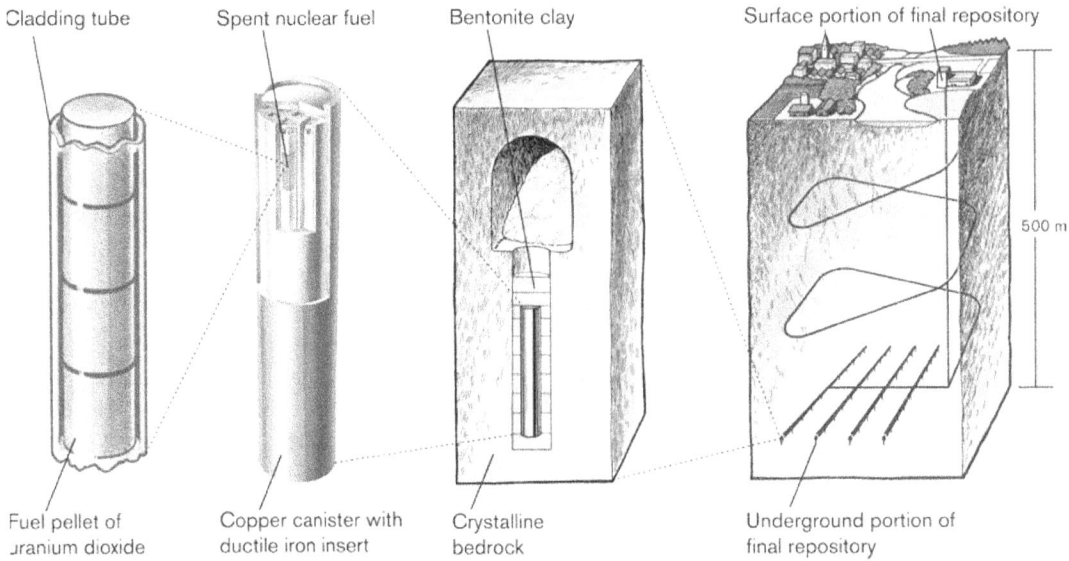

Cladding tube | Spent nuclear fuel | Bentonite clay | Surface portion of final repository

500 m

Fuel pellet of uranium dioxide | Copper canister with ductile iron insert | Crystalline bedrock | Underground portion of final repository

This paper *i*) gives an overview of the probabilistic dose calculations of the safety assessment in the license application and *ii*) presents some new results related to the probabilistic calculations obtained after the completion of the assessment.

2. OVERVIEW OF PROBABILISTIC DOSE CALCULATIONS

2.1. Assessment Strategy

2.1.1. General

The overall approach for <u>achieving</u> safety in the KBS-3 repository is to *i*) locate the repository in an environment that is expected to be stable in the long time frames during which the spent nuclear fuel poses a hazard to man and the environment, and *ii*) to use materials that are compatible with that environment in the construction of protecting barriers in the repository.

A central element in the <u>demonstration</u> of safety, i.e. in the safety assessment, is, therefore, to assess how the repository environment and the engineered barriers evolve in time after closure of the repository. A reference evolution spanning approximately 120,000 years is studied as a starting point. This time frame is the period of future glacial cycles that largely determine the boundary conditions for the evolution of the repository. In the reference evolution, a repetition of the last glacial cycle is assumed. This is a starting point, based on which other scenarios are considered and analyzed.

2.1.2. Scenario Disaggregation

In principle, the product of dose consequences and likelihoods of all possible future evolutions of the repository should be weighed together and presented as a time-dependent risk. The spectrum of possible evolutions is, however, very wide and cannot be captured in a detailed sense.

The usual approach taken in safety assessments, and also in the SR-Site assessment, is to work with scenarios and variants that are designed to capture the broad features of a number of representative possible future evolutions. Together, these are intended to give a reasonable coverage of possible future exposure situations.

The scenarios are selected based on a number of specified safety functions of the repository. For example, the canisters should withstand isostatic loads in the repository and they are designed to withstand loads of 45 MPa. In order to critically analyze the potential to withstand isostatic loads, a dedicated scenario is constructed, in which all factors of importance for the isostatic load are considered. Such factors are the thickness of future glaciers above the repository determining hydrostatic pressures on the canisters and the likelihood of manufacturing flaws rendering a canister more vulnerable to isostatic loads.

Each scenario may be subdivided into several variants and each variant may, in the probabilistic consequence calculations, be represented by several calculation cases.

2.1.3. Overestimation of Risk

The formulation of scenarios, variants and calculation cases, and the subsequent weighing together of these to give a total risk aims at an over prediction of risk. SSM's regulation requires that the annual risk should be less than 10^{-6}. There are a number of uncertainties that cannot be managed quantitatively in any other rigorous manner from the point of view of demonstrating compliance than by pessimistic assumptions.

Another situation in which risk has to be overestimated concerns scenario probabilities. Regarding e.g. future climate, both repetitions of conditions reconstructed for the past 120,000 year glacial cycle and an alternative where this development is considerably perturbed by a global warming effect can be envisaged. Although the two are mutually exclusive, both must be regarded as possible. In the risk summation, the logical position is adopted that the summed consequence of a set of mutually exclusive scenarios can, at any point in time, never exceed the maximum of the individual scenario consequences. For scenarios and variants where defensible probabilities are difficult to derive, a scenario or variant giving high consequences can pessimistically be assigned unit probability and other scenarios and variants yielding lower dose impacts can be "subsumed" under the one with the more severe consequences.

Although the primary aim with risk calculations is to demonstrate compliance, there is also the clear ambition of clarifying the sensitivities of the calculation results. For this aim, the calculation cases should be, in principle, as realistic as possible in capturing uncertainty. One quantitative tool for this is the use of probabilistic evaluations of calculation cases followed by sensitivity analyses of the results. For this reason, *i*) pessimistic simplifications are avoided where a sound scientific basis exists for a quantitative treatment and *ii*) pessimistically neglected features of the system are included in discussions of sensitivities.

2.2. The Structure of Scenarios and Calculation Cases

For each scenario (or scenario variant), the containment potential of the repository, i.e. the potential of the canisters to completely contain the spent fuel, is analyzed in a first step. If the system evolution is found to lead to failure of the canisters, the extent of canister failures and the failure times are quantified. This information is used as input to probabilistic radionuclide transport and dose calculations. For each scenario, a number of probabilistic calculation cases are formulated. The assessment strategy, when implemented, thus leads to a structure of scenarios (or scenario variants), each with a number of probabilistic calculation cases. Figure 2 shows an example for the scenario that yielded the highest risk contributions in the SR-Site assessment, the so called corrosion scenario, where canister failures due to corrosion are analyzed. The first two subdivisions of calculation cases arise due to two factors that are of importance for the scenario, and that can not readily be treated probabilistically: the hydrogeological model of the repository (three model variants, blue text in Figure 2) and the model used to quantify long-term loss of the protecting clay buffer around the canisters (three different models, red text). Together the two factors give rise to 3×3 calculation cases. Additional factors related to radionuclide transport and dose calculations give rise to a further multitude of calculation cases. A central corrosion case, based on reasonable assumptions regarding all the mentioned factors, is used as a starting point. The factors related to radionuclide transport and dose were studied in a first step, and this led to the conclusion that the central assumptions regarding these do not yield significantly different results than the alternative assumptions. The remaining $3\times3-1$ combinations of hydrogeological and buffer loss cases were therefore analyzed only for the central assumptions regarding radionuclide and dose factors.

Figure 2. Overview of calculation cases for the corrosion scenario.

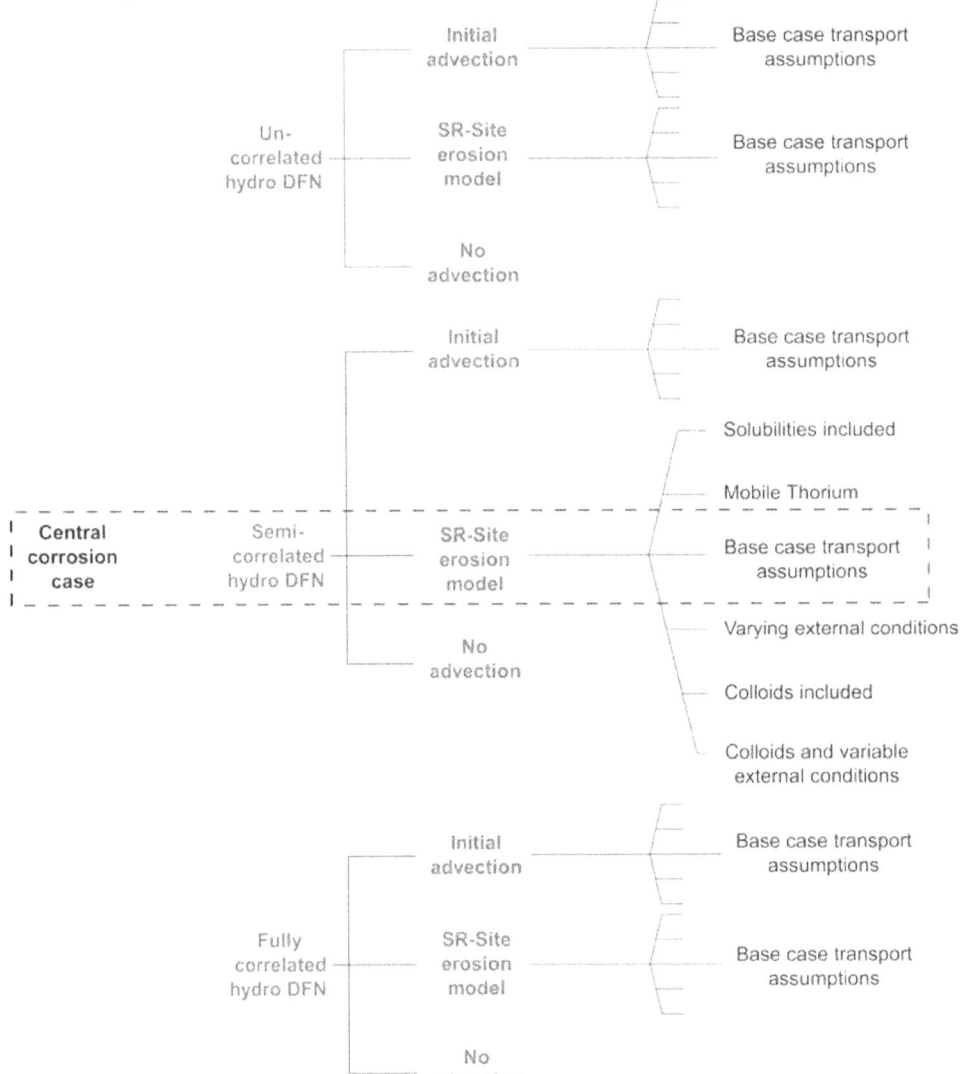

2.3. Key Probabilistic Results

2.3.1 Input Data Compilation

A protocol for the compilation of input distributions for the probabilistic calculations was established early in the project. The protocol requires, among other things and for each input data entity, critical evaluation of data sources, discussion of correlations between input parameters, and a qualification of an input probability density function, PDF. This is briefly described in chapter 9 of the safety assessment report [1], and in more detail in a dedicated Data report [2].

2.3.2. Key results

The result of the probabilistic calculation of the central corrosion case is shown in Figure 3. The figure shows the mean annual effective dose as a function of time, obtained with 10,000 realizations and using Latin Hypercube Sampling. The increase in dose over time is essentially caused by the increasing likelihood for canister failure as corrosion proceeds over time.

Figure 3. Mean annual effective dose equivalent release for a probabilistic calculation of the central corrosion case. The average number of failed canisters is 0.12. The peak mean annual effective dose over one million years in µSv is given in brackets in the legends.

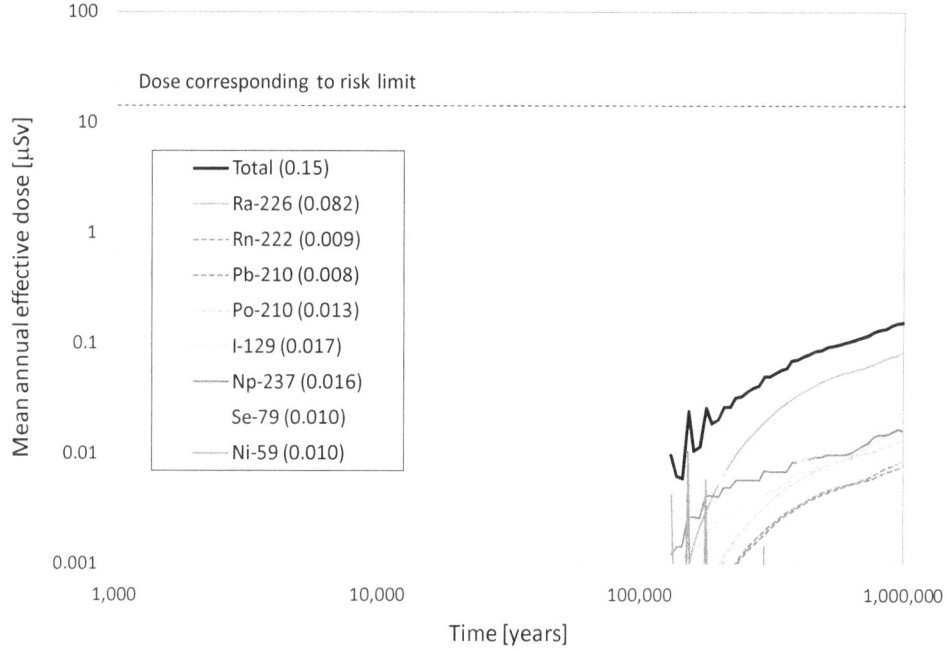

The results of all 3×3 main corrosion cases are shown in Figure 4. Since the three "no advection" cases, representing the cases where the protective buffer remains intact, yield no canister failures and thus zero releases, the nine cases generate only six release curves.

The corrosion scenario yields the highest risk contributions in the safety assessment. The cases shown in Figure 4 are argued to, together, cover uncertainties in the quantification of the corrosion scenario and to each be an upper bound on the cases they represent. Since the calculated mean doses are all below the dose corresponding to the regulatory risk limit over the entire one million year assessment period, these results are key elements in arguing that the proposed repository fulfils the regulatory risk limit at the Forsmark site.

Figure 4. Summary of far-field mean annual effective dose for probabilistic calculations of six variants of the corrosion scenario. The peak mean annual effective doses, all occurring at the end of the assessment period, are given in parentheses in μSv.

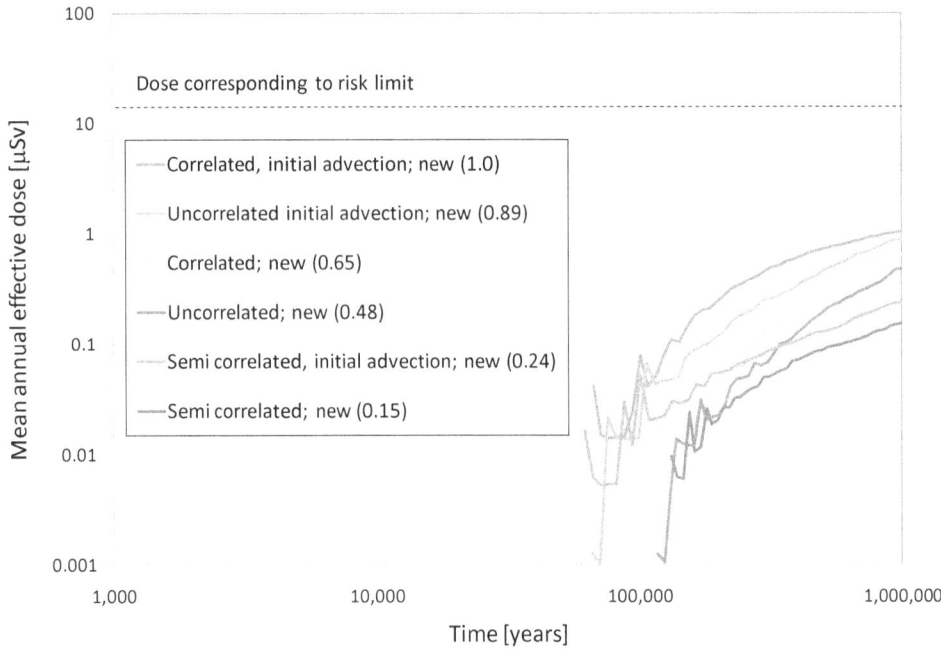

2.5. Sensitivity Analyses

Regarding sensitivities, it is of interest to determine *i*) the input parameters that correlate most strongly with the dose over the entire dose range and *ii*) the input parameter values that are related to high and low doses. Ra-226 dominates the dose in most of the realizations in the dose dominating corrosion scenario and it is thus of particular interest to clarify sensitivities of the Ra-226 dose to input parameters. The first purpose is achieved with two methods: *i*) determination of standardized rank regression coefficients and *ii*) determination of variance based sensitivity indices. The former is presented in the safety assessment report [1], whereas the latter has been applied after the completion of the report [3].

The second purpose is also achieved with two methods: *i*) the calculation of conditional mean values and *ii*) the application of so called cobweb plots. Again, the former is presented in the safety assessment report [1], whereas the latter has been applied after the completion of the report [3].

Finally, the results of the above methods are further explained by the use of a tailored regression model [1] that demonstrates how the variability in the output can be explained with analytic expressions derived from the conceptual understanding of the transport processes involved in the dose calculations. Based on the understanding and mathematical formulation of the transport processes involved in the radionuclide transport calculations, a number of tailored regression models that include successively more input variables were constructed for the peak dose of Ra-226, the radionuclide that dominates most of the numerical model realization. The highest order model yields an R^2-value of 0.99 when regressed on the results of the numerical transport models, see Figure 5 that shows both the regression model expressions as an insert, and the regression results.

Figure 5. Four tailored regression models, based on successively more variables, for the Ra-226 peak dose. For example, the black dots show the good agreement of the results of a five-parameter tailored regression model with those of the numerical transport model.

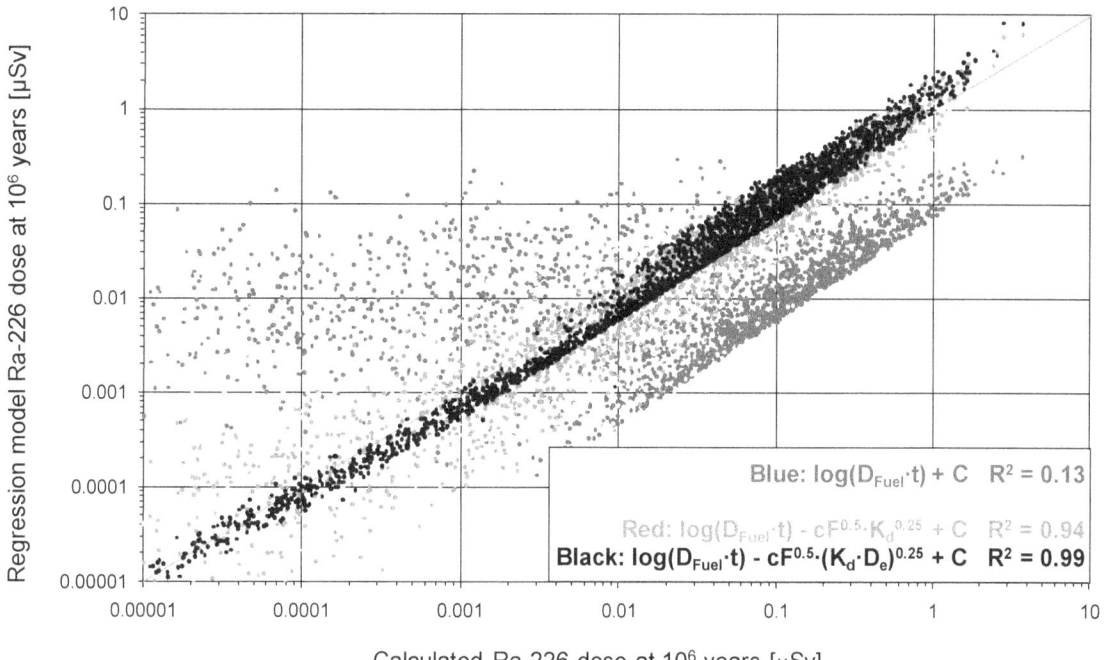

All the applied methods point to two input parameters to which the results are most sensitive: The rate at which the fuel matrix is converted, D_{Fuel}, with the associated release of radionuclides, and the transport resistance in the geosphere, F, which is closely related to the flow conditions along the transport path from the failing canister to the biosphere. This is further discussed in section 13.5.11 of the safety assessment [1] and in Reference [3].

3. ADDITIONAL RESULTS

What follows are new results in two areas, obtained after the completion of the safety assessment SR-Site.

3.1. Determination of the Variability of the Probabilistic Results

The probabilistic dose assessments were set up to obtain a statistically correct mean value, whereas the determination of output variability was simplified regarding some aspects of the problem. Specifically, in the probabilistic calculations, it was postulated that one canister fails, the mean dose of this case was calculated and subsequently multiplied by the mean number of failed canister for a particular calculation case.

To analyze the impact of this simplification, a more elaborated set-up of the calculation cases is applied and the variability is determined from these. Each case in Figure 2 is based on one of three hydrogeological model variants. The central corrosion case is e.g. based on the so called semi-correlated hydrogeological variant. In this variant, as in the other two, the network of rock fractures is described statistically. In a particular realization of a model variant, a network of fractures is generated in accordance with the statistical model related to the hydrogeological variant. This, in turn, will determine the flow condition at each of the 6,000 canister positions in the repository for the realization in question. Furthermore, the groundwater concentration of the main canister corroding agent, sulfide, is sampled for each canister position from a distribution determined from the present and future hydrogeochemical conditions at the site. These two stochastic factors, the groundwater flow and the sulfide concentration, will basically determine which (if any) of the 6,000 canisters that will fail in the corrosion scenario. The number of failed canisters, their failure times, their positions in the repository, and the flow related transport properties for radionuclides at these positions will thus vary between realizations of the hydrogeological model variant.

The statistically correct way of determining the variability of dose results this variation gives rise to is to determine the number of failed canisters, failure times and transport properties in each fracture network realizations, and then run a number of radionuclide transport and dose realizations for each of these. The former will essentially cover aleatory uncertainty related to the natural variability of the host rock conditions whereas the latter is essentially related to epistemic uncertainty where the fuel matrix conversion rate is a main contributor.

In summary, the following procedure was followed to evaluate one of the calculation cases:
1. Generate a realization of the network of water conducting fractures in accordance with the statistical model of the hydrogeological model variant. This yields a flow rate and transport related rock properties for each of the 6,000 canister positions.
2. Sample a groundwater sulfide concentration for each of the 6,000 positions. Determine, based on the flow rates and the sulfide concentrations, which (if any) of the 6,000 canisters will fail due to corrosion during the one million year assessment period; and the associated failure times. Repeat this step 1000 times to capture the variability caused by the distribution of sulfide concentrations, yielding 1000 sets of canister failure times and associated transport conditions.
3. For each of the 1000 sets generated in step 2, determine the dose consequences by running 1000 realizations of the radionuclide transport and dose model, sampling the additional parameters required by the model.
4. Repeat steps 1 through 3 for a number of realizations of the network of fractures. There are typically between 5 and 10 such realizations.

The result in terms of the number of failed canisters of this considerably more elaborate calculation scheme is shown in Figure 6 for the 5 realizations of the fully correlated hydrogeological model variant (see Figure 2). This variant was chosen since it gave rise to the highest doses in the safety assessment. Figure 6 shows the distribution of the number of failed canisters for each of the rock realizations when the sulfide concentration for the 6,000 positions is sampled 1,000 times. Also the results when combining all realizations is shown. As seen, there is a considerable probability that no canister will fail, and 6 canisters fail in the sulfide realization yielding the highest failure count. The mean number of failed canisters is 0.733, which is in good agreement with the theoretically correct 0.729 obtained with the simplified calculation scheme.

Figure 6. Distribution of the number of failed canisters for each of the five realizations of the fully correlated hydrogeological model.

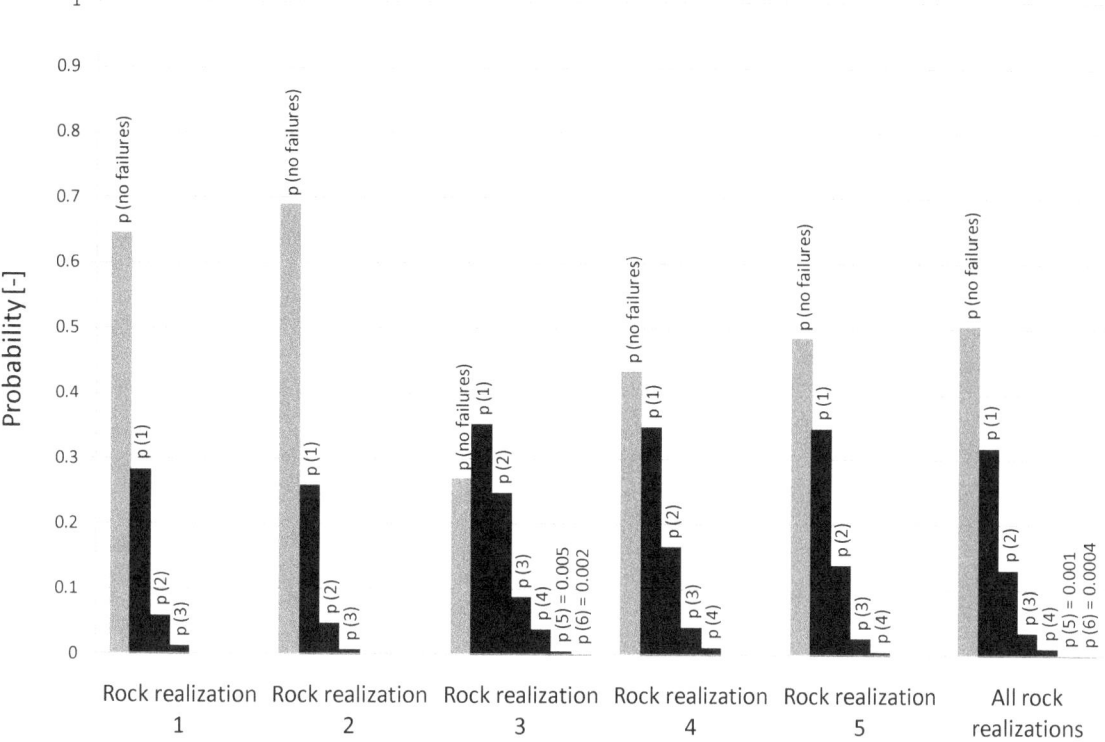

The distribution of the dose at 10^6 years for each rock realization is shown in Figure 7. These distributions were obtained by running each of the 1,000 canister failure realizations (for a particular rock realization) 1,000 times, sampling the parameters related to radionuclide release and transport. Note that the zero dose outcomes are not shown in the distributions. Hence the areas under the distributions are not unity, but reduced in accordance with the probability of a zero dose outcome.

Figure 7 also shows the distribution obtained with the simplified scheme aiming at calculating a correct mean value. As seen, the mean value calculated with that scheme, 1.01 µSv, is in good agreement with that calculated from all the realizations of the more elaborate scheme, 1.02 µSv, confirming that the simplified scheme yields a correct mean value. As expected, the variability of the results obtained with the elaborate scheme is higher, but the difference is not dramatic. Note that the area under the PDF curve of the simplified scheme is 1, since this distribution, by the nature of the calculation scheme, has no zero doses. Note also that the regulatory risk limit relates to the mean value and that the mean values of all the distributions are well below the risk limit of 14 µSv.

Figure 8 shows the distribution of peak dose taken over the one million year assessment period. The results are similar to those in Figure 7 showing the results at 10^6 year, which is expected since the peak dose in general occurs at the end of the 10^6 years assessment period.

Figure 9, finally shows the distribution of peak release from the near field over the one million year assessment period, converted to dose using the same biosphere dose conversion factors as for the results in Figures 7 and 8. It is noted that the near field doses are higher than the far field doses by a factor of about 8. The retention in the geosphere thus reduces the doses by less than an order of magnitude. Also the standard deviation is higher by about a factor of 8, meaning that the variability is not increased substantially by the transport of radionuclides through the geosphere. This is to a large extent explained by the fact that there is a two order of magnitude variability in the release rate of radionuclides from the spent fuel, controlling the variability of near field releases and playing a dominant role for the variability also after radionuclide transport through the geosphere.

Figure 7. Probability density function of annual effective dose at 10^6 years for each of the five realizations, R1-R5, of the fully correlated hydrogeological model.

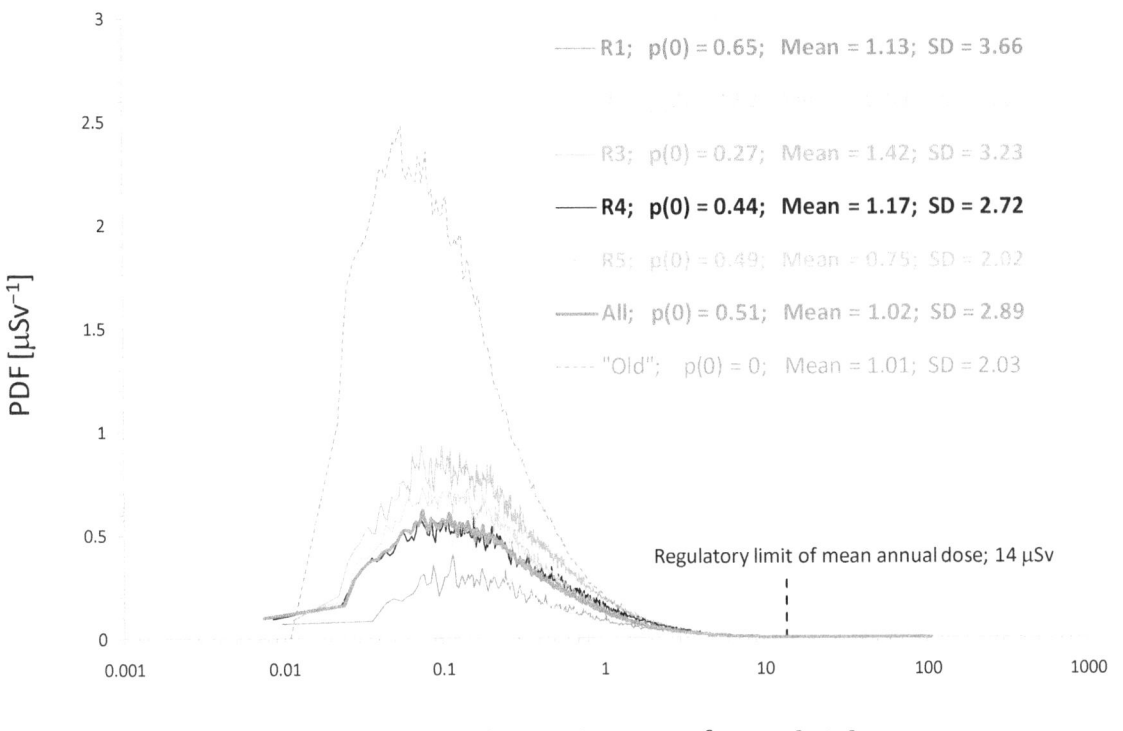

Annual effective dose at 10^6 years [μSv]

Figure 8. Probability density function of peak annual effective dose over the 10^6 year assessment period for each of the five realizations of the fully correlated hydrogeological model.

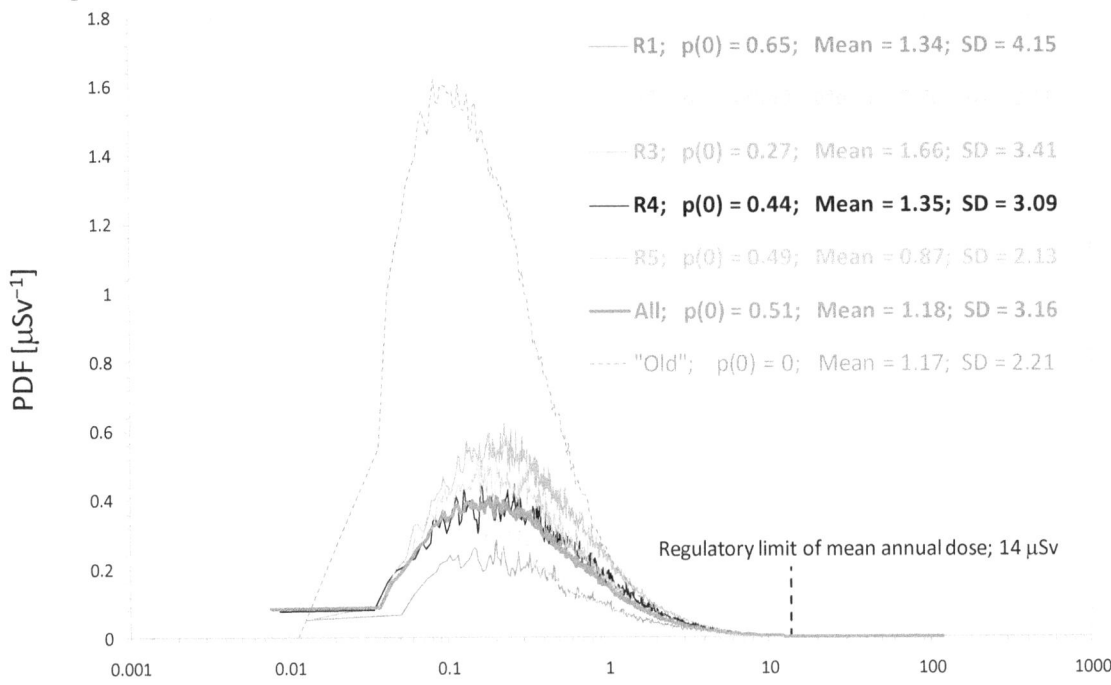

Peak annual effective dos over 10^6 years [μSv]

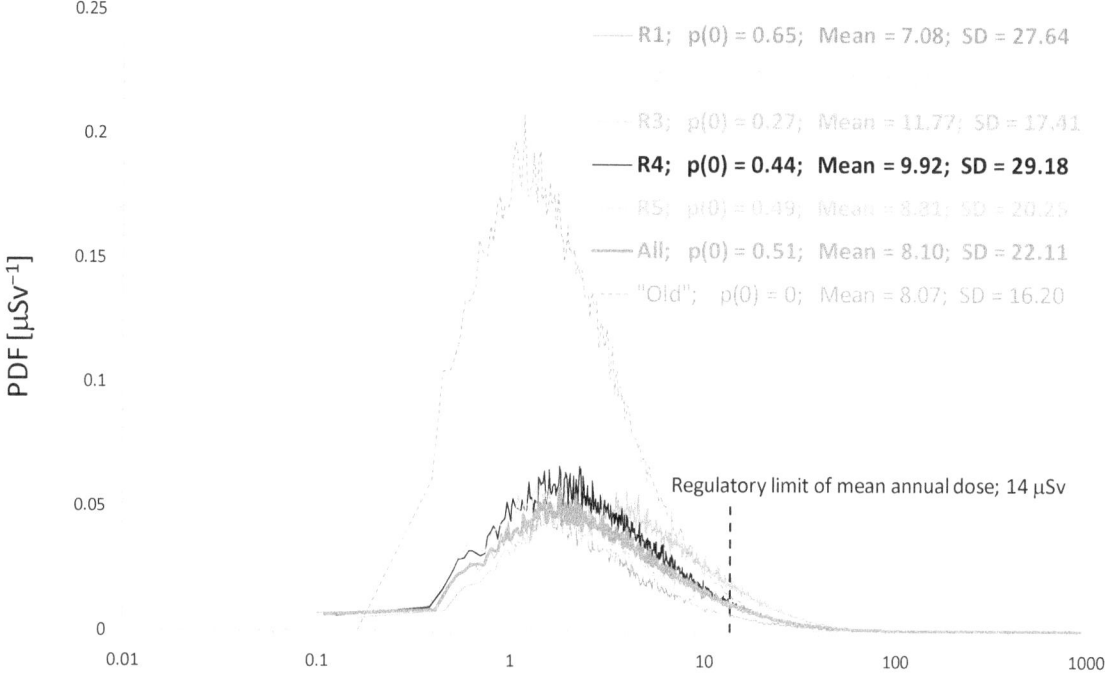

Figure 9. Probability density function of peak annual near field releases over the 10^6 year assessment period converted to effective dose for each of the five realizations of the fully correlated hydrogeological model.

Peak annual near field releses over 10^6 years expressed as effective dose [μSv]

3.2. Impact on Calculated Risk of Lifting Key Pessimistic Assumptions

The safety assessment in support of SKB's license applications is based on several pessimistic assumptions. In the following, a brief discussion of the potential impact of replacing some of these with more realistic assumptions is given.

First, it is noted that three of the cases in Figure 2 yield zero releases; the three variants with "no advection". These represent the situation where no advective conditions arise in the clay buffer protecting the canister. This is a possible outcome of the evolution of the buffer sub-system. The other two variants, advective conditions arising according to the SR-Site erosion model and the bounding case of initially advective conditions in the buffer, represent more pessimistic interpretations of the knowledge regarding the buffer erosion process and of those aspects of the system evolution related to the conditions required for erosion to occur, mainly low groundwater salinity and high flow rates.

Furthermore, for all cases calculated for the corrosion scenario it is assumed that a certain sulphide concentration, sampled from a specified distribution, prevails throughout the assessment period. It is only the highest sulphide concentrations in combination with the highest groundwater flow rates that yield canister failures. If, instead, the mean value of the sulphide concentrations is assumed throughout the assessment period for all canisters, reflecting a varying sulphide concentration over time, no canisters would fail. Periods with high concentrations would be followed by periods with low concentrations such that the mean value would not be sufficient to cause failures even in the deposition position with the highest flow rates.

Additionally, there is experimental evidence [4] that the oxidative fuel dissolution rate would in fact be near zero for the conditions present in the fuel tens of thousands of years into the future when the first canister failures occur. This would have as a consequence that the releases would be limited to only the fraction of radionuclides that is not embedded in the fuel matrix. This would lead to a reduction in calculated doses by several orders of magnitude for most of the assessment time, whereas doses caused by more rapidly released radionuclides, occurring during limited times after canister failures, would be unaffected.

Finally, in the safety assessment, it is assumed that the fuel matrix is converted according to a specified rate distribution. Uranium released in this conversion process is assumed to get reduced back to $UO_2(s)$ on iron and its corrosion products in the canister and there lead to the production of U-238 daughter nuclides such as Ra-226. All such daughter nuclides are assumed to be readily accessible for release, whereas a more reasonable assumption would be that the released uranium would re-crystallize and that the subsequently generated daughter nuclides would be contained in the UO_2 grains rather than being released. An assumed particle size of only one micrometer would lead to a reduction in the release rate of e.g. Ra-226 and its daughter products Rn-222, Pb-210 and Po-210, see Figure 3, by about a factor of 1000. The reduction in calculated total dose from all nuclides would be around one order of magnitude.

4. CONCLUSION

Key elements in the probabilistic radionuclide transport and dose calculations in support of SKB's license application for a spent nuclear fuel repository in Sweden have been summarized. The results demonstrate that a safe repository of the proposed KBS-3 type can be built at the Forsmark site in Sweden.

This paper has also confirmed that the simplified probabilistic calculation scheme adopted in the SR-Site assessment yields correct mean values of the calculated dose, the entity of interest regarding compliance with Swedish regulations. The variability is, as expected, somewhat reduced compared to that obtained with the more elaborate scheme used here, yielding a correct variability.

The paper also points to several factors treated pessimistically in the probabilistic assessment and gives a rough estimate of the potential reduction in dose that could result if more realistic assumptions were assumed. All these factors are assessed to yield reductions of at least an order of magnitude.

5. REFERENCES

[1]   "Long-term Safety for the Final Repository for Spent Nuclear Fuel at Forsmark. Main Report of the SR-Site project", SKB Technical Report TR-11-01, Swedish Nuclear Fuel and Waste Management Co, 2011, Stockholm. Available at www.skb.se
[2]   "Data Report for the Safety Assessment SR-Site", SKB Technical Report TR-10-52, Swedish Nuclear Fuel and Waste Management Co, 2011, Stockholm. Available at www.skb.se
[3]   A. Hedin. "Sensitivity Analysis of Probabilistic Dose Results in SKB's License Application", Proc. of 14th IHLRWM conference, Albuquerque, 28 April - 2 May 2013. American Nuclear Society.
[4]   M. Trummer and M. Jonsson. "Resolving the H_2 effect on radiation induced dissolution of UO_2-based spent fuel", Journal of Nuclear Materials, 396, pp.163-169, (2010).

Current Research in Storage and Transportation of Used Nuclear Fuel and High-Level Radioactive Waste

Sylvia J. Saltzstein

Sandia National Laboratories, Albuquerque, New Mexico, USA

Abstract: Through the Department of Energy (DOE)/ Office of Nuclear Energy (NE), Used Fuel Disposition Campaign (UFDC), numerous institutions are working to address issues associated with the extended storage and transportation of used nuclear fuel. In 2012, this group published a technical analysis which identified technical gaps that could be addressed to better support the technical basis for the extended storage and transportation of used nuclear fuel. This paper summarizes some of the current work being performed to close some of those high priority gaps. The areas discussed include: 1. developing thermal profiles of waste storage packages, 2. investigating the stresses experienced by fuel cladding and how that might affect cladding integrity, 3. understanding real environmental conditions that could lead to cask stress corrosion cracking, 4. quantifying the stress and strain fuel assemblies experience during normal truck transport and 5. performing a full-scale ten-year confirmatory demonstration of dry cask storage. Data from these R&D activities will reduce important technical gaps and allow us to better assess the risks associated with extended storage and transportation of used nuclear fuel.

Keywords: Used Nuclear Fuel, Storage, Transportation, dry cask storage

1. INTRODUCTION

The United States of America currently has 100 nuclear power reactors in 72 different nuclear power plants that supply about 19% of the country's electricity. The production of this energy creates "used nuclear fuel" which must be safely managed, contained, and disposed. These processes are often called the "back end" of the nuclear fuel cycle. Once the nuclear fuel no longer creates heat effectively in the reactor it becomes "used nuclear fuel" and is moved to a "spent fuel pool" at the nuclear power plant. As the pools near capacity, utilities move some of the older (typically 5 years) used fuel into "dry cask" storage. (U.S. NRC, 2013)

Dry cask storage allows used fuel to be surrounded by inert gas inside a container called a cask. The casks are typically steel cylinders that are either welded or bolted closed. The steel cylinder cask provides containment of the used fuel. Each cask is surrounded by additional steel, concrete, or other material to provide radiation shielding to workers and members of the public. (U.S. NRC, 2013) The casks can either be stored vertically or horizontally on the dry cask storage pad.

Figure 1: Horizontal Dry Storage (left) and Vertical Dry Storage (right)

There is currently about 70,000 metric tons of commercial used fuel accumulated in storage in the United States-about 80% of that is in used fuel pools and about 20% is stored in dry casks. This total increases by about 2000 metric tons annually. (U.S. NRC, 2013) Over time, nuclear power plants have found ways to use the fuel efficiently for a longer time in the reactor. Fuel that has been irradiated to more than 45 gigawatt-days per metric ton of uranium (GWd/MTU) is defined as high burn-up fuel. This high burnup fuel is comprising a larger percentage of the used fuel inventory. Due to the length of storage time and increased amount of high-burn-up fuel in the commercial used nuclear fuel inventory (UNF), the current infrastructure for the storage and transportation of commercial used nuclear fuel must provide both a safe and secure storage of our waste, but it must also maintain it in a condition so that the waste can withstand the stresses and strains of repackaging and transportation to a consolidated facility and permanent repository. While still believed to be safe, the R&D community does not have the same amount of data on the material characteristic of this fuel and therefore the DOE and NRC are supporting research in this area.

Current storage sites are shown below. Source: U.S. NRC website, downloaded 5/10/2012.

Figure 2. U.S. Locations where Used Fuel is currently stored.

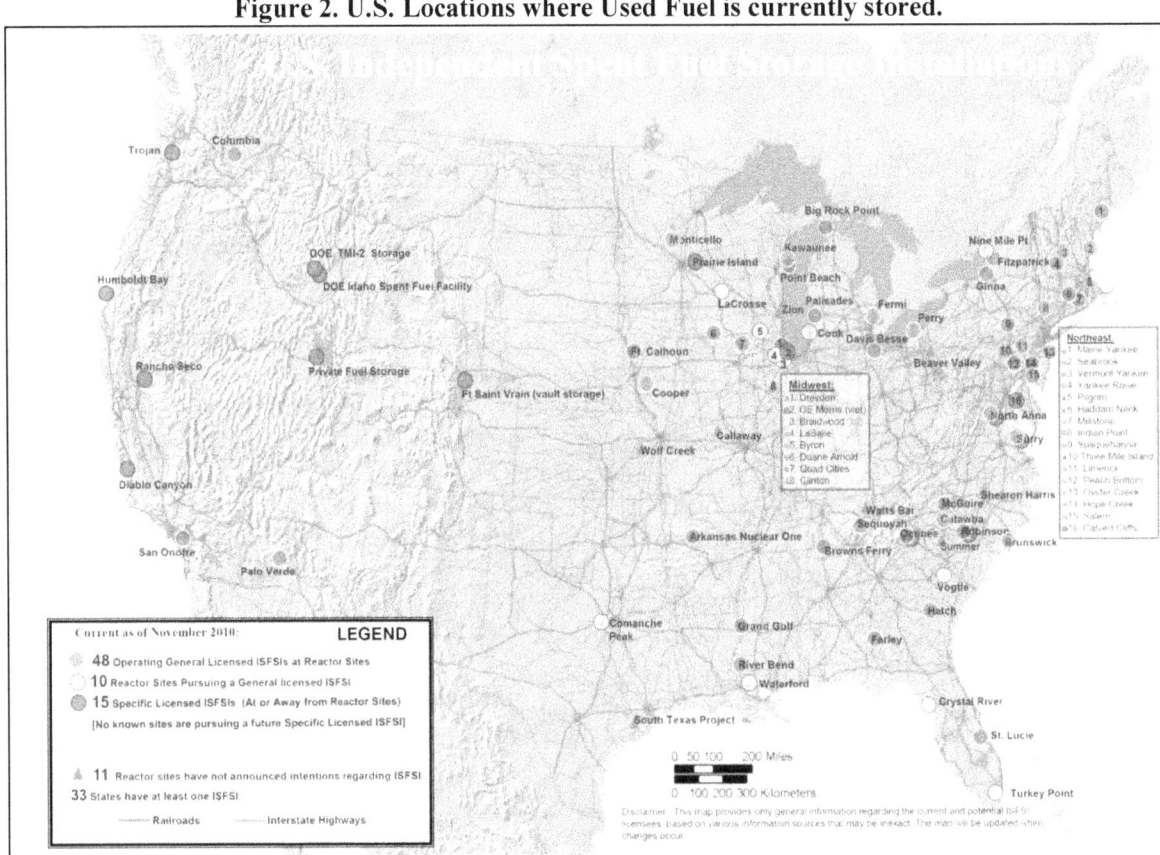

The U.S. Department of Energy supports the development of the technical basis for storage and transportation because they are responsible for the used fuel once it leaves the reactor site for further disposition. The U.S. NRC is tasked with regulating the storage, transportation and disposal. In 2009, the Department of Energy (DOE)/ Office of Nuclear Energy (NE) established the Used Fuel Disposition Campaign (UFDC) as part of the Nuclear Energy Fuel Cycles Technologies Program that supports overall research and development focused on issues associated with the nuclear fuel cycle. Establishment of the Used Fuel Disposition Campaign was recognition that important issues associated with the backend of the fuel cycle needed to be addressed on a national scale. The Used Fuel Disposition Campaign focuses on two principal areas; 1) R&D associated with the storage and transportation of used nuclear fuel and 2) research on permanent geologic disposal. This paper focuses on the R&D associated with storage and transportation and relates directly to the issue of extended storage with an added focus on high burnup used nuclear fuel. The UFDC used fuel storage and transportation R&D campaign has three main objectives:

1. Develop the technical bases to support the continued safe and secure storage of used fuel for extended periods.
2. Develop the technical bases for retrieval of used fuel after extended storage.
3. Develop the technical bases for the transport of low and high burnup fuel after extended periods of dry storage.

One of the first products of the UFDC Storage and Transportation R&D area was to identify data gaps that could be addressed to better support the technical basis for extended storage and transportation. This work was compared with other independent gap analyses to benchmark results. This comparison showed that there was good consensus with the technical gaps. The *very high* priority gaps are listed by rank order in Table 1.

Table 1. UFDC Top Priority Gaps Sorted by Rank ((Used Fuel Disposition Campaign, 2012)**)**

	Rank	Priority
Thermal profiles	1	Very High
Stress profiles	1	Very High
Monitoring – External	2	Very High
Welded canister – Atmospheric corrosion	2	Very High
Fuel Transfer Options	3	Very High
Monitoring – Internal	4	Very High
Welded canister – Aqueous corrosion	5	Very High
Bolted casks - Fatigue of seals & bolts	5	Very High
Bolted casks - Atmospheric corrosion	5	Very High
Bolted casks - Aqueous corrosion	5	Very High
Drying issues	6	Very High

Filling these gaps is a multi-year project. This report will focus on the work currently in progress to address some of those technical gaps.

2. Cladding Integrity Investigations (Gaps: Stress Profiles and Drying Issues)

A fuel assembly is comprised of numerous fuel rods held together in a spacer grid. Each fuel rod contains many nuclear fuel pellets. See Figure 2 for a picture of the assembly, spacer grid, fuel rod, and nuclear fuel pellet. The cladding is the outer layer of the fuel rod and is usually made of zirconium alloys, which are used because of their low absorption cross-section of thermal neutrons, high hardness, ductility, and corrosion resistance. Because the cladding is the first line of defense for fuel integrity, efforts are underway to better understand and predict the strength and ductility during drying, storage, and transportation conditions. When the fuel assembly is removed from the spent fuel pool and placed in a cask for dry storage, the temperature of the fuel and cladding peaks initially and then slowly cools. This peak temperature can cause numerous effects that need to be better understood for high burnup fuel such as: ductile to brittle transition temperature, hydride precipitation and re-orientation, and fuel pellet-to-pellet and pellet-to-clad interaction. Our goal is to be able to predict what happens to high-burnup cladding in different temperature and pressure conditions so that we can determine the best ways to manage the aging of the fuel cladding, and how it will perform during transport.

Figure 2: Diagram of a nuclear fuel assembly, fuel rods with outer cladding, and fuel pellets inside the rod.

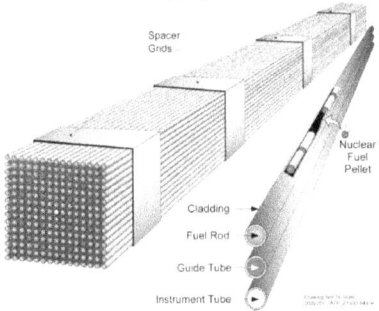

2.1 Ductile to Brittle Transition Temperature (DBTT): In order to maintain the integrity of the fuel and cladding, is important to understand where the fuel is in the ductile to brittle transition temperature curve as a function of time, burnup, and hydrogen up-take . Argonne National Lab has been working to determine the DBTT on the most common cladding types. Recent ring compression test on high burnup fuel cladding indicate that the cladding may not drop below the ductile to brittle transition temperature at pressures typically seen in high burnup fuel. See data for M5 cladding in Figure 3. (Billone, 2014) Work in this area will continue to verify these results and perform further tests at lower temperatures.

Figure 3: DBTT for High-Burnup M5 cladding. All at 400°C but at different different pressures. (Billone, 2014)

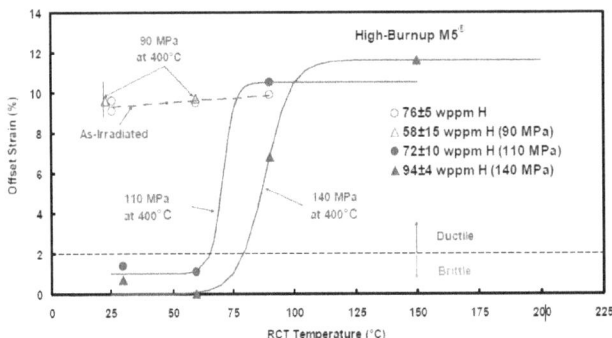

2.2 Hydride Formation: As the fuel is burned in the reactor for longer periods of time, hydrogen uptake by the cladding is also increased where it form hydrides in the circumferential direction. When the fuel is removed from the pool and begins the drying process, the fuel experiences a temperature spike and then subsequent cooling. During that time, the hydrides can dissolve and then precipitate again, but this time in the radial direction. Radial hydrides reduce ductility and can potentially lead to cracks in the cladding. Sandia National Labs is developing computer models to show how these hydrides form and be able to predict the conditions under which they form, as seen in Figure 4. This will allow us to better understand how to prevent radial hydride formation in fuel cladding.

Figure 4: Computer model showing the hydride growth in cladding. (Tikare, Hernandez, & Weck, 2013)

2.3 Fuel Pellet-To-Pellet And Pellet-To-Clad Interaction: With funding from the NRC, Oak Ridge National Labs has been using finite element analysis to develop models working on understanding the vibration integrity and flexural rigidity of the cladding of high burnup fuel. A surrogate used fuel rodlet containing surrogate fuel pellets was continually vibrated back and forth in ta U-frame bending fatigue testing system (see diagram in Figure 5). Results show that pellet-to-pellet and pellet-to-clad contact is very important to the strength of the fuel rod. When there is good pellet-to-pellet contact and pellet-to-cladding contact, the stresses are absorbed by the pellet, and much less stress is absorbed by the cladding. When there is poor cohesion at the pellet-to-pellet and pellet-to-clad interface, the embedded pellets can no longer provide effective structural support which can cause the rod to loose flexural rigidity, especially at the pellet-to-pellet interfaces. (Jiang & Wang, 2014) An example of the stress distribution with poor pellet-to-pellet cohesion is shown in Figure 6. The red color indicates areas of higher stress, the blue color show areas of lower stress.

Figure 5: U-frame Tool used to repeatedly flex the fuel rodlet to obtain data on rod vibration integrity and flexural rigidity. (Jiang & Wang, 2014)

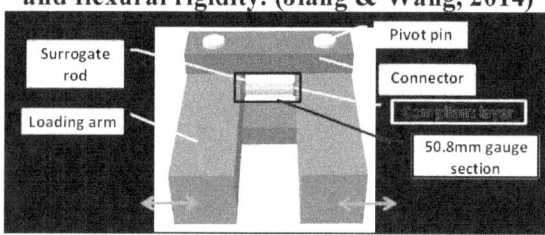

Figure 6: Example of FEA showing stress distribution with poor Pellet-to-Pellet contact (Jiang & Wang, 2014)

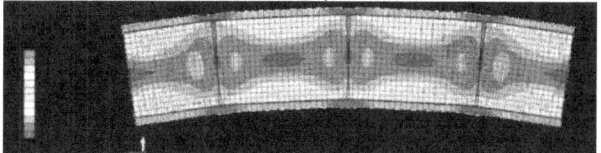

3.0 DEGRADATION OF THE METAL CANISTERS AND CASKS (Gaps: Atmospheric and Aqueous Corrosion): Additional work is being performed by DOE and its partners to understand environmental degradation of the metal casks. This research area is focused on the potential for stress corrosion cracking resulting from exposure of the cask to different inland and marine environments in combination with the residual stresses from the canister welds.

Figure 7: Dust accumulation on the surface of a horizontally stored cask. (Enos, Bryan, & Norman, 2013)

Figure 8: Dust particles collected from the outer surface of an in-use SNF storage container. (Enos, Bryan, & Norman, 2013)

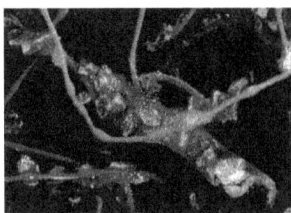

Dust samples from dry storage casks collected in both east and west coast marine environments are being analyzed to better understand when and if conditions exist for initiation and growth of stress corrosion cracking. See Figure 7 and Figure 8 for sources of the dust samples and an enlarged picture of a dust sample. Initial results indicate that the salts do not appear to have much sodium or chlorine, but are high in calcium and SO_4. The remainder of the sediment appears to be mostly comprised of pollens. (Enos, Bryan, & Norman, 2013) The initial analysis of this dust composition does not appear to be very corrosive to stainless steel.

Our goal in this research area is to determine what conditions could lead to canister corrosion and potential cracking, if those conditions exist where casks are stored, and how to potentially mitigate those conditions.

4.0 UNDERSTANDING THE TEMPERATURE GRADIENTS AND AIR FLOW DUIRNG STORAGE (Gap: Thermal Profiles): Dry canisters are typically stored on a concrete pad adjacent to the nuclear power plant. They can be stored horizontally or vertically in storage overpacks and are cooled by natural convection through vent openings in the concrete overpack. The Pacific Northwest National Laboratory is currently working to develop models that predict the thermal profiles of the canisters as they cool. Data to validate this modeling comes from canister temperature measurements and fuel radioactive decay calculations. Being able to predict the temperatures and thermal conditions provides needed data for understanding the ductility of the fuel and cladding, the rate of chemical reactions on the cask surface, and the pressures within the fuel. These data and tools will allow us to predict when fuel temperatures may drop below the Ductile to Brittle Transition Temperature, and when salts may deliquesce onto a cask surface. Knowing this information may affect how the fuel is handled during storage and transportation.

Figure 9: PNNL Analysis of temperature on cask surface. (Adkins)

Figure 10: PNNL Analysis of airflow through a horizontal storage module. (Adkins)

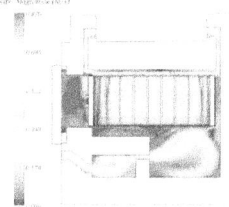

5.0 FORCES EXPERIENCED DURING NORMAL TRANSPORT (Gap: Stress Profiles): While we have a good understanding of the ability for a waste package to withstand numerous accident scenarios. We do not have a complete understanding of the loadings that fuel within a waste package would experience during normal road or rail transport. Currently, investigations are underway using a surrogate fuel assembly (Figure 11) and subjecting it to shock and vibration on a shaker table (Figure 12) similar to the shocks and vibrations obtained from routine truck transport. Results indicate that the stresses and strains experienced by the assembly during normal truck transportation conditions are much less than the levels thought to create damage to used fuel cladding --200 microstrain during the test vs. 9400 microstrain for high burnup used fuel. (Figure 14). Due to limitations of the shaker table testing, further analysis incorporating low frequencies and more degrees of freedom are needed. (McConnell, et al., 2013) On February 25, 2014, Sandia National Labs tested the same surrogate assembly on a truck over a 38 mile distance in Albuquerque, New Mexico. (Figure 15)This test was designed to incorporate the low frequencies and numerous degrees of freedom that the shaker table could not incorporate. The truck driving route consisted of dirt road, interstate highway, and city roads. Initial data indicate that stresses and strains are again much lower than the levels thought to create damage to used fuel (~400 microstrain vs 9400 microstrain for high burnup fuel). These results suggest that failure of the rods is unlikely during normal conditions of transport due to stress or strain. In addition, data from these tests is used to validate fuel assembly models (Figure 13).

Data for rail transport have recently been obtained, and testing using this data is planned within the next year. The data indicates that truck transport subjects the fuel to greater stresses and strains than rail, but because it is estimated that the majority of shipments will be over rail, it is still important to obtain and analyze this rail data. The goal of this work is to determine if high burnup used fuel can safely withstand truck and rail transport and to validate models being developed to better understand the structural performance of fuel under normal conditions of transport.

Figure 11: The surrogate fuel assembly instrumented with accelerometers and strain gauges. (McConnell, et al., 2013)

Figure 12: Surrogate Assembly on Shaker Table to be subjected to the vibrations seen on a 700 mile Truck Transport. (McConnell, et al., 2013)

Figure 13: Fuel Assembly Submodel (Adkins, et al., 2013)

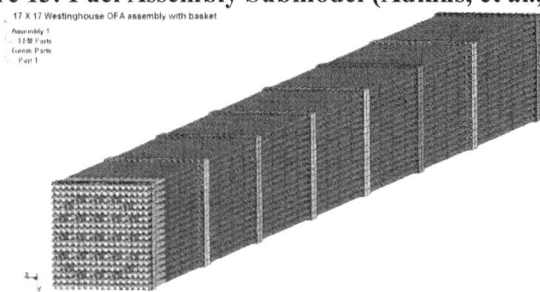

Figure 14: Elastic portion of stress—Shaker test shows that stresses and strains during normal conditions of transport are well below levels believed to create damage. (Klymyshyn, Sanborn, Adkins, & Hanson, 2013)

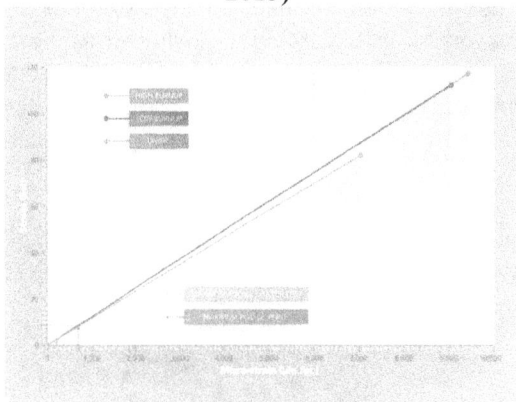

Figure 15: Surrogate assembly on 38 mile Truck transport. The assembly is bolted and strapped to concrete blocks to simulate the weight of a transport canister. (photo: Saltzstein)

6.0 FULL-SCALE DEMONSTRATION (Gaps: Internal and External Monitoring and Drying Issues). In order to obtain confirmatory information on drying and long-term storage conditions of high burnup fuel, DOE, EPRI, and Dominion Power are starting a ten-year project to obtain data from a cask of high burnup fuel as it goes through the typical drying process and then cools for many years. Analysis if similar rods with similar pedigrees and burnup history (sister rods) will be tested before cooling to obtain time-zero data. Data from periodic inspection of rods in the cask will provide information on how the high burnup fuel degrades over time. Thesedata can be compared to the data from the sister rods. Environmental data collected from the cask, such as temperature and fission gas, will provide information on the environment around the rods and potentially detect any failures. This data will be used to assist in the extension of storage and transportation licenses, validate our current predictive models, and better recommend the best conditions to for safe storage and aging.

7.0 CONCLUSION

In the United States, two unrelated factors have focused attention on the integrity of high burnup spent fuel (i.e., > 45 GWD/MTU) during storage and transportation operations: the need for longer-term storage at the independent fuel storage sites and the higher fuel burnups related to efficiencies in US reactor operations. A strong technical basis is being developed to demonstrate our understanding of high burnup fuel response during storage and transport. This requires validation of the integrity of high burnup spent fuel through a combination of testing and analysis to demonstrate actual performance in the areas of loads, applied strains, material properties, and pellet-clad interaction. Through the national labs and universities, the US DOE Storage and Transportation R&D Program is developing this technical basis.

8.0 ACKNOWLEDGEMENTS

Sandia National Laboratories is a multi-program laboratory managed and operated by Sandia Corporation, a wholly owned subsidiary of Lockheed Martin Corporation, for the U.S. Department of Energy's National Nuclear Security Administration under contract DE-AC04-94AL85000. Approved for public release; further dissemination unlimited.

9.0 REFERENCES

Adkins, H., Geelhood, K., Koeppel, B., Coleman, J., Bignell, J., Klymyshyn, N., et al. (2013). *Used Nuclear Fuel Loading and Structural Performance Under Normal Conditions of Transport – Draft FCRD-UFD-2013-000289.* U.S. Department of Energy Fuel Cycle Research and Development.

Billone, M. a. (2014). UQ for Hydride Reorientation and Embrittlement in High-Burnup PWR Cladding. *Uncertainty Quantification Workshop.* Las Vegas, NV.

Enos, D., Bryan, C., & Norman, K. (2013). *Data Report on Corrosion Testing of Stainless Steel SNF Storage Canisters, FCRD-UFD-2013-000324.* U.S. Department of Energy, Used Fuel Disposition Campaign.

Jiang, H., & Wang, J.-A. W. (2014). Potential Impact of Interfacial Bonding Efficiency on High-Burnup Spent Nuclear Fuel Vibration Integrity during Normal Transportation. *Waste Management 2014 Conference.* Phoenix, Arizona.

Klymyshyn, N., Sanborn, S., Adkins, H., & Hanson, B. (2013). *Fuel Assembly Shaker Test Simulation, FCRD-UFD-2013-000168.* Pacific Northwest National Laboratory: U.S. Department of Energy, Used Fuel Research and Development.

McConnell, P., Flores, G., Wauneka, R., Keonig, G., Ammerman, D., Bignell, J., et al. (2013). *Fuel Assembly Shaker Test, FCRD-UFD-2013-000190.* U.S. Department of Energy Fuel Cycle Research and Development.

Tikare, V., Hernandez, E., & Weck, P. (2013). Simulation of Hydride Reorientation on Zr-Based Claddings During Dry Storage. *International High-Level Radioactive Waste Management Conference.* Albuquerque,.

Modified-LOPA; a Pre-Processing Approach for Nuclear Power Plants Safety Assessment

Seyed Mohsen Gheyasi[*][a], Mohammad Pourgol-Mohammad[†][a]
[a] Sahand University of technology, Tabriz, Iran

Abstract: Risk and safety assessment are important subjects in modern industries. Different methods have been proposed for safety and risk evaluation of high hazardous facilities. The risk assessments methods are classified in three main groups of qualitative, semi-quantitative and quantitative. The methodology is selected depending on scope and objective, level of details and requirements. Nuclear facilities regulations require more detailed assessment of system safety. Regulatory body requires utilization of probabilistic risk assessments (PRA) for appraisal of design, modifications and operation of nuclear power plants. This method usually is very complicate, expensive and time consuming. Significant amount of resources are needed for a PRA project completion which in some cases for preliminary safety evaluation are not justified. Simpler methods would be used for preliminary evaluation as a pre-processor to quick find out of the situations (especially in operational nuclear power plant). Layer of Protection Analysis (LOPA) is one of the powerful risk analysis methods. It is a semi-quantitative approach widely used in chemical process industry. This method is not a competitive alternative to full quantitative methods of risk analysis for nuclear facilities like PRA. However, it is simpler and less expensive methodology comparing to full probabilistic risk assessment methods. It evaluates the probability of failure per demand for the safety system failures and the resulting consequences. It is introduced here as a practical technique for early and quick risk assessing in many other industries. But if LOPA has been selected as a risk evaluator pre-processor in nuclear systems it requires some modifications in methodology structure.

This research examines utilization of LOPA method for nuclear systems as an order of magnitude evaluation of the safety status. Conventional LOPA method requires some essential modifications in methodology to prepare it as a suitable approach for nuclear systems, especially in its scenario development and quantitative calculations. The so-called modified layer of protection analysis (Modified-LOPA) methodology is based on improvement of some features of conventional LOPA. Some changes are proposed to the classic LOPA method by using event tree method and Bayesian logic. Since LOPA and event-tree methods use definition of scenarios to represent the paths of the accidents, therefore scenario development is completed in modified method by using event-tree method. Then initiating event frequency and probability of failure on demand (PFD) of independent protection layers (IPLs) estimations are updated by Bayesian approach which increases the reliability of results by combination of plant specific data with generic data from other similar industries.

In this paper "Modified-LOPA" method is proposed as a primary tool for quick hazard analysis, risk assessment and risk based decision in nuclear systems. This method is more accurate comparing to conventional LOPA. However it is not a complete substitute for Full PRA in nuclear systems. A simple example of a fire protection system shows application of this method and the results are compared with the results of a PRA approach.

Keywords: LOPA, Risk, Scenario, Event Tree, Initiating Event, Independent Protection Layer (IPL), Probability of Failure on Demand (PFD), Frequency, Bayesian, Nuclear Power Plant.

1. INTRODUCTION

Nowadays, different methods have been proposed for safety evaluation of high hazardous facilities. The methods include Probabilistic Risk Assessment (PRA), Failure Mode and Effects Analysis (FMEA), Hazard and Operability (HAZOP), Reliability Block Diagram (RBD) and Layer of Protection Analysis (LOPA) [1]. Depending on scope and objective, level of details and requirements, the methodology is selected.

[*] Gheyasi.Mohsen@gmail.com
[†] mpourgol@gmail.com (Author of Corresponding)

Nuclear facilities regulations require more detailed assessment of system safety. Regulatory body requires utilization of PRA for assessment of design, modifications and operation of nuclear power plants. Although it's essential to meet all safety criteria in nuclear systems, the extent and complexity of analysis usually make it difficult to reach the results in a limited time. Significant amount of resources are needed for a PRA project completion which in some cases for preliminary safety evaluation are not justified. Simpler methods would be used for preliminary evaluation as a pre-processor to quick find out of the situations (especially in operational nuclear power plant). If the finding concludes the need for a more detail assessment then PRA is recalled.

LOPA is one of the powerful risk methods which is a semi-quantitative approach. It's widely used in chemical process industry. Actually, this method is not a competitive alternative to the full quantitative methods of risk analysis for nuclear facilities, but the characteristics of this method makes it capable of performing a preliminary nuclear safety and risk analysis. They include:

- Being systematic and straight-forward.
- Expression of results as semi-quantitative.
- Affordable cost, time and effort requirement compared to Full PRA method.
- Capability of focus risk reduction efforts on impact events with high severity and high probability [2].
- Capability of quick system design weakness identification for improvement and modification.

In spite of the fact that LOPA has got many positive features, some difficulties in scenario identification and in using statistical quantities, has led many researchers to modify this method. In this paper "Modified-LOPA" method is proposed as a primary tool for quick hazard analysis, risk assessment and risk based decision in nuclear systems. This method is more accurate comparing to traditional LOPA which gets help from Event Tree structure for developing scenarios and Bayesian logic to update the failure data. However it is not a complete substitute for Full PRA in nuclear systems, but the highly reliable results justify the using of Modified-LOPA as a nuclear risk pre-processing method.

NOMENCLATURE

FMEA	Failure Mode and Effects Analysis
HAZOP	Hazard and Operability
I.E.	Initiating Event
IPL	Independent Protection Layer
LOPA	Layer of Protection Analysis
NPP	Nuclear Power Plant
OFS	Off-Site Fire Protection System
ONS	On-Site Fire Protection System
PFD	Probability of Failure on Demand
PRA	Probabilistic Risk Assessment
RBD	Reliability Block Diagram
SIL	Safety Integrity Level
SIS	Safety Instrumented System

2. LITERATURE REVIEW

LOPA has been presented in several works and the results of its successful implementation have been reported in the various literature. LOPA is used in [2] to evaluate a highly reactive process and illustrates the benefit of risk assessment to follow a HAZOP hazard analysis. Hydroxylamine production facility has been evaluated as a practical case study in this paper. LOPA has been described in [3] for determining the requirements for Safety Integrity Level (SIL) of a Safety Instrumented System (SIS). Summers [4] briefly described LOPA as a powerful analytical tool for assessing the adequacy of protection layers used to mitigate process hazards.

An overview has been provided in [5] that mainly discusses the commercially available explosion prevention and mitigation systems applicable to gas, dust, mist and hybrid (gas-aerosol) explosions, including basic principles and proper application for single and combined systems and their

Limitations. Another research [6] attempts to explain the principles of LOPA and the means by which it can be used within the accidental risk assessment methodology for industries.

In some articles it's attempted to develop LOPA in methodology. Yun and Mannan [7] presented a Bayesian–LOPA methodology which studied a LNG importation terminal as a case study to demonstrate application of the method. It proposes that the Bayesian–LOPA method is a powerful tool for risk assessment of not only the LNG facilities but also in other industries, such as petrochemical, nuclear, and aerospace.

LOPA is presented in [8] as an approach that may include human harm and is independent of the analyst. It also provided how to identify and evaluate scenarios for LOPA and briefly describes the contribution of human errors in accidents. Markowski and Kotynia [9] applied including an uncertainty aspect in LOPA to the risk assessment of a hazardous substance release. It has been provided by a "bow-tie" approach being a composition of fault and event tree. The quantitative application of the "bow-tie" model has been proposed in the methodology of LOPA.

Summers et al. [10] improved the frequency and risk reduction tables in the estimate of the hazardous event frequency, and how consequence severity tables can significantly increase confidence in the severity estimate have been showed.

A mixed integer nonlinear programming model is presented in [11] to improve the computational use of LOPA. The human role and activities is reviewed in [12] as potential initiating events and human performance within independent protection layers in LOPA methodology.

3. MODIFIED-LOPA DESCRIPTION

As told before, layer of protection analysis is a semi-quantitative approach to evaluate the risk of potential incidents and to provide guidance on the adequacy of independent protection layers (IPLs) to lower the risk. LOPA typically uses order of magnitude categories for initiating event frequencies and for the probabilities of failure of IPLs, which can mitigate the frequency or reduce the consequence of an incident [2].

LOPA focuses risk reduction efforts on impact events with high severity and high probability, so its primary requirement is to determine these sever events. As a result, LOPA often follows a qualitative risk analysis performed as part of a HAZOP, check list, etc. to identify and characterize hazards.

LOPA methodology typically builds on the information developed during a qualitative hazard evaluation. Then, layers of protection are intended to independently comply with three main functions: Prevention, protection and mitigation. To be considered as independent protection layers (IPL's), safeguards need to satisfy some characteristics: independence, specificity, dependability and auditability [4].

The methodology typically uses order of magnitude to express the initial event frequency, the probability of failure on demand of the independent protection layers and the magnitude of the consequence. This way of expression provides good achievement to simple comparison and calculation.

It's expected that the results of LOPA be accompanied by [13]:
- Providing rational, semi-quantitative, risk-based answers
- Reducing emotionalism
- Providing clarity and consistency
- Documenting the basis of the decision
- Facilitating understanding among plant personnel

According to the literature review, especially the research of Yun et al. (2009) and its proposal for using the Bayesian–LOPA method to risk assessment in nuclear systems, this decision was made to recommend a modification in LOPA method for this purpose. So Modified-LOPA method has been recommended which uses Event-tree method for better scenario development, and Bayesian probabilistic method for updating data and calculating uncertainty of results.

4. METHODOLOGY STEPS

Modified-LOPA is based on improvement of some features of conventional LOPA. Since LOPA and event-tree methods are using definition of scenarios to represent the paths of the accidents, therefore scenario development in modified method is completed by using event-tree method. Initiating event frequency and Probability of Failure on Demand (PFD) of Independent Protection Layers (IPLs) data are updated by Bayesian approach. Figure 1 demonstrates the flowchart and steps for Modified-LOPA method. This flowchart is adapted based on the previous researches done in [2,7]. The basic steps of this approach are described below:

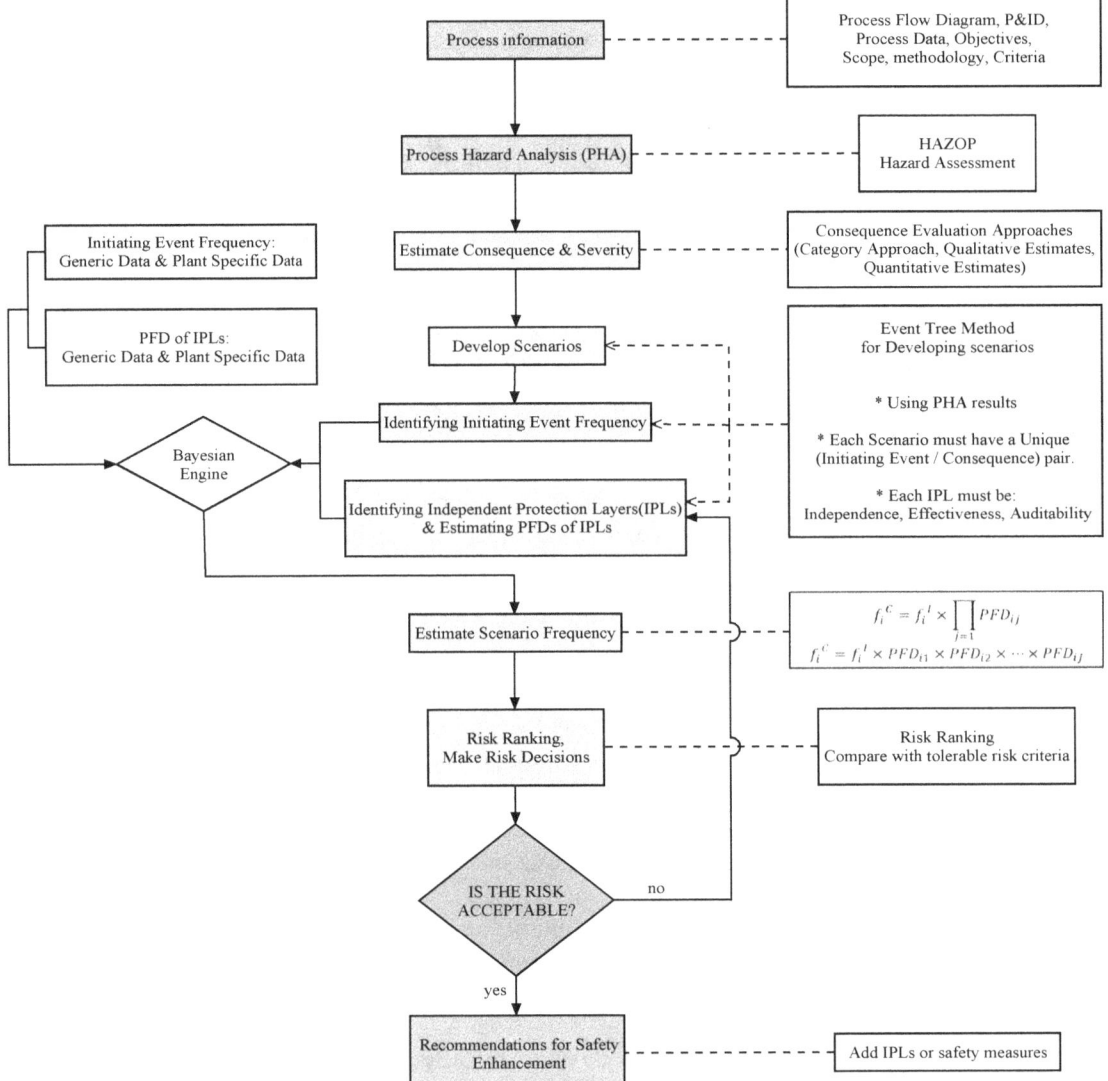

Figure 1. Modified-LOPA Process Flowchart

Step1: Process Information
First, a system should be completely identified. The piping and instrumentation diagram, process flow diagram, process data, objectives, scope, methodology, criteria and every data of maintenance and failures should be studied. It's recommended that the criteria be provided in this step to determine the endpoint of accidents. Some examples for considering the endpoint of accidents are: overpressure, leak of toxic and flammable fluids, fire or explosion, etc.

Step 2: Process Hazards Analyses (PHA)
Qualitative hazard analysis is a fundamental step for identification consequences of events in LOPA, which is usually done by HAZOP which usually is used to identify all probable events. In a HAZOP study, the severity of events can be categorized and it helps the analyzers to opt just the critical scenarios for LOPA.

Step 3: Estimate Consequence and Severity

Consequences are the undesirable outcomes of accident scenarios. One of the first decisions of an organization must make when choosing to implement LOPA is how to define the consequence endpoint. Since the consequences must be categorized, special attention to primary steps of LOPA is very important.

There are some approaches for this purpose include [13]:

Method 1: Category Approach and using matrices.

Method 2: Qualitative Estimates; that use the final impact on humans as the consequence of interest.

Method 3: Quantitative Estimates with Human harm; which uses mathematical models.

Step 4: Developing Scenarios

A scenario is an unplanned event or sequence of events that results in an undesirable consequence. Each scenario consists of at least two elements which show the beginning and the end of an event. These elements are:

- An initiating event
- A consequence

A scenario in its perfect form, should illustrate the pathway of an event. Each scenario must have a unique Initiating Event-Consequence pair. Since the definition of scenarios in LOPA is similar to the Event-Tree method and both approaches are based on analyzing scenarios, it's considered to use the tree structures to clarify the principles of LOPA scenarios. However process risk assessment in nuclear systems requires all spectrums of possible accidents that subsequently may exceed the specified risk tolerance level, analyzing the worse cases of events is useful a pre-processing. In order to obtain more appropriate and accurate analysis, the complete accident scenario model is developed.

In the traditional LOPA, an accident scenario is defined as a single cause–consequence pair using an event tree approach. Only one path of the accident scenario, which merely leads to a major hazard, is analyzed. For more complex scenarios, LOPA should be used several times for each initiating event (IE) separately. Another limitation of LOPA is the fact that there is no separation of top event or loss event. As mentioned in literature review, Markowski and Kotynia [9] suggested bow-tie method, which is composed of a fault tree which identifies the causes of the top event or loss event, usually representing unwanted release of the substance and an event-tree showing what are the consequences of such a release. In the "bow-tie" model all connections between initiating events, loss event and outcome events are fully identified.

Although this method is a very comprehensive approach for improving accuracy of LOPA, but increasing the computation size, makes LOPA exceed a semi-quantitative method. For this reason it has been assumed that event tree method is approximately enough for developing scenarios in LOPA. Using event tree causes the accident path to be identified exactly, it also causes the IPLs be more transparent, and capability of linkage the results to Full PRA method be increased. Also existence of a schematic of a scenario facilitates its understanding.

Step 5: Event Tree Method

The event tree is a logical structure in the form of a tree branch that maps out the different pathways by which the bad event can come about. All of the paths that cause an adverse outcome must be included and analysts routinely rely on the experience of subject matter experts to know which events to include. The tree structure enables the analyst to order events (usually chronologically), to separate clusters of events from each other, and to show whether or not events are important. The branching structure shows how an initiating event that starts a sequence at the left side of the tree may lead to the bad event that is shown at the far right side. Events or options that depend on other events are shown to the right of those events on which they depend [14]. As mentioned before, both LOPA and Event-Tree method utilize the scenario concepts.

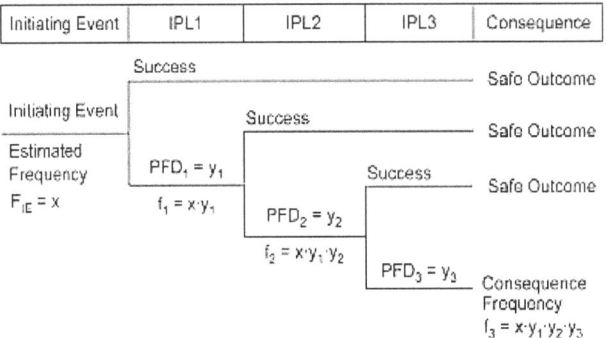

Figure 2. LOPA Event Tree Example [15]

Step 6: Identifying Initiating Event Frequency

For LOPA, each scenario has a single initiating event. The frequency of the initiating event is normally expressed in events per year. Some sources use other units, such as events per 106 hours [13]. LOPA uses order-of-magnitude to express the frequency of initiating events. In general, initiating events are divided in three main categories: external events, human errors and equipment failures. A HAZOP study should represent these initiators.

Step 7: Identifying IPLs and Estimating PFDs

An IPL is a device, system, or action that is capable of preventing a scenario from proceeding to its undesired consequence independent of the initiating event or the action of any other layer of protection associated with the scenario. An IPL must be:

- Effective
- Independent
- Auditable

Like the previous step, a HAZOP study should be able to illustrate the safeguards which are considered against the final consequence. Then LOPA analyzers have to separate the IPLs from other safeguards. For LOPA calculations, it is essential to know the failure rate or probability of failure of system's IPLs. Thus, the concept of probability of failure on demand is introduced. PFD for an IPL is the probability that, when demanded, it will not perform the required task [13]. PFD is a complement to availability and also is a probabilistic value.

Step 8: Bayesian Probabilistic Method

Bayesian estimation incorporates degree of belief and information beyond that contained in the data sample, forming the practical difference from classical estimation. The subjective interpretation of probability forms the philosophical difference from classical methods. Bayesian estimation is comprised of two main steps. The first step involves using available information to fit a prior distribution to a parameter, such as frequency of an IPL. The second step of Bayesian estimation involves using additional or new data to update the prior distribution. This step is often referred to as "Bayesian Updating" [16]. The generalized form of Bayes' theorem for discrete variables is:

$$\Pr\left(A_j | E\right) = \frac{\Pr\left(A_j\right).\Pr\left(E | A_j\right)}{\sum_{i=1}^{n} \Pr\left(A_i\right).\Pr\left(E | A_i\right)} \tag{1}$$

The terms of this equation are:

$\Pr(A_j|E)$: The posterior probability of event Aj given event E or updated probability of event A_j

$\Pr(A_j)$: The Prior probability of event A_j

$\Pr(E|A_j)$Likelihood function based on sample data

$\sum_{i=1}^{n} \Pr\left(A_i\right).\Pr\left(E | A_i\right) = \Pr\left(E\right)$: Total probability

The above equation means that probability data can be updated by combining the prior probability (from previous information or generic data) and the relative likelihood (from plant-specific data).

Typically, the selection of the prior distribution is somewhat subjective, so a selection of a conjugate prior from the same family of distributions as the posterior can make the choice more objective for easier computation of the posterior parameters [7].

Since using Bayesian equation in its primary form is difficult in some cases, it's recommended to use conjugated distributions. In these cases, for example gamma and poison distributions are conjugated. If there is a prior distribution in the form of gamma and a likelihood distribution in poison, the Bayesian calculation will result in a gamma posterior distribution.

Step 9: Estimate scenarios frequency

After updating data, scenarios frequency is estimated. The following is the general procedure for calculating the frequency for a release scenario with a specific consequence endpoint. For this scenario, the initiating event frequency from step 5 is multiplied by the product of the IPL and PFDs from step 6 [13].

$$f_i^C = f_i^I \times \prod_{j=1}^{J} PFD_{ij}$$
$$= f_i^I \times PFD_{i1} \times PFD_{i2} \times \ldots \times PFD_{iJ} \qquad (2)$$

Step 10: Calculating Risk

In this step, the severity of categorized consequences from step 3 is multiplied by scenarios frequency from step 7.

$$R_k^C = C_k \times f_k^C \qquad (3)$$

Step 11: Make risk decision

The calculated risk is compared with risk tolerance criteria for the decision-making. If, however, the calculated risk exceeds the risk criteria, the scenario is judged to require additional (or stronger) mitigation (IPLs), or to require changes in the design to make the process inherently safer, thus reducing scenario frequency or consequence, or (preferably) eliminating the scenario [13]. This change in accident path should be considered in event tree if other IPLs are needed to be added. Also the Bayesian calculations and scenarios frequency estimations must be repeated considering the effect of new changes.

Step 12: Safety Management

Risk management must be applied to the all levels of system including design, operation, monitoring, test, maintenance, etc. It's important to mention that LOPA do not suggest any way to control the risks, but it clarifies the way of decision making to help the management team.

5. Application OF MODIFIED-LOPA on NPP Fire Protection System

The methodology is applied on a fire protection system for a typical Nuclear Power Plant (NPP) which is designed to extinguish fires in this facility. Fire protection is considered a mitigation IPL as it attempts to prevent a larger consequence subsequent to an event that has already occurred. If a company can demonstrate that it meets the requirements of an IPL for a given scenario it may be used [13].

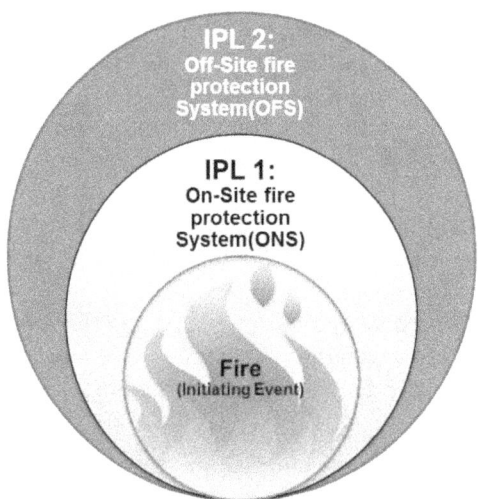

Figure 3. Layers of Protection Against a Fire Scenario.

Description of facility

This fire protection facility includes two separate systems, On-Site fire protection system (ONS) and Off-Site fire protection system (OFS). Each of the systems is a set of different components such as sensors, alarms, tanks, valves, pumps, etc. but it's supposed that ONS and OFS systems will meet all characteristics of the independent protection system.

Analysis by PRA

This system has been completely studied by PRA approach [1]. This method has considered all subsystems and equipment in order to get the most accurate results. Three different scenarios are defined, Fig. 4 shows the event-tree of these scenarios. PRA calculations shows the frequency of fire event in such plant is 7.1E-4, and PFD of ONS system is 2.8E-3, also PFD of OFS system is 1E-4 [1].

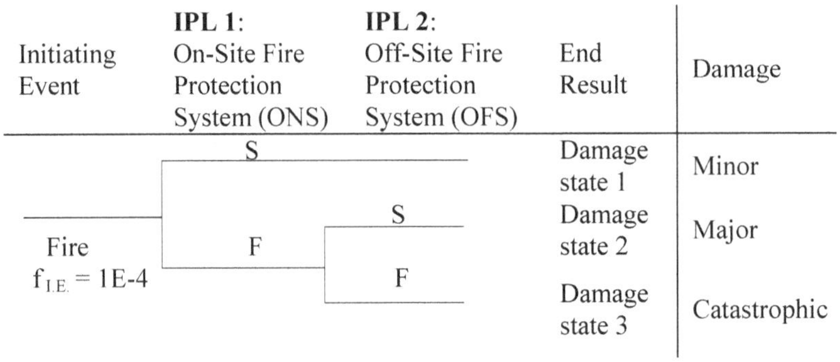

Initiating Event	**IPL 1:** On-Site Fire Protection System (ONS)	**IPL 2:** Off-Site Fire Protection System (OFS)	End Result	Damage
	S		Damage state 1	Minor
Fire $f_{I.E.} = $ 1E-4	F	S	Damage state 2	Major
		F	Damage state 3	Catastrophic

Figure 4. Scenario of Events Following a Fire Using the Event-Tree Method [1]

Modified-LOPA

Consider a change in components of primary ONS system. If this system be a fire protection of an in operation nuclear power plant, a new risk assessment will be needed. But PRA method requires a huge amount of calculations and resources. So Modified-LOPA would be used as a simpler method and a pre-processor. The pre-assumptions considered to analyze this system by Modified-LOPA are as followed:

- The same defined scenarios in Figure 4. Are also considered.
- The calculated results of PRA are used as mean values of prior data.
- Frequency of fire accident and PFD of OFS system are same as before, because of no change.
- New ONS system was tested in similar facilities for 1000 demands, and 3 failures were observed.

- The prior data will be updated by Bayesian method.

Bayesian Updating

The simplest type of prior distribution from the standpoint of the mathematics of Bayesian inference is a so-called conjugate prior, in which the prior and posterior distribution are of the same functional type (e.g., beta, gamma), and the integration needed to obtain the normalizing constant in Bayes' Theorem is effectively circumvented [17].

For prior distribution of PFD of ONS system, the mean value is 2.8E-3. And from engineering judgment its standard deviation is estimated 1.4E-3 and a Beta distribution has been assigned for it.

For the binomial distribution, the conjugate prior is a beta distribution. So in this case, the likelihood function is modeled by a Binomial distribution. Elements of Bayesian updating from Eq. (1) for updating PFD of ONS system are shown in Table 1 from Appendix A.

Risk Calculation

Table 2 in Appendix A, shows the results of risk calculation by Modified-LOPA and compares it to the result of PRA approaches. As observed in Table 2, a little increase in PFD of ONS system, the calculated risk of scenarios leads a larger discrepancy of the results between two approaches. Common cause failure is considered in the calculation of PRA but the events are considered independent in LOPA. This is another reason for the discrepancy between two methods. Another reason for this difference is due to the updated value of PFD using Bayesian formula.

6. CONCOLUDING REMARKS

 The research demonstrated the application of modified LOPA methodology on safety evaluation of nuclear facilities. In classic LOPA, only the most severe consequences are often considered. However, Modified-LOPA considers all probable scenarios with assistance of Event Tree method. Bayesian updating makes estimation of the frequencies and PFD more accurate by utilization of historic and field data. In this paper Modified-LOPA is represented as a powerful pre-processing method in nuclear power plants. The example shows good agreement of its result in comparison with a full PRA approach.

The effect of using Event Tree structure could be better demonstrated if the studies consist a large scope system with very complicate components. Besides, for new designed systems with lack of failure data or in case of unreliable collected failure data, using Bayesian logic which gives analysers the ability of updating the plant specific data with generic data from other similar systems, can lead to more reliable results. Modified-LOPA will be known as the most comprehensive semi-quantitative method if economic survey be added to it.

7. REFERENCES

[1] Modarres, M., *Risk Analysis in Engineering: Techniques, Tools, and Trends*. 2006, New York: CRC Press, Taylor & Francis Group. 85-111.

[2] Wei, C., W.J. Rogers, and M.S. Mannan, *Layer of protection analysis for reactive chemical risk assessment*. Journal of Hazardous Materials, 2008. **159**(1): p. 19-24.

[3] Dowell Iii, A.M., *Layer of protection analysis for determining safety integrity level*. ISA Transactions, 1998. **37**(3): p. 155-165.

[4] Summers, A.E., *Introduction to layers of protection analysis*. Journal of Hazardous Materials, 2003. **104**(1–3): p. 163-168.

[5] Pekalski, A.A., et al., *A Review of Explosion Prevention and Protection Systems Suitable as Ultimate Layer of Protection in Chemical Process Installations*. Process Safety and Environmental Protection, 2005. **83**(1): p. 1-17.

[6] Gowland, R., *The accidental risk assessment methodology for industries (ARAMIS)/layer of protection analysis (LOPA) methodology: A step forward towards convergent practices in risk assessment?* Journal of Hazardous Materials, 2006. **130**(3): p. 307-310.

[7] Yun, G., W.J. Rogers, and M.S. Mannan, *Risk assessment of LNG importation terminals using the Bayesian–LOPA methodology*. Journal of Loss Prevention in the Process Industries, 2009. **22**(1): p. 91-96.

[8] First, K., *Scenario identification and evaluation for layers of protection analysis.* Journal of Loss Prevention in the Process Industries, 2010. **23**(6): p. 705-718.

[9] Markowski, A.S. and A. Kotynia, *"Bow-tie" model in layer of protection analysis.* Process Safety and Environmental Protection, 2011. **89**(4): p. 205-213.

[10] Summers, A., W. Vogtmann, and S. Smolen, *Improving PHA/LOPA by consistent consequence severity estimation.* Journal of Loss Prevention in the Process Industries, 2011. **24**(6): p. 879-885.

[11] Ramírez-Marengo, C., et al., *A formulation to optimize the risk reduction process based on LOPA.* Journal of Loss Prevention in the Process Industries, 2012(0).

[12] Myers, P.M., *Layer of Protection Analysis – Quantifying human performance in initiating events and independent protection layers.* Journal of Loss Prevention in the Process Industries, 2012(0).

[13] CCPS, *Layer of Protection Analysis -Simplified Process Risk Assessment.* 2001, New York, NY: Center for Chemical Process Safety. 1-258.

[14] J Wreathall, C.N., *Assessing risk: the role of probabilistic risk assessment (PRA) in patient safety improvement.* Qual Saf Health Care, 2004. **13**: p. 206–212.

[15] Markowski, A.S. and M.S. Mannan, *Fuzzy logic for piping risk assessment (pfLOPA).* Journal of Loss Prevention in the Process Industries, 2009. **22**(6): p. 921-927.

[16] Michael Stamatelatos, H.D., *Probabilistic Risk Assessment Procedures Guide for NASA Managers and Practitioners.* Second ed. 2011, Washington, DC: NASA Headquarters.

[17] Dana Kelly, C.S., *Bayesian Inference for Probabilistic Risk Assessment; A Practitioner's Guidebook.* 2011, London: Springer.

[18] O'Connor, A.N., *Probability Distributions Used in Reliability Engineering.* 2011, College Park, Maryland: Reliability Information Analysis Center (RIAC).

APPENDIX A

Table 1. Bayesian Approach for Updating PFD of ONS System [17-18]

Prior	Likelihood	Posterior
Beta Distribution $f(p) = \dfrac{\Gamma(\alpha+\beta)}{\Gamma(\alpha)\,\Gamma(\beta)}\, p^{\alpha-1}(1-p)^{\beta-1}$	**Binomial Distribution** $Pr(X = x) = \binom{n}{x} \cdot p^{x}(1-p)^{n-x}$	**Beta Distribution**
$\alpha_{prior} = \mu_{prior}\left[\dfrac{\mu_{prior}(1-\mu_{prior})}{Var_{prior}} - 1\right]$ $\beta_{prior} = (1-\mu_{prior})\left[\dfrac{\mu_{prior}(1-\mu_{prior})}{Var_{prior}} - 1\right]$	x : No. of failures n : No. of demands	$\alpha_{post} = \alpha_{prior} + x$ $\beta_{post} = \beta_{prior} + (n-x)$ $\mu_{post} = \dfrac{\alpha_{post}}{\alpha_{post} + \beta_{post}}$ $Var_{post} = \dfrac{\alpha_{post}\cdot\beta_{post}}{(1+\alpha_{post}+\beta_{post})(\alpha_{post}+\beta_{post})^{2}}$
$\mu_{prior} = 2.8\,E-3 \qquad \alpha_{prior} = 3.986$ $Var_{prior} = 1.96\,E-6 \quad \Rightarrow \quad \beta_{prior} = 1419.585$	$x = 3$ $n = 1000$	$\alpha_{post} = 6.986 \qquad \mu = 2.88\,E-3$ $\beta_{post} = 2416.585 \quad \Rightarrow \quad Var_{post} = 1.18\,E-6$

Table 2. Modified-LOPA for NPP Fire Protection System

Scenario No.	Economic consequence severity	Category	frequency of I.E.	IPL(s)	Updated PFD	Calculating Risk from Updated Data (Modified-LOPA)	Calculated Risk from Primary Data (PRA)
1	$1,000,000	Minor	7.10 E - 4	$\overline{ONS}$ *	2.88 E - 3	$\approx$ $ 708.0	$ 710.0
2	$92,000,000	Major	7.10 E - 4	ONS , $\overline{OFS}$	1.40 E - 4	$\approx$ $ 188.0	$ 230.0
3	$210,000,000	Catastrophic	7.10 E - 4	ONS , OFS	2.88 E-3 * 1.40 E-4 = 4.03E-7	$\approx$ $ 0.060	$ 0.018
* The sign "‾" shows successful operation of an IPL.							

Uncertainty Analysis for Target SIL Determination
in the Offshore Industry

Sungteak Kim[a*], Kwangpil Chang[a], Younghun Kim[a], and Eunhyun Park[a]

[a] Hyundai Heavy Industries, Yongin, Korea

Abstract: The requirements on design of SIS (Safety Instrumented System) based on SIL (Safety Integrity Level) has been developed continuously in the offshore industry. Especially, IEC 61508 and IEC 61511 illustrates various methodologies to determine a target SIL for specified safety function such as risk graph, hazard matrix, etc. These methods could derive different target SILs for the identical safety function. Model uncertainty might be the main cause of the result. In addition, since various methods require many input parameters, parameter uncertainties contribute to a target SIL with variance, either. In the offshore industry, engineers usually utilize two or even more methods to assess target SILs for the same function simultaneously and determine the more conservative value as the target SILs from the results. The conservatism would keep the system safe, but sometimes it could be too safe by installing excessive safety systems. For better decision-making, this article identifies the uncertainty factors in determining target SILs and evaluates the effects of the uncertainties on target SILs. Case studies have been performed for the practical systems used in the offshore industry.

Keywords: SIL determination, Uncertainty analysis, Offshore industry.

1. INTRODUCTION

In the offshore industry, safety instrumented systems (SIS) are installed for reducing risks to allowable level by detecting hazardous events and taking actions to prevent them from developing into further accidents. Elements of SIS consist of initiators, a logic solver, and final elements, as illustrated in Figure 1. According to the safety lifecycle approach in Figure 2, based on the international standards IEC 61508 [1] and IEC 61511 [2], the series of activities should be conducted: identifying safety instrumented functions (SIFs), assessing target safety integrity level (SIL) for each SIF, designing SIS, calculating the achieved SIL and verifying by comparison to target SIL, and putting SIS into the operation phases. Defined target SIL would affect whole SIS lifecycles including design and operation since these target values draw the upper limit of the reliability performance. From this point of view, target SIL should be derived carefully in order to not only satisfy the required risk reduction but also to get rid of unnecessary additional SISs upon existing safety systems which are already in place.

Figure 1: Safety Instrumented System

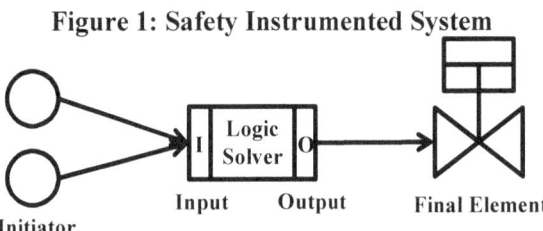

According to IEC 61511 [2], SIL requirement for each identified SIF would be determined using the following methods: semi-quantitative method, qualitative method, and semi-qualitative method. Choosing the proper one among the methods follows the criteria as follows: complexity, regulatory authorities, experience, accessibility of information on the input parameters, and whether the required risk reduction is given in a quantitative or qualitative form.

---

* *Contact author: sungteak.kim@gmail.com*

Probabilistic Safety Assessment and Management PSAM 12, June 2014, Honolulu, Hawaii

Figure 2: Safety Lifecycle Approach regarding SIL

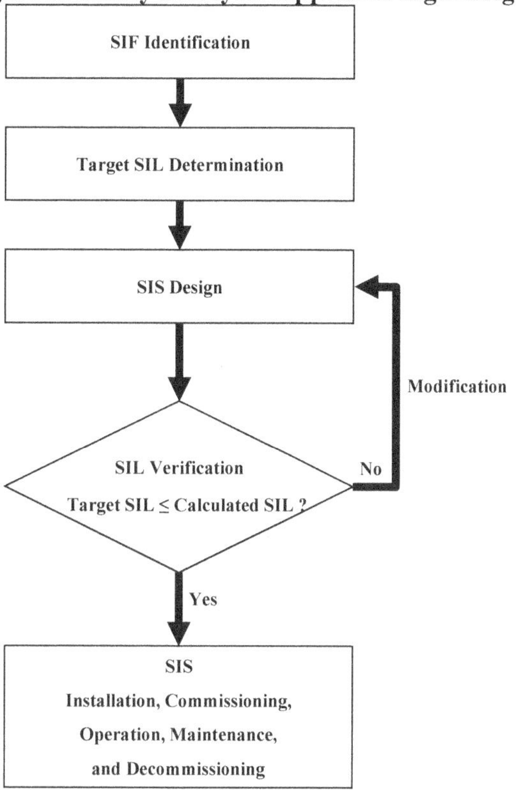

Abovementioned methods have both advantages and disadvantages with respect to rigor and effort, fit with SIL lifecycle, inputs required, etc. Above all, however, the methods are required to show consistency in target SIL determination. For example, if a SIF is identified from SIL workshop, the target SIL for this SIF should be the same regardless of the persons who take part in the SIL determination. Subjectivity is the main contributor of inconsistency in target SIL [3].

In practice, to avoid the problem of inconsistency, engineers and analysts usually utilize at least two methods simultaneously and apply the more conservative result as target SIL. This approach of conservatism would help the system have sufficient safeguards and keep it safe. However, there could be an argument of excessive safety system design resulting in increasing cost in terms of CAPEX as well as OPEX due to more frequent maintenance.

Assessment of target SIL always involves uncertainties due to its nature – decision-making based on either qualitative experts' opinion or quantitative statistical data with variance. Managing these uncertainties might lead the system to be optimized via setting the appropriate level of system reliability target. Uncertainty analysis for target SIL determination would be the starting point of effective SIL lifecycle management by checking the current uncertainty level in target SIL and identifying dominant factors which contribute to output uncertainties.

The objectives of this article are to (i) explain the characteristics of SIL determination methods, and (ii) propose the relevant application of the results obtained from uncertainty analysis.

The article is organized as follows: introduction of methods for target SIL determination, which are popular in the offshore industry in Section 2, identification of uncertainty factors and applicable proposed procedures for uncertainty analysis in SIL determination phase in Section 3, case studies performed to show the effects of the uncertainty analysis in Section 4, and conclusion with some future works in Section 5.

2. TARGET SIL DETERMINATION FREQUENTLY USED IN THE OFFSHORE INDUSTRY

2.1. Risk Graph

Risk graph is the one of popular methods used for target SIL determination in the offshore industry. Hazard matrix, another frequently used one, is known as the similar method to the risk graph. The common thing of two methods is that some parameters are combined to present the level of unmitigated risks based on decisions made by experts. However, the risk graphs consider likelihood (or demand rate), consequence, occupancy and probability of personnel avoiding hazard while the hazard matrices consider only likelihood and consequence [4]. This means that the risk graphs enable engineers to model more detailed situation.

Still, the risk graph method has a limitation. The method is suitable for assessing target SIL of SIS with defined equipment under control (EUC), for instance a pressure vessel, which is defined as local safety functions. On the other hand, safety functions where the whole platform is the EUC, e.g. emergency shutdown and fire & gas safety functions, defined as global safety functions, are not easily assessed through risk graphs [3].

2.2. Minimum SIL Requirements in OLF 070

Since IEC 61508 and IEC 61511 provide a variety of methods for SIL requirement determination, but without the specified guideline of which method to be used, it is difficult to choose the proper method. Additionally, it is known that the risk graphs and/or hazard matrices can result in non-consistent SIL requirement [3].

In this respect, OLF 070 suggests the use of minimum SIL requirements for target SIL. Minimum SIL requirements are developed for the most typically used safety functions in the oil and gas production plants. The table 7.1 in OLF 070 [3] contains the description of each safety function, functional boundaries and minimum SIL requirements. The goal of the minimum SIL requirements is for checking the minimum safety level of frequently used safety functions, simplifying calculation and documentation, and thus encouraging the standardization of target SIL determination in the industries [4].

Minimum SIL requirements are based on the typical loop assumption and estimated using the industrially verified component reliability data. Since the minimum SIL values are literally the minimum requirements, it is possible to establish stricter requirements where overall risk levels are much higher, which result from quantitative risk assessment (QRA) using the minimum SIL values as input data.

However, because of plant specific conditions and technological improvements, deviations from the defined minimum SIL requirements may be identified. To handle the deviations, OLF 070 Appendix C [3] suggests compensating methods using the tabulated minimum SIL requirements, whereas practical oil and gas projects go back to the original approach using IEC 61508 and IEC 61511 methodologies [4].

3. UNCERTAINTY ANALYSIS IN DETERMINING TARGET SIL

3.1. Concept of Uncertainty

Uncertainty is incomplete knowledge and information about a system as well as inaccuracy of the behaviour of systems [5]. Based on its nature, uncertainty is classified into two categories; epistemic or aleatory uncertainty. Epistemic uncertainty stems from the lack of knowledge. Accordingly, this kind of uncertainty can be reduced or controlled if additional knowledge becomes available. Aleatory uncertainty arises from inherent and natural randomness and variability. In this respect, aleatory uncertainty may be associated with observable quantities while epistemic uncertainty with unobservable quantities such as a failure rate [6].

In practice, a system cannot be characterized exactly due to epistemic uncertainties in both values of the model parameters and assumptions supporting the model itself [7]. The former and the latter are called parameter uncertainty and model uncertainty, respectively. An uncertainty analysis aims at identifying uncertainty factors and presenting uncertainties in analysis results for better decision making in terms of parameter and model uncertainty.

3.2. Parameter Uncertainty

Parameter uncertainty is about uncertainty in quantitative parameter values [5]. In this paper, concerning parameters are failure rates of components, beta-factors for common cause failure, proof or diagnostic test coverage factors, etc. Main influencing factors to parameter uncertainty are relevance and amount of generic data for a specific application and environment.

In terms of epistemic or aleatory uncertainty, parameter uncertainty can be deemed epistemic, aleatory, or both [10]. Regarding epistemic uncertainty, parameter uncertainty comes from imperfect knowledge about distribution types and values of the parameters. For aleatory uncertainty, the distribution of parameters represents its inherent variability. In the same context, the distribution includes combined effect of aleatory and epistemic uncertainty.

The effect of parameter uncertainty can be analyzed by observing uncertainty propagation [8]. Uncertainty propagation results in the distribution of uncertainty measures of interest (in our case, average of PFD or SIL). The techniques used for uncertainty propagation are Monte Carlo simulation, moment propagation, or discrete probability distribution [9]. It is recommended that sensitive analysis can be utilized to rank the importance of parameters in addition to uncertainty propagation [6].

3.3. Model Uncertainty

Model uncertainty mainly concerns the validity of model assumptions [11]. A model is the interpretation of real world. To design and develop a model, a lot of assumptions and hypotheses have to be defined. Even if these assumptions are well-defined logically, validation should be taken into account in order to check how much the model reflects real world.

Moreover, selection of models contributes to model uncertainty. Since the interpretations can vary, it is necessary that the models have gaps. Various models show the differences especially in the number and the kind of parameters used. Model selection is also influenced by regulations, standards, guidelines, and internal company policies. Further, the more the model deals with detailed level, the much time and effort are needed [11].

Assuming that the models have similar level of validity, model uncertainty can be reviewed by comparing the results of various models. However, it is sometimes hard to decide which model should be selected for the analysis where validation cannot be performed. In this case, a consensus model [8], which has been publicly published, peer reviewed, and widely adopted by stakeholders, is recommended to be used.

3.4. Uncertainty Analysis for Target SIL Determination

Technically, there are two representative methods for analysis of uncertainty, fuzzy set approach and probabilistic approach [12]. The fuzzy set approach is a set of mathematical principles for knowledge representation as degrees of belief using membership functions. It reflects how people think and attempts to model sense of words and intent of decision making [13]. The structure of a fuzzy set approach consists of three main components: a fuzzifier, which converts parameters into membership functions; an inference engine, based on a set of rules that reflects experts' opinions about how to link input to output; and a defuzzifier that re-convert the obtained output parameters into scalar value, in this case the target SILs.

Probabilistic uncertainty analysis is mainly based on sampling techniques: Monte Carlo sampling or Latin Hypercube sampling [12]. These methods generally involve the generation of random samples of input random variables, the deterministic evaluations of the performance function at these samples, and the post-processing to extract the probabilistic characteristic (statistical moments, reliability, and PDF) of the performance function.

Considering qualitative features of the risk graph methodology into uncertainty modeling, the fuzzy set approach should be utilized since experts' knowledge and consensus are key factors for target SIL determination [14]. On the other hand, sampling-based method is applicable to quantitative SIL determination methods such as OLF 070 minimum SIL requirement. Input parameters can be modeled as assumed probability distributions and this makes the basis of sampling technique.

4. CASE STUDIES

4.1. System Description

An example study has been illustrated for the local safety function, MEG Subsea Injection Pump Discharge PSHH. This protection function is to prevent overpressure in discharge of MEG injection pump, which is positive displacement type. Any obstruction at the user point or no MEG injection due to process shutdown might lead to this hazard. To prevent this hazard, MEG injection pump should be stopped on high-high pressure detected at the pump discharge. There will be not only MEG spill as an environmental consequence but loss of containment with very high pressure in terms of personnel risk. The dangerous undetected failure includes all possible modes of failure leading to any of the following effects: the transmitter failing to signal high pressure on demand, the logic solver failing to initiate pump stop, the circuit breaker failing to stop the pump motor on demand. For this reason, the MEG pump is not included in the reliability calculation. Figure 3 shows the configuration of this SIF.

Figure 3: MEG Subsea Injection Pump Discharge PSHH Configuration

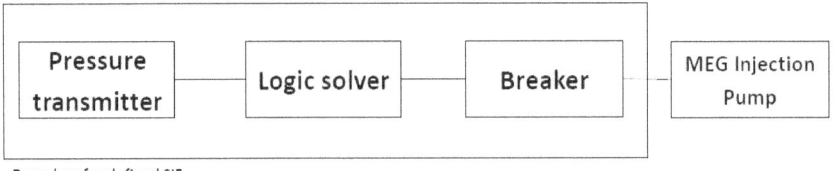

Boundary for defined SIF

It is assumed that there are already existing protective measures, so-called non-SIS. One measure is two pressure safety valves (PSVs) provided on the MEG injection pump discharge, sized for blocked outlet condition. Another is the valve that will be open to maintain the pressure in the header. However, if obstruction is sudden, pressure control may not act.

4.2. Effects of the Uncertainties on Target SIL

4.2.1. Target SIL obtained from risk graph using fuzzy set approach

To investigate the effects of the uncertainties on determined target SIL, uncertainty analysis, using either fuzzy set approach or sampling-based method, has been performed for two different SIL determination methods; risk graph and OLF 070 minimum SIL requirement.

Previously, in section 3.4, the fuzzy set approach is appropriate for uncertainty analysis of risk graph model. For this case study, the calibrated risk graph has been used as shown in Figure 4, which conveys the result of SIL assessment for the SIF in deterministic way. The first step of the fuzzy set approach is to fuzzify the parameters into membership functions. Parameters used here are listed in

Figure 5 with corresponding membership functions, respectively. Membership functions are modelled using both trapezoidal and triangular shaped functions. For parameter P, the mark with a star indicates as follows: P_A should be selected if only all the following are true; a) facilities are provided to alert the operator that the SIS has failed, b) independent facilities are provided to shutdown such that the hazard can be avoided or which enable all persons to escape to a safe area, c) the time between the operator being alerted and a hazardous event occurring exceeds 1 hour or is definitely sufficient for the necessary actions.

After the fuzzification, the fuzzy inference system is modelled using 'If-then rule', for example, If (C is Medium_high) and (F is Low) and (P is High) and (W is Medium_high) then (SIL is SIL 2). In this case study, total 52 rules are generated.

The last stage is to defuzzify the results obtained back into the scalar value. There are several methods of defuzzification such as Center-of-Maximum (CoM), Mean-of-Maximum (MoM), Center-of-Area (CoA), etc. For this case, CoA has been used because this method can produce more accurate results [15] and the result is shown in Figure 6.

Figure 4: Calibrated Risk Graph and Determined Target SIL for the SIF

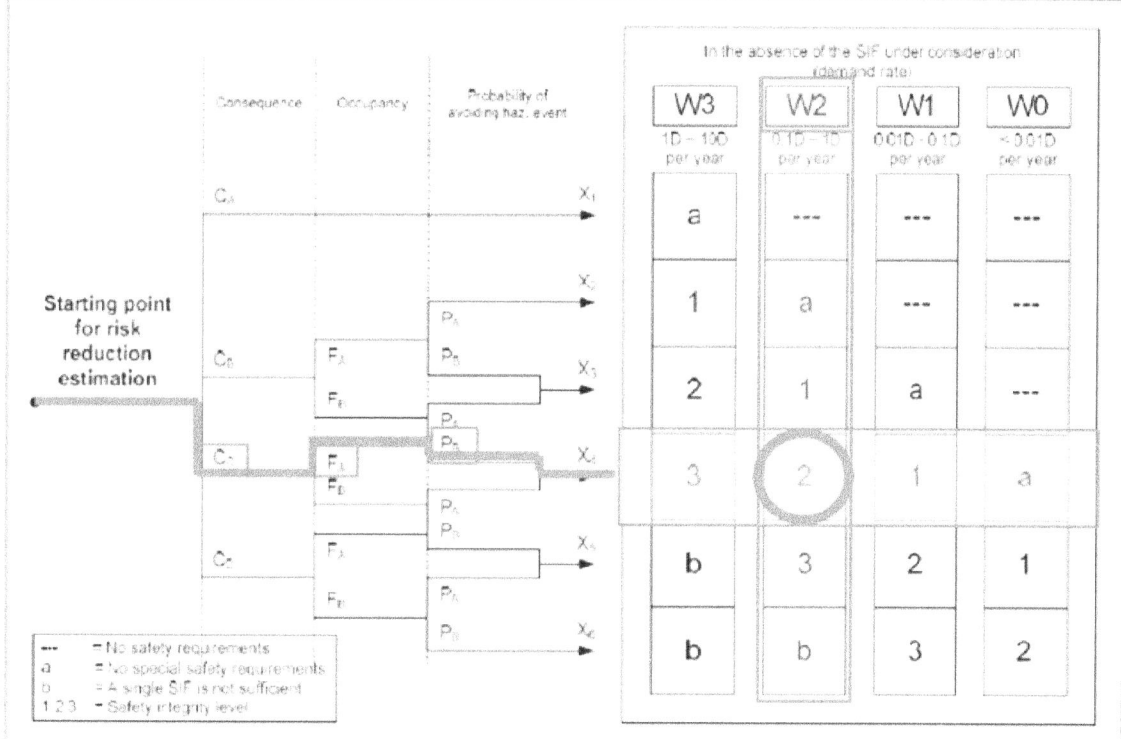

Figure 5: Membership Functions for Risk Graph Parameters

C	C_A	C_B	C_C	C_D
Classification	No risk to personnel or minor injury Range of C = ~ 0.01	Moderate injury Range of C = 0.01 ~ 0.1	Serious or permanent injury or potential for single fatality Range of C = 0.1 ~ 1	Multiple fatalities Range of C = 1 ~
Linguistic value in membership function	Low	Medium_low	Medium_high	High
Membership function	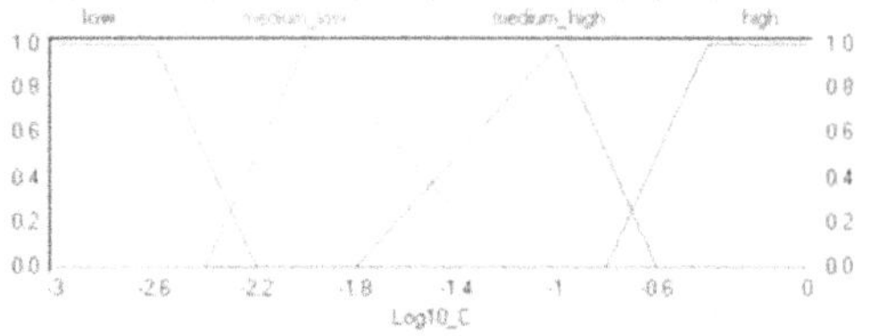			

F	F_A	F_B	P	P_A	P_B
Classification	Rare to more frequent exposure in the hazardous zone Range of F = 0's ~ 10's	Frequent to permanent exposure in the hazardous zone Range of F = 10's ~ 100's	Classification	Use parameter if all conditions* are satisfied	Use parameter if all conditions* are not satisfied
Linguistic value in membership function	Low	High	Linguistic value in membership function	Low	High
Membership function	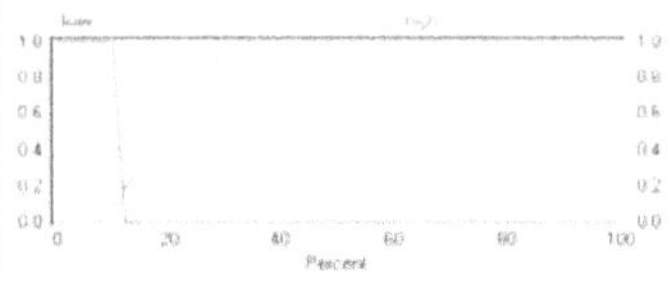		Membership function	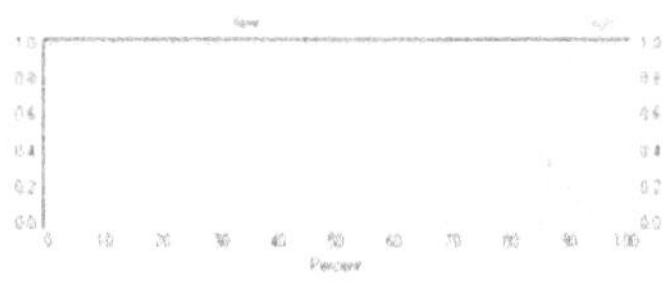	

W	W_0	W_1	W_2	W_3
Classification	Demand rate (= of demands / year) Range of W = ~ 0.01	Demand rate (= of demands / year) Range of W = 0.01 ~ 0.1	Demand rate (= of demands / year) Range of W = 0.1 ~ 1	Demand rate (= of demands / year) Range of W = 1 ~ 10
Linguistic value in membership function	Low	Medium_low	Medium_high	High
Membership function	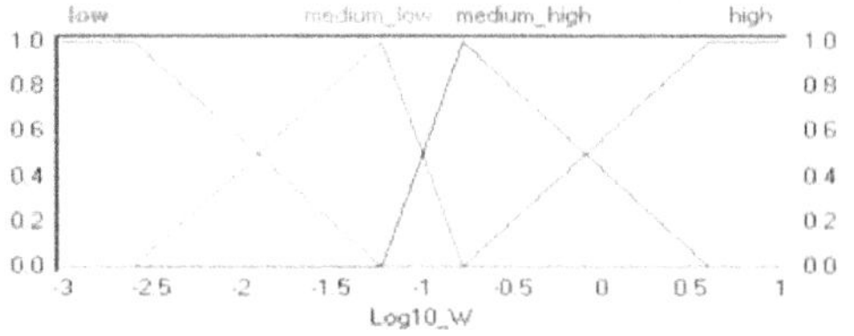			

SIL	---	SIL a	SIL 1	SIL 2	SIL 3	SIL b
Classification	No safety requirements Range of PFD = N A	No special safety requirements Range of PFD = 0.1 ~ 1	Range of PFD = 0.1 ~ 0.01	Range of PFD = 0.01 ~ 0.001	Range of PFD = 0.001 ~ 0.0001	A single SIF is not sufficient Range of PFD = 0.0001 ~ 0.00001
Linguistic value in membership function	Very_low	Low	Medium_low	Medium_high	High	Very_high
Membership function	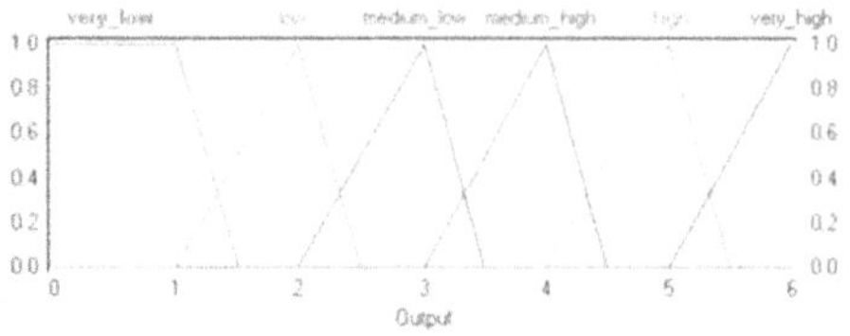					

Figure 6: Target SIL Obtained using the Fuzzy Set Approach

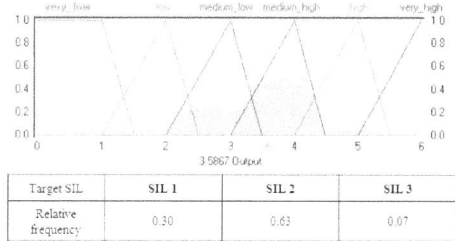

Target SIL	SIL 1	SIL 2	SIL 3
Relative frequency	0.30	0.63	0.07

4.2.2. Target SIL obtained from OLF 070 minimum SIL requirement using sampling method

As mentioned in section 2.2, minimum SIL requirements in OLF 070 are derived based on the typical loop assumption and PFD estimation using industrially verified component reliability data. Thus, the uncertainty analysis, applied to determination of target SIL using the method from OLF 070, would be performed by the sampling methods to investigate the effect of uncertainty propagation.

To calculate the target SIL of the SIF, PDS method [16] has been used for maintaining consistency with OLF 070. Since every component has simple configuration, 1oo1, the average of PFD follows the formula (1) without consideration of common cause failure. It should be noted that the probability of so-called test independent failure (TIF) can be added to the PFD to reflect the effect of incomplete testing. OLF 070 takes P_{TIF} into consideration when calculating PFD. The values for P_{TIF} come from the PDS data handbook [17].

$$PFD_A = \lambda_{DU} \cdot \tau/2 + P_{TIF} \qquad (1)$$

In addition to the abovementioned model (1), another model has been used in the case study for the purpose of comparison. In order to replace the effect of imperfect testing with P_{TIF}, proof test coverage (PTC) is added to the input parameters [16] and the PFD model is modified as follows:

$$PFD_B = PTC \cdot \lambda_{DU} \cdot \tau/2 + (1 - PTC) \cdot \lambda_{DU} \cdot T/2 \qquad (2)$$

T is the assumed interval of complete testing that the residual failure modes will be detected. If some failure modes are not able to be tested for, then T should be taken as the lifetime of the equipment. In this case, T is assumed to be 5 years, the periodic overhaul duration of the offshore plant where the SIF would be installed.

Table 1 shows the reliability data used for the calculation of minimum SIL requirement of the SIF. Among the parameters, λ_{DU} and PTC are assumed to be random variables due to uncertainties from incompleteness of data. The uncertainty of the DU failure rate is given by a lognormal distribution with median equal to the values in Table 1. The error factors are assumed to be 3 [11]. The PTC and P_{TIF} are given by a uniform distribution with the intervals shown in Table 1. In regard to proof test coverage, this assumption is due to lack of accumulated data from generic databases in the offshore industry. Also, P_{TIF} has certain amount of uncertainty because its value is determined by experts' opinion. On the other hand, number of components and τ are assumed to be constant since the uncertainties of configurations and proof test intervals can be controlled [8].

Table 1: Reliability Data for the SIF Components

Component	No. of Components	λ_{DU} (/hour)	τ (hours)	PTC (%)	P_{TIF}
Pressure transmitter	1	$0.3 \cdot 10^{-6}$	8760	$80 \sim 99$	$4.0 \cdot 10^{-4} \sim 6.0 \cdot 10^{-4}$
Logic solver	1	$1.0 \cdot 10^{-6}$	8760	$80 \sim 99$	$3.0 \cdot 10^{-5} \sim 7.0 \cdot 10^{-5}$
Circuit breaker	1	$0.2 \cdot 10^{-6}$	17520	$80 \sim 99$	$3.0 \cdot 10^{-5} \sim 7.0 \cdot 10^{-5}$

The uncertainty propagation has been studied by Monte Carlo simulation. For each simulation run, random values for each uncertain parameter have been obtained and then used as an input to calculate target SIL_A and SIL_B based on PFD_A and PFD_B, respectively. 50,000 simulation runs are performed for the precision of results. The target SIL distributions are shown in Figure 7 with input parameter distributions. Also, the statistics of target SIL simulation results are arranged in Table 2, where P_α represents $\alpha\%$ percentile of each output value.

Figure 7: Target SIL Distribution using Monte Carlo Simulation

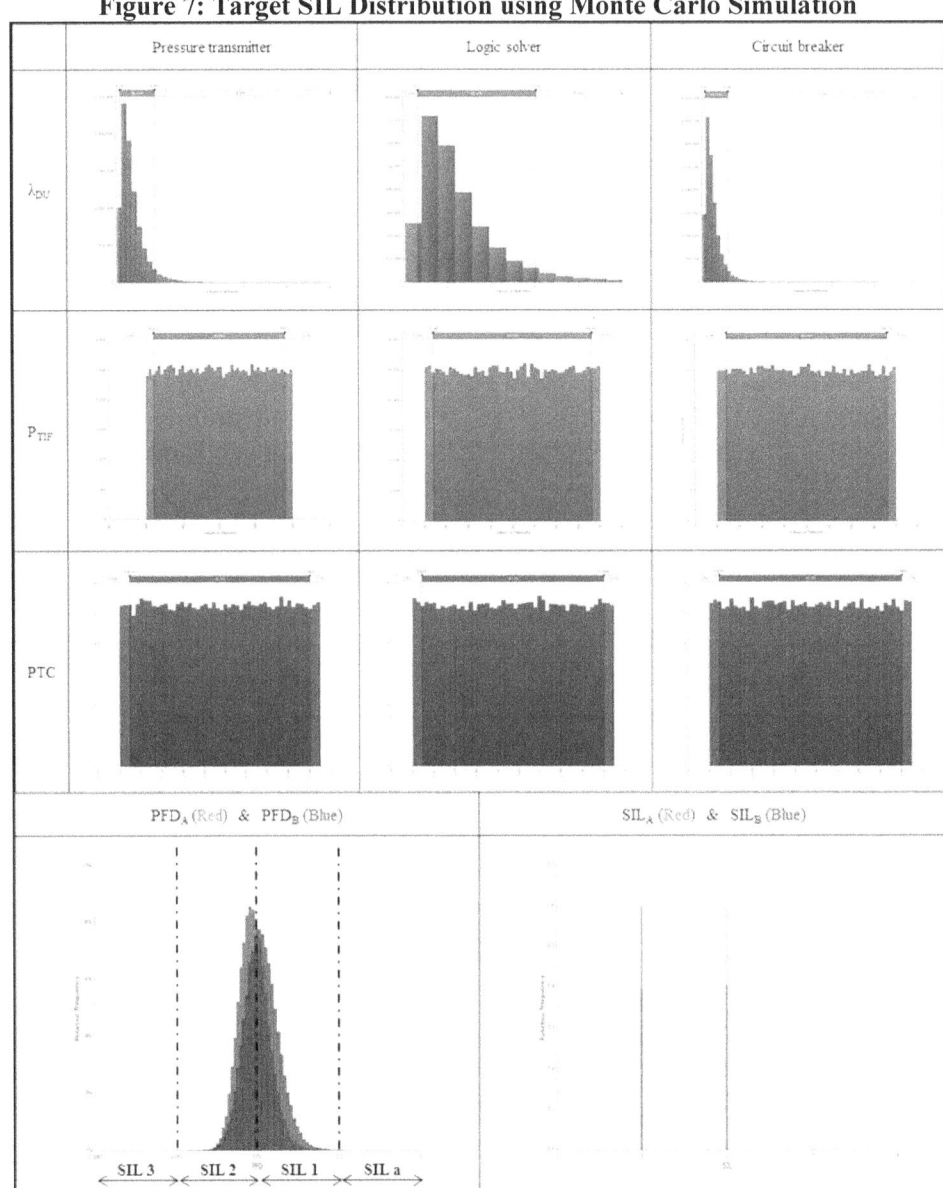

Table 2: Statistics of Target SIL Simulation Results

Output	Mean	Standard Deviation	Coefficient of Variation	P_{10}	P_{50}	P_{90}
PFD_A	9.94×10^{-3}	4.64×10^{-3}	4.67×10^{-1}	5.42×10^{-3}	8.93×10^{-3}	1.56×10^{-2}
PFD_B	1.27×10^{-2}	6.65×10^{-3}	5.24×10^{-1}	6.35×10^{-3}	1.12×10^{-2}	2.07×10^{-2}
SIL_A	1.61	4.88×10^{-1}	3.03×10^{-1}	1.00	2.00	2.00
SIL_B	1.40	4.91×10^{-1}	3.51×10^{-1}	1.00	1.00	2.00

4.3. Discussions

Without any uncertainty taken into consideration, i.e. in deterministic way, SIL 2 requirement has been derived when using calibrated risk graph as shown in Figure 4. Moreover, OLF 070 [3] refers that the PSD function for PAHH is required to satisfy SIL 2. The function is defined to start with the pressure sensor and terminates with closing of the critical valve. On the minimum SIL table, it is noted that the final element of this function could be different from a valve, e.g. a pump which must be stopped. From the viewpoint of OLF 070, MEG Subsea Injection Pump Discharge PSHH is also classified in the PSD function for PAHH and SIL 2 requirement can be applied.

Unlike the target SILs determined by deterministic way, the results show differences when considering underlying uncertainties together. For the same SIF, the fuzzy risk graph result and sampling-based PFD_A show similar tendency that SIL 2 is dominant than SIL 1. However, sampling-based PFD_B results in SIL 1 as a target SIL since the mean value of PFD is slightly over the 1.00×10^{-2}, which is the boundary of SIL 1 and SIL 2. Although the distributions of PFD_A and PFD_B approximate each other in terms of both location of center and amount of relative variation, the final outputs are distinguished. It can be guessed that the reason of the difference in target SILs mainly comes from the difference in models between PFD_A and PFD_B. Since PTC shows much sensitivity than other parameters [4], the PFD_B model including PTC can be vulnerable to the parameter uncertainty. Even if P_{TIF} is exposed to a certain level of uncertainties since the values are general based on expert judgment [16], trustworthy and reliable database dealing with PTC does not exist yet. Thus, as of now, PFD_A model has much, even a little, robustness that PFD_B model.

The fuzzy risk graph, in Figure 6, shows that target SIL values range from SIL 1 to SIL 3. It is not necessary to take the result into account seriously because the portion of SIL 3 is small, 7% of all results. However, a situation is likely to occur when there is no dominant target SIL value. For instance, target SIL results have values from SIL 1 to SIL 3 with relative frequencies of 35%, 35%, 30%, respectively. In this case, some decision-makers can think and act cautiously by choosing SIL 3 as the target SIL. Still, other decision-makers consider the result as the average of PFD or SIL. From this point of view, SIL 2 is determined since the median and/or mean value is located in SIL 2 range. It is not easy to judge which decision is more reasonable. The same problem can also occur when using sampling-based method.

5. CONCLUSION

This article has given overall understandings of uncertainty analysis for target SIL determination phase. Especially, risk graph method and minimum SIL requirement in OLF 070 have been introduced and studied as subjects for uncertainty analysis since these are popular methods in offshore industry.

Both model and parameter uncertainty contribute to uncertainties of determined target SIL values when using either risk graph or minimum SIL requirement. To investigate the effect of uncertainties, the fuzzy set approach and sampling-based simulation have been used for risk graph and OLF 070 minimum SIL requirement, respectively.

The case studies have been performed on the SIF, MEG subsea injection pump discharge PSHH. When applying deterministic approach, SIL 2 is derived as the target SIL by risk graph and OLF 070 minimum SIL requirement. Similarly, the fuzzy set approach and the sampling-based method for PFD_A model show the same result, SIL 2 on average. PFD_B model, however, shows the difference which results in SIL 1. The main reason was due to the high sensitivity parameter, PTC.

In conclusion, uncertainty analysis can provide broader possibility of having various output values when determining target SIL. When making decisions on target SILs based on the results obtained from the uncertainty analysis, balance is the most considerable issue between sufficient safety margins and economic feasibility.

Acknowledgements

This work was supported by the Korea Evaluation Institute of Industrial Technology (10042430, Development of 500 MPa URF & SIL 3 Manifold and Subsea System Engineering for Deepsea Field) funded by the Ministry of Trade, Industry & Energy (MI, Korea).

References

[1]   IEC 61508. *"Functional safety of electrical / electronic / programmable electronic safety-related systems,"* International Electrotechnical Commission, 2010, Geneva.

[2]   IEC 61511. *"Functional safety—safety instrumented systems for the process industry,"* International Electrotechnical Commission, 2010, Geneva.

[3]   OLF-070. *"Application of IEC 61508 and IEC 61511 in the Norwegian petroleum industry,"* Technical report. The Norwegian Oil Industry Association, 2004, Stavanger, Norway.

[4]   S. Kim, K. Chang, and Y. Kim. *"Risk-based design for implementation of SIS functional safety in the offshore industry"*, In: Steenbergen RDJM, van Gelder PHAJM, Miraglia S, Vrouwenvelder ACWM, editors. Amsterdam, The Netherlands, 29 September-2 October 2013 Safety, reliability and risk analysis: Beyond the horizon. Proceedings of the European safety and reliability conference 2013 (ESREL 2013), CRC Press, pp. 1875-80, (2013), London.

[5]   M. Rausand. *"Reliability of safety-critical systems,"* John Wiley & Sons, Inc., 2014, Hoboken, New Jersey.

[6]   M.A. Lundteigen. *"Safety instrumented systems in the oil and gas industry: concepts and methods for safety and reliability assessments in design and operation"*, Doctoral thesis, Norwegian University of Science and Technology (NTNU), 2009, Trondheim, Norway.

[7]   T. Aven and E. Zio. *"Some considerations on the treatment of uncertainties in risk assessment for practical decision making,"* The Journal of Reliability Engineering and System Safety, volume 96, pp.64-74, (2011).

[8]   H. Jin, M.A. Lundteigen, and M. Rausand. *"Uncertainty assessment of reliability estimates for safety instrumented systems"*, In: Berenguer C, Grall A, Guedes Soares C, editors. Troyes, France, 18-22 September 2011 Advances in safety, reliability and risk management. Proceedings of the European safety and reliability conference 2011 (ESREL 2011), CRC Press, pp. 2213-21, (2011), London.

[9]   NASA, *"Probabilistic risk assessment procedures guide for NASA managers and practitioners,"* Technical report, NASA Office of Safety and Mission Assurance, 2002, Washington, DC.

[10]   Abrahamsson, M. *"Uncertainty in quantitative risk analysis – characterization and methods of treatment,"* Doctoral thesis, Lund University, 2002, Lund, Sweden.

[11]   A.F. Janbu. *"Treatment of uncertainties in reliability assessment of safety instrumented systems"*, Master's thesis, Norwegian University of Science and Technology (NTNU), 2009, Trondheim, Norway.

[12]   G. Rausand. *"Uncertainty management in reliability analyses"*, Master's thesis. Norwegian University of Science and Technology (NTNU), 2005, Trondheim, Norway.

[13]   O.Y. Abul-Haggag, and W. Barakat. *"Application of fuzzy logic for the determination of safety integrity in light of IEC 61508 & 61511 standards"*, International Journal of Emerging Technology and Advanced Engineering, volume 3, pp.41-48, (2013).

[14]   C. Simon, M. Sallak, and J.F. Aubry. *"SIL allocation of SIS by aggregation of experts' opinions"*, Proceedings of the European safety and reliability conference 2007 (ESREL 2007), 2007, Stavanger, Norway.

[15]   I. Elaamvazuthi, P. Vasant, and J. Webb, *"The application of Mamdani fuzzy model for auto zoom function of a digital camera,"* International Journal of Computer Science and Information Security, volume 6, pp.244-249, (2009).

[16]   Sintef. *"Reliability prediction methods for safety instrumented systems, PDS method handbook,"* Sintef, 2013, Trondheim, Norway.

[17]   Sintef. *"Reliability data for safety instrumented systems, PDS data handbook,"* Sintef, 2013, Trondheim, Norway.

Using Fault Trees to Analyze Safety-Instrumented Systems

Joseph R. Belland[*]
Isograph, Inc., Irvine, USA

Abstract: Safety-instrumented systems are protection functions frequently seen in automotive, chemical processing, and oil and gas refining systems. These functions are designed to engage in case a hazardous condition arises and mitigate any potentially catastrophic consequences. Because of the potential for loss of life or other safety-related risks related to these systems, safety-instrumented systems usually have a very strict reliability requirement.

Fault Tree analysis is a method of analyzing a system to determine its reliability and identify weak points. This method uses a qualitative and quantitative approach that graphically shows how component failures logically combine to create system failures, and quantifies the system failure probability using failure rate data from component failures.

Due to its powerful and flexible nature, Fault Tree analysis is an ideal method for analyzing safety-instrumented systems to determine if they are meeting their reliability goals, to find weak points in the design, or for focusing maintenance efforts. Fault Trees may also be used to determine the spurious trip rate of the safety system, that is, how frequently the safety system will engage unnecessarily. This paper will provide a guide to using Fault Tree analysis software for these purposes.

Keywords: Fault Tree Analysis, Safety-Instrumented Systems, Safety Integrity Level, Automotive Safety Integrity Level

1. INTRODUCTION

1.1. Fault Tree Analysis

Fault Tree analysis was first developed in 1961 at Bell Laboratories to evaluate the launch control systems of ICBMs [1]. Since then, it has become widely used in many different industries to effectively model potential causes of system failures.

Fault Tree analysis is a deductive failure analysis which focuses on one particular undesired event and provides a method for determining causes of this event. This undesired event, usually a hazard or catastrophic failure, constitutes the top event in a fault tree diagram. This TOP event is connected to basic events through intermediate logic gates. These logic gates indicate the combination of failures or occurrences that will lead to the TOP event. In this way a fault tree is a qualitative analysis.

The basic events typically represent component failures or other hazards or events that can contribute to the TOP event hazard. If probability values for the base events are known, Boolean algebra and probability laws can be applied to calculate a probability value for the TOP event. In this way, Fault Tree analysis is also quantitative.

---

[*] jbelland@isograph.com

Figure 1: An Example Fault Tree

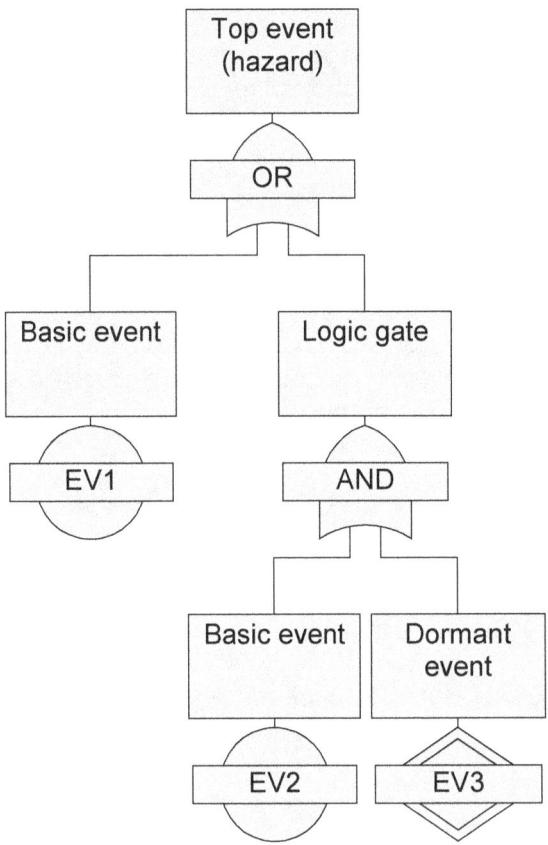

1.2. Safety-Instrumented Systems

Critical processes or systems appear in many different industries. These processes or systems, if they are not properly maintained or controlled, can malfunction is such a way as to cause significant risk to safety, environmental impact, or financial loss. Examples of such critical systems may be found in the process industry, nuclear, railway, automotive, and many others. Examples include a chemical reactor in a refinery, a nuclear power generator, or an airbag on an automobile.

Safety-instrumented systems (SIS) are systems that are designed to lower the risk of critical processes or systems. The SIS operates in such a way that, if the critical equipment malfunctions, the SIS will restore the system to a safe state.

A typical SIS consists of three elements, inputs, a logic solver, and final elements [2]. The inputs monitor the critical system, examining for unsafe or potentially unsafe conditions. The logic solver interprets the inputs from the sensors. The final elements are typically actuators of some sort, such as control valves, whose function is to either halt the system or process or bring it to a failsafe state.

Let us consider an example of a safety-instrumented system.

1.2.1. High-Integrity Pressure Protection System

A high-integrity pressure protection system (HIPPS) is a kind of SIS frequently seen in chemical plants and oil refineries. It is designed to prevent an over-pressurization event of a fluid line or vessel by shutting off the input of the fluid. Since the over-pressurization event may lead to a rupture or explosion, this SIS is utilized in such a way to mitigate the risk of the potential hazardous consequences.

The example HIPPS shown in Figure 2 consists of three pressure transmitters (PT) (the inputs), a logic solver, and two block valves (the final elements). The PTs monitor the fluid flow, and send this information to the logic solver. If two out of three PTs read an unsafe pressure, then the logic solver will signal both block valves to close, shutting off the flow into the downstream equipment. The 2oo3 voting in this case is for two reasons: it provides redundancy, so a single PT dangerous failure will not negate the entire SIS; and it prevents spurious trips or accidental engagements of the SIS when it is not needed due to a single PT safe failure.

Figure 2: An Example HIPPS

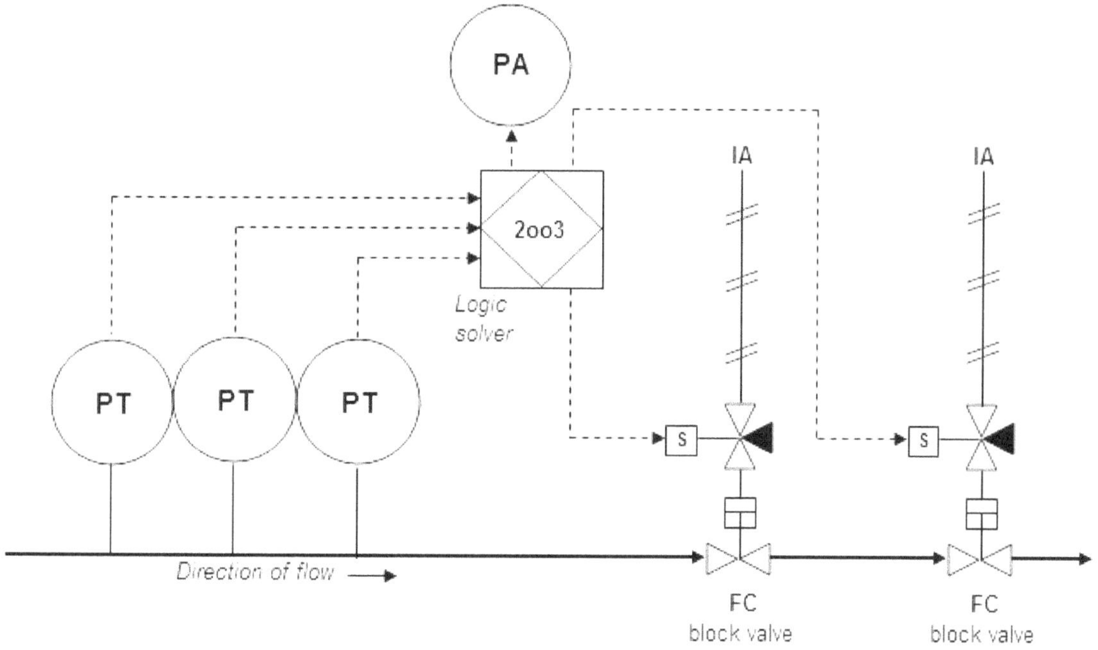

2. EVALUATING THE PFD OF A SIS USING FAULT TREE ANALYSIS

Due to their usage in critical applications, safety-instrumented systems have a very stringent probability of failure on demand (PFD) requirement. This requirement is usually determined by industry standards, such as the safety integrity level (SIL) rankings defined in the IEC 61508 standard, or the automotive safety integrity level (ASIL) rankings defined in ISO 26262 [3]. Fault Tree analysis can calculate a PFD for a SIS, and therefore determine what SIL ranking applies to the function of the SIS.

For the purposes of this paper, we will not explore Fault Tree calculations in mathematical rigor. There are many useful resources—such as NUREG-0492, the U.S. Nuclear Regulatory Commission's Fault Tree Handbook, or IEC 61025—that describe the equations needed to solve a Fault Tree by hand. There are also many Fault Tree analysis computer software tools that can be used to create and evaluate a Fault Tree. For these reasons, we shall only focus on the general techniques of representing a SIS with a Fault Tree, and leave the mathematical analysis to the computers.

2.1. Fault Tree Construction

To begin constructing the Fault Tree representation of our SIS, we must first understand the logic symbols used in a Fault Tree diagram. Table 1 shows commonly-used Fault Tree logic symbols.

Symbol	Name	Logic
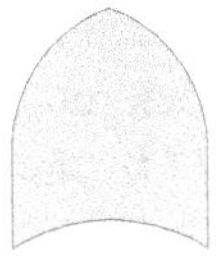	OR	TRUE if any input is TRUE
	AND	TRUE if all inputs are TRUE
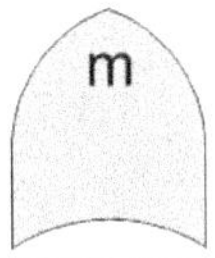 m	VOTE	TRUE if m inputs are TRUE

One primary consideration is that when a Fault Tree is used to evaluate the PFD of a SIS, it is generally constructed using failure logic. That is, the basic events and gates represent dangerous failures of the components and systems. This is the opposite of the logical expressions normally used in SIS design. For instance, a 2oo4 vote arrangement in a SIS means that the system will trip if two of the four elements meet the trip criteria. In PFD Fault Tree, a 2oo4 VOTE gate means the system will fail to trip if two of the four elements fail to meet the trip criteria.

Once we understand the logic gates that appear in a Fault Tree, we can construct the representative tree of our example SISs. We will start from a template. Since SISs consist of three basic subsystems—inputs, logic solver, final elements—and the failure of any one of these three subsystems will cause a failure of the SIS, the basic template will have an OR gate as the TOP gate, with each of those three subsystems as inputs, as in Figure 3.

Figure 3: A Generic SIS Fault Tree

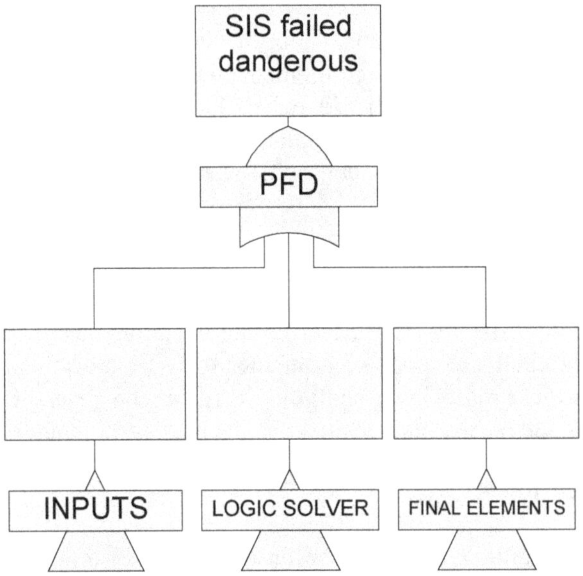

2.1.1. HIPPS PFD Fault Tree

In the HIPPS system from section 1.2.1, there are three pressure transmitters serving as the inputs to the system. They have a 2 out of 3 vote configuration, meaning if two of the three PTs register a high pressure, the SIS will trip. For the PFD Fault Tree, we interpret this to mean if two of the three transmitters fail to register the high pressure (fail low), then the SIS will not engage.

The two block valves provide redundancy to each other. In fault tree terminology this is represented by an AND gate, meaning the block valve system fails if both block valves fail.

Figure 4: A Fault Tree Representation of the HIPPS PFD

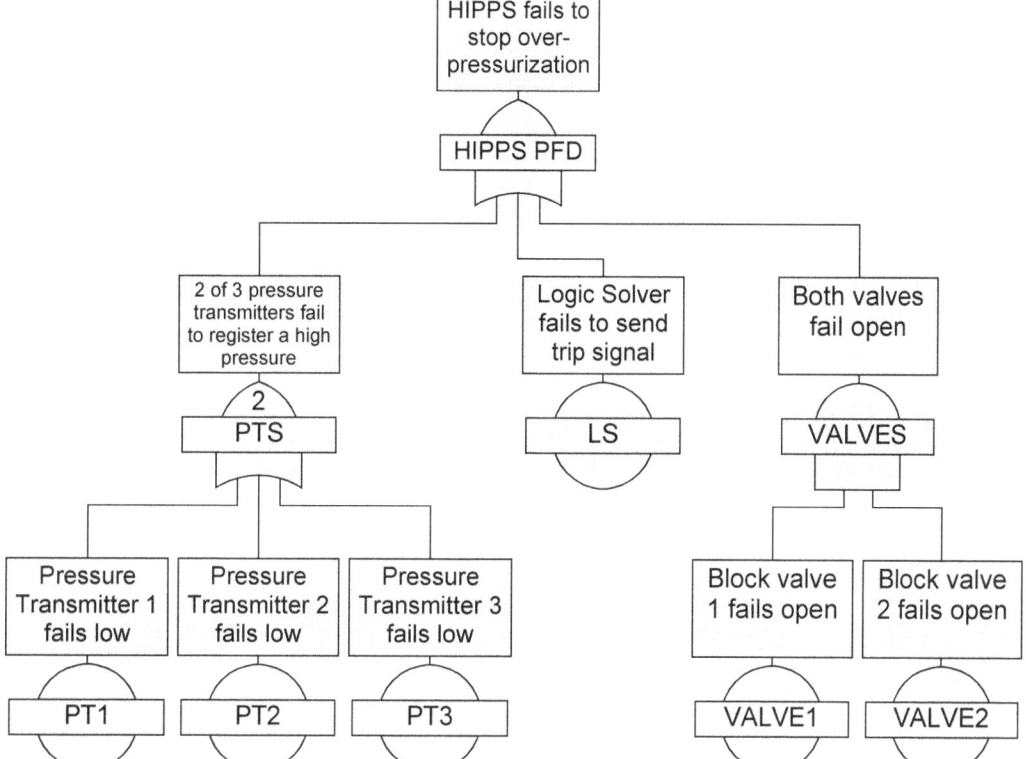

2.2. Failure Data

The failures of components in a SIS are typically divided into four modes: safe detected, safe undetected, dangerous detected, and dangerous undetected [2].

Graph 1: Failure Mode Classifications

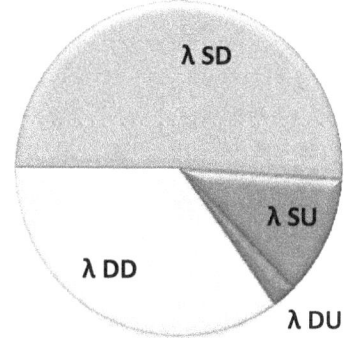

Since Fault Trees are constructed to evaluate one hazard, the basic event failure data must describe only failures that contribute to that hazard. If the TOP event hazard being evaluated is the demand failure of the SIS, then only dangerous failure rates should be included for the basic events. If the spurious trip of the SIS is being evaluated, then only safe failures should be included.

Some Fault Tree computer programs may require that the detected and undetected failure modes be modeled with separate basic events. The inputs to each basic event would be the failure rate appropriate to the failure mode it represents, e.g., λ_{DD} for the dangerous detected failure component event and λ_{DU} for the dangerous undetected failure component event.

If imperfect proof testing is used, then it may be important to also account for the proof test coverage. An imperfect proof test essentially splits the undetected failures into two categories: those that are corrected after testing (λ_{DUC}) and those that remain uncorrected after testing (λ_{DUU}). Failures that are uncorrected after proof tests will only be corrected by a replacement of the component.

Figure 5 provides an example of modeling each component failure mode with a separate basic event.

Figure 5: Failure Modes Modeled with Separate Basic Events

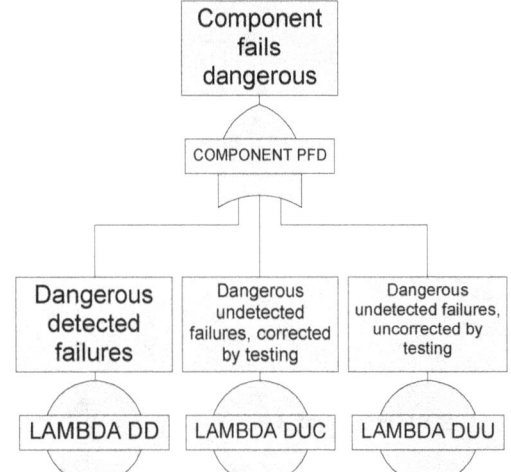

2.2.1. HIPPS Failure Data

For the HIPPS Fault Tree, we will use the data found in Table 2.

Table 2: HIPPS Failure Data

Component	λ (per hour)	λ Safe %	DC % (Safe)	DC % (Dang.)	MTTR (hours)	τ (months)	PTC%	θ (years)
Pressure Transmitter	1.1E-6	37.5	0	60	0.001	48	90	6
Logic Solver	6.1E-6	99.84	100	0	8	48	99	10
Block Valve	4.6E-6	0	0	0	0.001	24	99.5	20

Key:
λ – failure rate
λ Safe % – Percent of failures that are safe detected or safe undetected
DC % (Safe) – Diagnostic Coverage percent for safe failures; the percent of safe failures that are detected by diagnostics.
DC % (Dang.) – Diagnostic Coverage percent for dangerous failures; the percent of dangerous failures detected by diagnostics.
MTTR – Mean Time to Repair
τ – Test interval
PTC% – Proof Test Coverage percent; the percentage of undetected failures that are uncovered and corrected during proof testing.
θ – Replacement interval

This is characteristic of the type of data available for SIS components. However, some translation may be needed to enter this into a Fault Tree computer program. We can use the following equations to convert these values to data usable in a Fault Tree software tool.

$$\lambda_S = \lambda \frac{\lambda_{SAFE\%}}{100} \tag{1}$$

$$\lambda_D = \lambda \left(1 - \frac{\lambda_{SAFE\%}}{100}\right) \tag{2}$$

$$\lambda_{DD} = \lambda_D \frac{DC_{Dang.}}{100} \tag{3}$$

$$\lambda_{DUC} = \lambda_D \left(1 - \frac{DC_{Dang.}}{100}\right)\left(\frac{PTC}{100}\right) \tag{4}$$

$$\lambda_{DUU} = \lambda_D \left(1 - \frac{DC_{Dang.}}{100}\right)\left(1 - \frac{PTC}{100}\right) \tag{5}$$

Entering this information into a Fault Tree computer program gives us the following PFD values for the individual components of the HIPPS system:

Table 3: HIPPS Component PFD Values

Component	PFD$_{avg}$
Pressure Transmitter	5.1E-3
Logic Solver	1.7E-4
Block Valve	4.2E-2

2.3. Common Cause Failures

One concern when modeling safety-instrumented systems is the occurrence of common cause failures (CCF). A common cause failure is the failure of more than one component due to a single cause. This single-point failure affecting multiple components can be due to a variety of issues, such as environmental stresses (temperature, humidity), improper maintenance and testing, manufacturing defects, incorrect installation or calibration, or other similar causes.

The effect of CCFs on a SIS is to negate some of the benefits of redundancy. Redundant components are intended to provide extra protection against failures. However, if redundant components carry the risk of simultaneous failure, this protection is reduced.

Generally, Fault Tree methodology assumes independence amongst the basic events. That is, the assumption of a Fault Tree model is that the occurrence of one basic is independent of the occurrence of any other. This is opposite to how CCFs affect the system.

For this reason, CCFs must be explicitly accounted for when constructing a Fault Tree model of a SIS. The standard way to accomplish this is to add a new basic event representing the common cause failure of the components to the Fault Tree. See Figure 6 for an example.

Some Fault Tree software tools may allow implicit inclusion of CCFs. In these programs, the user would flag certain basic events as belonging to a common cause group. The Fault Tree program would then automatically perform the correct calculations, accounting for the loss of redundancy due to CCFs.

Figure 7 demonstrates how this might be accomplished in a Fault Tree tool. In this example, COMP1 and COMP2 are both flagged as sharing a CCF event.

Quantitatively, input failure data for the CCF basic event must also be accounted for. The most common method of accomplishing this is the beta factor model. The beta factor essentially represents the percentage of failures of the components that are due to common causes. So for example, a beta factor of 0.05 would mean that 5% of the component failures are due to common causes.

We can calculate the independent and common cause failure rates using equations (6) and (7). These values would then be used in equations (1) and (2) for calculating the failure mode rates.

$$\lambda_{IND} = \lambda \cdot (1 - \beta) \tag{6}$$

$$\lambda_{CCF} = \lambda \cdot \beta \tag{7}$$

Figure 6: A Fault Tree Model of a Common Cause Failure

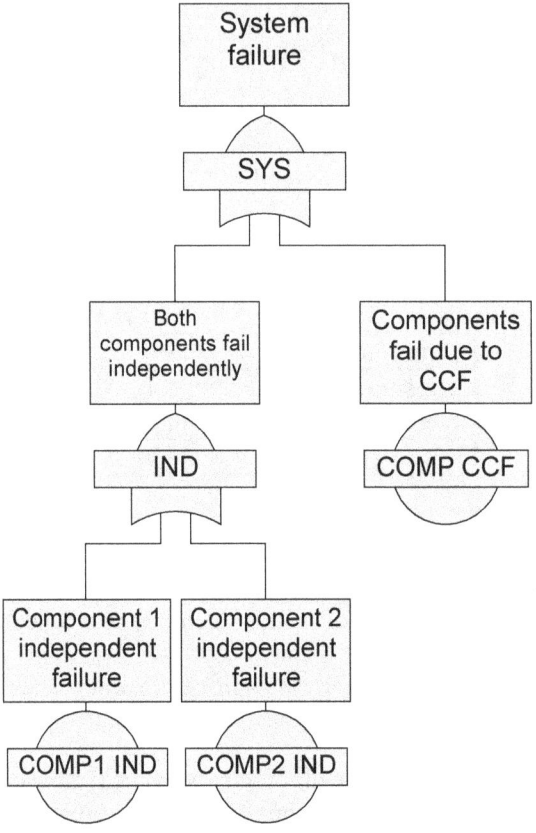

Figure 7: Implicit Inclusion of Common Cause Failures

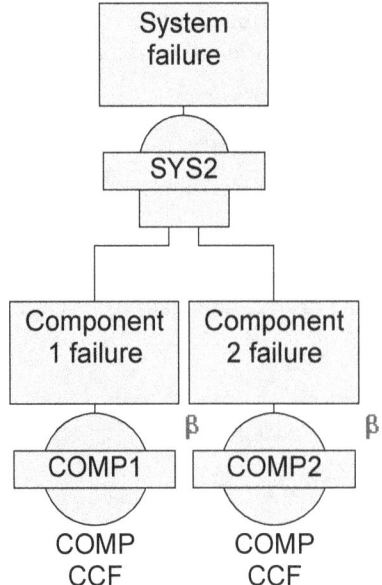

2.4. Logic before Average versus Average before Logic

One final consideration when using Fault Tree analysis methods to evaluate a SIS is the effect of averaging component PFD values before applying system logic. Standard Fault Tree methodology works by first calculating a PFD_{avg} for each basic event, then combining those averages using the multiplication and addition laws of probability. This can create a discrepancy with other SIS evaluation methods, which first apply the system logic, then calculate a PFD_{avg}.

For instance, the equation for calculating the PFD_{avg} of a 1 out of 2 voted configuration is given in the IEC 61508-6 standard as [4]:

$$PFD_{avg} = 2([1 - \beta_D]\lambda_{DD} + [1 - \beta]\lambda_{DU})^2 t_{GE} t_{CE} + \beta_D \lambda_{DD} MTTR + \beta \lambda_{DU} \left(\frac{\tau}{2} + MTTR\right) \quad (8)$$

Where

$$t_{GE} = \frac{\lambda_{DU}}{\lambda_D}\left(\frac{\tau}{3} + MTTR\right) + \frac{\lambda_{DD}}{\lambda_D} MTTR \quad (9)$$

$$t_{CE} = \frac{\lambda_{DU}}{\lambda_D}\left(\frac{\tau}{2} + MTTR\right) + \frac{\lambda_{DD}}{\lambda_D} MTTR \quad (10)$$

Using the failure data for the block valve from Table 2, and applying β and β_D values of 0.05, yields a PFD_{avg} of 3.969E-3. However, using standard Fault Tree methodology on the same data yields PFD_{avg} of 3.348E-3, which is only 84% of the IEC 61508-6 calculation. The reason for this discrepancy is because for a function $f(x)$,

$$\overline{f(x) \cdot f(x)} \neq \overline{f(x)} \cdot \overline{f(x)} \quad (11)$$

If a computer program is being used to evaluate the Fault Tree, then this discrepancy can be accounted for with sophisticated computer algorithms. For instance, that same Fault Tree, when evaluated in a program with a compensating algorithm, yields a PFD_{avg} of 3.913E-3, which differs from the IEC 61508-6 calculation by less than 1.5%.

2.5. HIPPS Final Analysis

Entering all this information into a Fault Tree computer program, using a beta factor of 5% for the pressure transmitters and block valves yields the final results shown in Table 4.

Table 4: HIPPS Reliability Metrics

PFD_{avg}	λ (/hour)	MTBF (hours)	RRF
4.7E-3	6.193E-7	1,622,000	212.8

This qualifies the system as a SIL-2 ranking, pending review of the architectural constraints.

3. EVALUATING THE SPURIOUS TRIP RATE OF A SIS USING FAULT TREE ANALYSIS

When evaluating a SIS, it is often important to also consider the safe failures of the system [2]. These safe failures are sometimes referred to as spurious trips, and occur when the safety system engages unnecessarily, when there was no hazardous condition for the safety system to mitigate. An unnecessary trip of the safety system may also carry a hazard risk with it. For instance, if the airbag on a vehicle deploys when no collision occurred, then it may cause the driver to lose control of the vehicle.

Usually, the relevant reliability metric for spurious trips is mean time to failure, or $MTTF_{spurious}$.

3.1. Construction

The logic used to construct a spurious trip evaluation Fault Tree is usually the reverse of that used in the PFD evaluation tree. This is because, if a system is built with redundancy such that 2 out of 2 failures must occur to cause the PFD hazard, then usually 1 of the 2 elements tripping spuriously will cause the system to engage. For this reason, all gates except the TOP gate are logic-swapped. AND gates become OR gate and vice versa. Generally, the reverse of m out of n vote logic is $n - m + 1$ out of n.

Likewise, instead of using the dangerous failure rates, safe failure rates are used to calculate component failures. This is accomplished by modifying equations (3),(4), and (5) with the safe failure rates.

3.1.1. HIPPS Spurious Trip Fault Tree

Applying the rules above, we can construct a Fault Tree to model the spurious trips of the HIPPS and obtain an $\text{MTTF}_{\text{spurious}}$.

Figure 8: HIPPS Spurious Trip Fault Tree

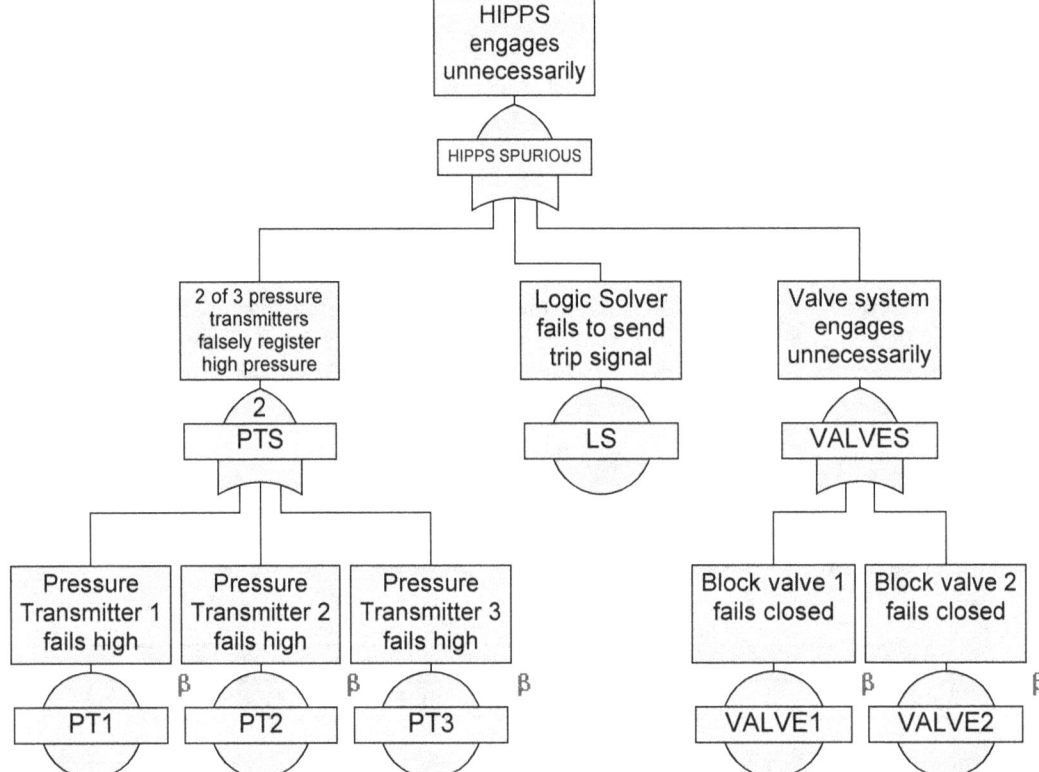

Since the only failure mode of the block valves is the dangerous failure—they cannot fail safe (closed)—the events could be removed from the tree.

Table 5: HIPPS Spurious Trip Calculations

Spurious trip rate (/hour)	$\text{MTTF}_{\text{spurious}}$ (hours)
6.165E-6	162,200

4. OPTIMIZATION

There are many advantages to using a computerized Fault Tree to analyze SISs. For instance, the computer can analyze the tree more quickly and the tree can easily be modified to consider alternative designs or maintenance plans. This can be very useful when trying to optimize the SIS.

4.1. Importance Analysis

One place to begin any attempt at optimization would be with an importance analysis. Importance analysis considers each basic event's contribution to the TOP gate hazard, and ranks the events by the percent of hazardous failures that are contributed to, at least in part, by each event. This is done quite simply by analyzing the tree, assuming that the event never occurs, and then comparing this result with the normal results. The importance results for the HIPPS PFD tree are given in Graph 2.

Graph 2: HIPPS Importance Rankings

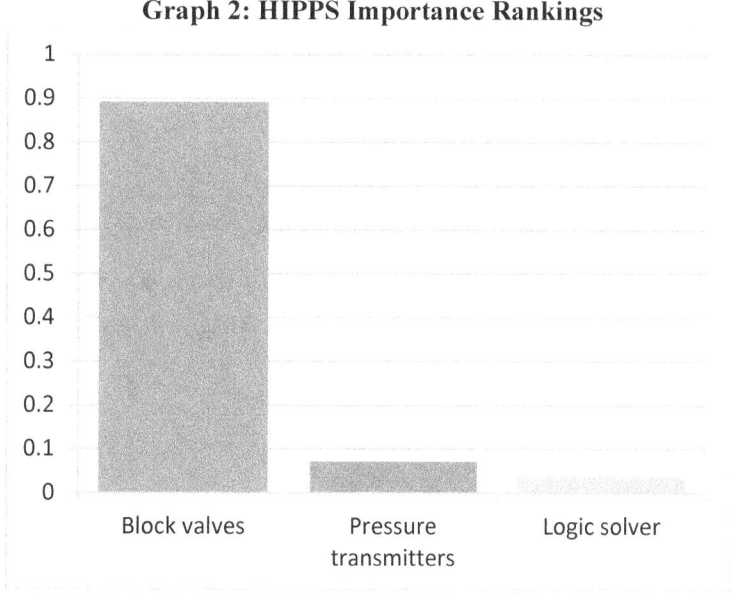

The block valve system accounts for 89% of the HIPPS demand failures, indicating that any attempt at improving the reliability should start there.

4.2. Sensitivity Analysis

Another method of optimizing a SIS that can easily be done by computer is sensitivity analysis. In this analysis method, a basic event input parameter, such as failure rate or test interval, is modified and the effects on the system reliability are recorded. This will show, for example, the impact of using a more reliable component or the effects of different test intervals.

Table 6 shows the impact on the HIPPS PFD with respect to changes in the test interval of the block valves. Using sensitivity analysis in this way, we could look for the test interval that would give us a target reliability goal.

Table 6: HIPPS PFD Sensitivity Analysis

τ (months)	4	6	8	12	18	24
PFD_{avg}	1.028E-3	1.274E-3	1.547E-3	2.174E-3	3.314E-3	4.700E-3

5. CONCLUSION

Fault Tree analysis is a useful tool in the reliability engineers tool belt to analyze safety-instrumented systems. It can help calculate the probability that the SIS will not perform its required function on demand, and thus determine the safety integrity level of the system. It can also be used to calculate how frequently the safety system will engage unnecessarily (the spurious trip rate). Many Fault Tree computer programs exist and using these can help the reliability engineer optimize the design or maintenance planning of the SIS. Those interested in more information about the quantitative cut set methodology used in Fault Tree analysis are encouraged to seek out additional resources or investigate software packages to see how this technique can meet their needs.

References:

[1] C. Ericson, "Fault Tree Analysis - A History," *Proceedings of the 17th International System Safety Conference,* 1999.

[2] International Society of Automation, *ISA-TR84.02-2002: Safety Instrumented Functions — Safety Integrity Level Evaluation Techniques,* North Carolina: ISA, 2002.

[3] International Organization for Standardization, *ISO 26262-5:2011: Road Vehicles — Functional Safety — Part 5: Product Development at the Hardware Level,* Geneva: ISO, 2011.

[4] International Electrotechnical Commission, *IEC 61508-6: Functional Safety of Electrical/Electronic/Programmable Electronic Safety-Related Systems,* Geneva: IEC, 2000.

Are Cognitive and organizational human factors missing from the blunt end in the oil and gas industry?

Stig O. Johnsen[a**]

[a] SINTEF, Trondheim, Norway

Abstract: The area of Human factors (HF) has been in focus to increase safety and quality of operation in the industry. HF covers three main domains, physical ergonomics, cognitive factors and organizational factors. HF has often been identified as root causes in accidents, i.e. 40-90% in different industries. The International Association of Oil & Gas producers (OGP) prioritized "more attention paid to HF" as one of four issues after the Macondo disaster in 2010. However in several projects, HF focus on cognitive factors and organizational factors is missing in the blunt end (i.e. early phase). HF is conceptualized as physical ergonomics (work environment and layout). In addition layout is often based on prior experiences and not on new was of operations. There is poor systematic HF education; few HF experts in user organizations and limited HF research to improve safety. We propose that HF must focus on cognitive factors and organizational factors early from the blunt end. The organizations should have local HF competence. External validation and verification from certified HF experts should be performed early. A simple set of HF guidelines and standards should be used to ensure HF focus, in addition to continued regulatory focus and review.

Keywords: Human Factors, Safety, Resilience, Crew Resource Management.

1. INTRODUCTION

In the last 60 years, Human factors (HF) have been developed as a discipline, covering a broad set of domains. Traditionally the main domains have been physical ergonomics (consisting of layout, work environment..), cognitive factors (perceptions, information processing, Human Machine Interface..) and organizational factors (communication, teamwork, Crew Resource Management..), see [1]. HF has an important influence on work design and execution, and thus it has a key relationship to safety. In many reviews of accidents, HF has been seen as a causal or contributing factor in 40 to 90 percent of accidents, depending on the industry, [2], [3], [4], [5]. Thus human factors has been key area in design and operations of safety critical equipment, this has been particular evident in aviation, where human factors experts has been a key part of the organization and in design teams. The oil and gas industry have a different tradition related to HF, in many instances HF has often been considered at later stages, and not from the blunt end i.e. early in the concept phase. Recent accidents and incidents have created more focus on HF in the oil and gas industry, i.e. a reactive approach. The Macondo blowout killed 11 workers, released 4.9 million barrels of oil, and expenses are 42Bn$ so far. More focus on HF, such as a better design (or improved human machine interface) giving warning of a blowout may have mitigated the accident. Such a design activity is a part of the area cognitive human factors. The International Association of Oil & Gas producers (OGP), has identified "more attention paid to HF" as one of four prioritized areas after Macondo, see [6]. In the oil and gas industry in Norway, there has been a focus on physical ergonomics i.e. layout and work environment, based on regulation, from PSA [7], the Facilities regulation §20 – focusing on ergonomic design, and §21, focusing on HMI. When designing a control center or a drilling cabin in the oil and gas industry, some human factors issues are usually explored and discussed. This has been done based on the use of recognized HF standards such as ISO 11064 "Ergonomic design of control centers", [8], and exploring alarm standards from EEMUA, [9].

---

[*] Stig.o.johnsen@sintef.no

Cognitive factors and organizational factors are not always explored sufficiently, and have sometimes been identified as areas of consideration or causes in incident investigations, [10] and [11]. A review of plans for new oil and gas fields (PUD), identified that mainly physical ergonomics was in focus at the early phases. Several cognitive and organizational issues in the drillers workplace was identified by the PSA, ref [12] [10] indicating insufficient focus on HF both in design and operations. Insufficient focus on human factors has also been identified in the design phase during implementation of new automated technology in drilling, ref [13]. Insufficient cognitive focus has also been identified in a design review of four new control centers, 2010 through 2012. The missing focus on cognitive issues and organizational issues may create weaknesses and holes in defenses as described in [14] and [15].

Based on the above factors, the research questions we would like to discuss are:
- Are key HF issues such as cognitive factors and organizational factors prioritized as a part of design of new control centers, i.e. from the blunt end?
- Why are cognitive and organizational human factors missing from the blunt end?
- Do the responsible organizations have sufficient knowledge to require, buy and involve the relevant HF experts at the right time?

2. METHOD AND RESEARCH

The research questions mentioned above have indicated the need for the following activities:
- A review of status of HF in critical areas, as documented by reviews by the authorities, and an exploration of key challenges related to actual implementation of HF issues by interviewing the HF actors involved in the design process
- A participatory review of HF focus in the recent design of control centers, i.e. an evaluation of the focus on physical ergonomics, cognitive factors and organizational factors. The reviews have especially been focused on new ways of operation based on remote operations and increased collaboration between onshore and offshore, i.e. if HMI has been designed on new responsibilities and if there has been focus on collaboration in distributed teams such as explored by crew resource management training (CRM).

The areas of exploration have been design of the control centers, based on recognized human factors standards, such as ISO 11064 [8], implying a task driven iterative design process.

The research approach when reviewing the control centers have been based on participatory action research (PAR); see [16]. PAR involves three basic elements – research, action and participation. PAR aims at creating a joint learning process between researchers and the various stakeholders holding interests in the problem under study, thus the findings has been discussed with the involved stakeholders, which have helped prioritize the findings. The study has focused on the blunt end, i.e. early design phases (feed phases) and during detailed design, when implementing new solutions or integrating several systems.

We have also explored the need for standards and guidelines through a Norwegian human factors network, consisting of around 400 stakeholders, having been involved in relevant projects where Human factors has been a key issue, see www.hfc.sintef.no.

3. RESULT POOR FOCUS ON COGNITIVE AND ORGANIZATIONAL FACTORS

In the following we have documented the results from our review and exploration.

3.1. Review of prior evaluations and incidents

We have performed a review of status based on open incident reports from the authorities and OGP.

From PSA, [12], the result of a survey of drillers work situation was presented, some of the results from the drillers were: many unnecessary alarms (reported from 50% of the drillers), the alarms gives no support during upsets (reported from 20% of the drillers), no support during critical situations (reported from ca 20% of drillers) no advance indication prior to upset/problem (reported from 20% of the drillers). There was too much information on the screens (reported from 50% of drillers), there was a mix of old and new systems, and 1/3 of drillers lose concentration and has problems keeping awake during operations and 1/3 of drillers are not aware of procedures when performing an operation.

From PSA, [10], there was a discussion of well-control incidents, and there was a need for improved systems to present safety critical information, improved alarms and physical ergonomics such as improved layout of drillers' cabin. It seems there had not been any notable improvement in the period from 2007. In several instances the systems used in drilling had weaknesses; with inadequate designs of displays, control panels, alarm and data systems. There is also room for improvement in HMIs outside the CCR, as an example touch screen HMIs is being used including alarm lists that cannot distinguish between alarms active unacknowledged, active acknowledged, and return to normal unacknowledged.

In the design phase, the system selection, it is often pointed out that the driller must use systems from different vendors with different user interfaces. As an example - from a review it was found that the same kind of graphs goes from top to bottom in one systems, while in another system the graphs goes from left to right. Thus there is poor coordination of HMI/cognitive factors across the different systems. This should have been addressed when the requirements for the systems were established, by specifying the need for common HMI design prior to the design phase.

HF in Crane cabins can also be improved, related to cognitive issues (Graphical displays, and use of Closed Circuit Television - CCTV) also connected to physical ergonomics (i.e. anthropometrics/ adjustability, integrated control in the sitting chair, quality of information display and glare).

In PSA report [11], discussing an incident related to stability, it was pointed out that "Several HMI shortcomings have been identified, especially with regards to legibility and to the way information of low operational value is emphasized on the safety system's HMIs." Thus the HMI interface was not optimal, and in combination with poor training, this can escalate an incident.

The research and development in the oil and gas area, including safety and HF, is handled by the Norwegian PETROMAKS program. In the 10 year period, documented by Research Council [17], 447 projects had been awarded grants – around 1%, i.e. 4 minor projects were related to Human Factors. This indicates poor HF focus related to research and exploration of new technology in the oil and gas industry.

The reports and cases indicate that there are challenges related to cognitive and organizational human factors, and support the view from OGP [6] –"more attention paid to HF", and it is suggested that HF should be more in focus in the blunt end in the early phases. Key issues are suggested to be:

- Requirements of common HF design, such as HMI and user interfaces should be performed as early as possible, to ensure consistent HF and HMI across different systems from different suppliers
- Improved design of Human Machine Interface (HMI) to present safety critical information, improved design of alarms, improved design to accommodate stressful situations
- Exploration of "worst case" scenarios prior to the design phases, in order to ensure that the systems at the workplace can accommodate worst-case scenarios even in a stressful situation

3.2. Varying focus on organizational HF and team collaboration, Crew Resource Management during the design phases

In the oil and gas industry it has been increased focus on collaboration between onshore and offshore through initiatives such as integrated operations (IO), as described in a white paper, [18]. The implementation of IO creates the need for improved collaboration and coordination between onshore and offshore, thus it was suggested in [19], that the Oil and Gas industry in Norway, should implement an adapted Crew Resource Management (CRM) training. The need for CRM training has been based on a 6th generation CRM concept, ref [20], identifying/preventing threats to safety at the earliest time and managing errors (i.e. Threat and Error Management). The CRM topics have been conceptualized as communication, situational awareness, teamwork, decision making, leadership and personal limitations (stress). Since 2005 the need for CRM training has also been a part of a HF validation method, called CRIOP, ref [21], The CRIOP validation method is often used in the oil and gas industry when building control rooms, control centers or driller cabins. The CRIOP method is an open, freely available method from the web (www.criop.sintef.no).

The need for CRM training among the crews involved in drilling and operations between onshore and offshore have been explored during the validation of design and operation of new control centers. The validation activities have been based on participatory action research (PAR) in the different phases of design. Meetings, discussions and prioritization of issues were conducted in a group setting, involving between 6and 26 participants in the different projects. Participants were HF experts, management, technical safety, work environment, automation, telecom and control room users. We have been involved in a limited set of validation activities, and we have performed analyses of 10 control centers as a part a verification and validation activity, and exploring two centers with collaboration between drillers, onshore support and expert centers more in depth.

The two analysis of collaboration between drillers, onshore support and expert centers were based on extended observations, interviews and discussions. Collaboration between the driller's onshore operation center and the different oil rigs were impacted by the variability of procedures and systems between rigs from the same operator. There was little standardization of procedures between the different rigs. Between the onshore support centers and onshore expert center, there were poor common perceptions missing communication and missing support of situational awareness.

In one operational setting (i.e. assessment of the control centre in operations) there were 15 operators involved in the control center. 30 % of the control room operators wanted to discuss problems and challenges in a team setting (needed access to experts locally or onshore) – and in one instance "13% - only 2 of 15 operators was confident that they could handle the CCR on their own (87%", i.e. 13 of 15 was not confident that they could handle incidents.) This also supported the need for systematic CRM training.
During the verification and validation activities, the participants prioritized issues and findings through expert judgments. In 6 of 10 projects, some sort of CRM focus and training was explicit suggested and prioritized (such as collaboration and communication) by the involved experts. The CRM focus or training was seldom suggested so early that it impacted the HMI design or layout.

The CRM issues discussed during validation and verification were related to design of procedures and systems for collaboration in a distributed team between onshore and offshore. Issues were clarity in responsibility, design for situational awareness in a distributed setting; how to support common mental models, common risk perceptions, Common HMI across different systems placed onshore and offshore and supporting communication in a distributed setting.

In Norway the need for CRM training has been suggested from 2005 at least. The need for CRM was prioritized in 6 of 10 projects we explored, thus the users have seen the need for CRM training. Based on accidents, such as Deepwater Horizon, OGP [23], has recommended implementing CRM training in a wide range of wells operational roles. A systematic approach has not yet been implemented in the oil and gas industry, thus there is a need to prioritize CRM in the industry. It may be a sign of complacency that it takes 10 years to implement necessary HF based training regime in the oil and gas industry.

3.3. Varying focus on HF during the design phases – poor focus on cognitive HF and organizational HF – when new ways of operation were explored

The focus on HF, to support safety and resilience has, been explored in four projects covering the design of control rooms (workplace design and HMI) involving remote operations and report support. The HF standard ISO 11064 [8] was used as a context for the work.

The activities has been based on participatory action research (PAR); exploring "best practices" guidelines and using a scenario approach in the early phases of design. Meetings, discussions and prioritization of issues were conducted in a participatory group setting, involving between 6 and 26 participants in the different projects. Participants were HF experts, management, technical safety, work environment, automation, telecom and control room users. The following common issues were identified between the projects:

- Physical ergonomics: Issues related to layout and work environment were minor. The quality of voice communication between distributed actors could be improved. However layout was often based on prior solutions – and not based on new ways of operation and new possibilities.
- Organizational ergonomics: Responsibility, work procedures and information between distributed actors should be clarified and had not been explored sufficiently. In addition, the expert teams involved in verification and validation prioritized increased focus on team collaboration and suggested adaptions of Crew Resource Management (CRM) training.
- Cognitive ergonomics: HMI development has been immature, and should be specified more clearly by the users, developed based on user requirements and tested in collaboration with users, suppliers and HF experts. There are many interfaces – and complexity due to missing consistency between different systems from different vendors. The role of humans as a safety barrier has not been explored sufficiently. We see the need for documentation of when the human operator – with sight, hearing and perception has been a safety barrier in operations. The extensive use of CCTV in remote operations and remote support has not been based on HF guidelines. During not normally manned operations (NNM) when humans are not present - how can the CCTV support the operator? What are the defined situations of hazard and danger that can be discovered on CCTV, and how can these situations or scenes be mitigated? These questions had not been explored sufficiently.

It was varying HF knowledge and awareness between the four different companies having the responsibility of the installation. Only one operator had a broad set of Human factors experts integrated in their organization. The other operators had outsourced Human Factors activities, and when discussing cognitive and organizational issues the operator used a physical ergonomics as a reference, that had scant knowledge of the area. It was varying (i.e. missing) Human Factors knowledge in the project teams. To ensure the right competence, there are several HF certification schemes internationally, such as Centre for Registration of European Ergonomists (CREE) and Board of Certification in Professional Ergonomics. However there is missing systematic certification of Human factors experts in Norway, so far there is one certified HF expert in Norway.

We have also seen instances of missing human factors focus during the initial phase of relevant projects even at this point in time, i.e. 2013. In [13], it was found that there was insufficient focus on human factors in a design phase of new automated technology, i.e. the design process of new drilling equipment.

4. DISCUSSION AND CONCLUSION

We have seen that there are varying practices related to the implementation of HF in the blunt end, prior to design. In some projects - cognitive human factors, such as human machine interfaces, responsibilities and procedures are not prioritized. When discussing the key theme of this article

"(Why) Are Cognitive human factors missing from the blunt end in the oil and gas industry?", our position is:

- Missing proactive focus on cognitive human factors and organizational factors. It is a great deal of variability, but in several projects in the oil and gas industry (cases from drilling), cognitive and organizational human factors is missing from the blunt end i.e. in the early project phases.
- Missing knowledge of the scope of human factors. Knowledge and awareness of Human factors seems poor in the responsible organizations. Human Factors are often conceptualized as physical ergonomics (layout and working environment) and necessary steps to perform cognitive analysis and organizational analysis are not done. Human factor knowledge is usually outsourced, and necessary key knowledge is not integrated in the responsible organizations. Training and certification seems wanting in Norway.
- Missing use of HF standards. There is missing knowledge of a required set of Human Factors tools, guidelines and theories. In some instances new projects has not been aware of a simple set of human factors guidelines, thus it seems important to focus on a selected set of standards such as ISO 11064 [8], HF guidelines for the use of CCTV, guidelines for team training as CRM (Crew Resource Management).

As discussed in [22], risk management is based on a complex relationship between regulators, organizations and actors thus HF must become more in focus through coordinated actions on many levels. Cognitive factors and organizational factors are important both to sustain safety and avoidance of major disasters, but also to sustain resilience and a positive work environment thus these elements should be prioritized. We propose that HF analyses and work also must focus on cognitive factors and organizational factors. The knowledge and focus of human factors should be explored in the "byer" organizations and environment (authorities, educational institutions and research council), and cognitive and organizational factors should be prioritized and validated/verified in the early phases of all projects (i.e. conceptual design and feed phase). External validation and verification from certified HF experts should be performed as early as possible. A simple set of HF guidelines and standards should be used to ensure early HF focus, in addition to continued regulatory focus and review.

HF focus should be proactive and not only reactive as a result of an accident investigation.

Acknowledgements

This research has been supported by the Human Factors in Control (HFC) network in Norway.

References

[1]   Karwowski, W. (2012). "*The disciplione of human factors and ergonomics*" In Salvendy, G. (2012). Handbook of human factors and ergonomics. John Wiley and Sons.

[2]   Rothblum, A. M., Wheal D., Withington S., Shappell S. A., Wiegmann D. A., Boehm W. and Chaderjian M. (2002). "*Human factors in incident investigation and analysis.*" 2nd international workshop on human factors in offshore operations (HFW2002).

[3]   Luxhøj, J. T. (2003). "*Probabilistic Causal Analysis for System Safety Risk Assessments in Commercial Air Transport.*" Workshop on Investigating and Reporting of Incidents and Accidents (IRIA). Williamsburg, Virginia, USA, NASA.

[4]   Wiegmann, D. A. and S. A. Shappell (2003). "*A human error approach to aviation accident analysis: the human factors analysis and classification system*". Aldershot, Ashgate.

[5]   DoD Department of Defence. (2005). "*Department of Defense Human Factors Analysis and Classification System.*" U.S. Navy, from www.uscg.mil/safety/docs/ergo_hfacs/hfacs.pdf

[6]   OGP (2013a) "*Offshore safety: Getting it right now and for the long term*" retrieved from www.ogp.org.uk/files/1513/6007/8310/Web_Post_GIRG_Final_300113.pdf

[7]   PSA (2014) "*The facilities regulation*" retrieved from ptil.no/activities/category399.html

[8]   ISO 11064 "*Ergonomic design of control centres*" (2013)

[9]   EEMUA 191 "*Alarm Systems - A Guide to Design, Management and Procurement*" (2013) (3rd edition) ISBN 0 85931 192 2

[10] PSA-RNNP (2011) "*Risk levels in Norwegian petroleum activities*".

[11] PSA (2012) "*Gransking Saipem - Ballasthendelse Scarabeo 8, 2012.09.04*"

[12] PSA (2007) "*Human Factors i bore- og brønnoperasjoner - borernes arbeidsituasjon*"

[13] Sætren G. (2013) "*Consequences of insufficient focus on human factors in a design phase of new automated technology*". SRA-E2013; 2013-06-17 - 2013-06-19

[14] Reason, J. (1997). "*Managing the risks of Organizational Accidents*" Ashgate.

[15] Reason, J. (2008). "*The Human contribution.*" (2008). Ashgate.

[16] Greenwood, D. J., & Levin, M. (1998). "*Introduction to action research: social research for social change*". Thousand Oaks, California: Sage Publications.

[17] Research council (2012) *Status – Petromaks* ("*Statusrapport Petromaks – 10 år*").

[18] *Stortingsmelding 38*, (2004) retrieved at 2009-12-03 from www.regjeringen.no/nb/dep/oed/dok/regpubl/stmeld/20032004/Stmeld-nr-38-2003-2004-.html?id=404848.

[19] Johnsen S. O., Lundteigen M. A., Albrechtsen E., Grøtan T. O. (2005) "*Trusler og muligheter knyttet til eDrift*" ISBN 92-14-03138-9, SINTEF STF38A04433

[20] Helmreich, R. L. & Merritt, A. C. (1998). "*Culture at work in aviation and medicine: National, organizational, and professional influences*". Aldershot, U.K.: Ashgate.

[21] Aas, A. L.; Johnsen, S. O.; Skramstad, T. "*CRIOP: A Human Factors Verification and Validation methodology that works in an industrial setting*". Lecture Notes in Computer Science 2009 ;Volum 5775. s. 243-256.

[22] Rasmussen, J. (1997). "*Risk management in a dynamic society: A modelling problem.*" Safety Science, 27, 183-213.

[23] OGP (2013b) Report 501 "*Well Operations Crew Resource Management – Guidance for developing WOCRM training syllabus*"

ACHIEVING A TOTAL SAFETY CULTURE THROUGH BEHAVIOR BASED SAFETY, ESTABLISHING AND MAINTAINING AN INJURY FREE CULTURE

Author, NJF van Loggerenberg

University of South Africa, Pretoria, South Africa

Abstract: Historically the focus in the industry has been on improving safety by addressing the work environment and eliminating, mitigating, and identifying hazards and risks. Industry has however reached a plateau regarding safety and safe procedures. It is therefore essential to focus on directives to establish a total safety culture through behavior based safety. The existence of a safety culture is evident when employees believe that safety is a value in life and the employer enhances employee safety ownership.

Positive and negative reinforcement can encourage safe behaviors. Behavior based safety aims at increasing safety in industry by positively influencing the behavior of all stakeholders involved. In focusing on a total safety culture, it is important to also focus on behavior and to understand the ABC model. Behavior based safety includes the four steps of the improvement process: define, observe, intervene and test, which will empower a shift in the employee safety culture from bad to good behavior.

An injury free culture requires the reconsideration of safety activities and the engagement of employees. Creating a workplace with an overall injury free culture that includes safety and empowers employees can only be done by establishing and maintaining a total safety culture.

Keywords: Safety culture, behavior based safety (BBS), safety ownership, ABC model, injury free culture, total safety culture

1. INTRODUCTION

Historically the focus in industry has been on improving safety by addressing the work environment and eliminating, mitigating, and identifying hazards and risks. To this end better tools and equipment have been provided and safe procedures enforced. Industry has, however, reached a plateau, regarding safety and safe procedures. It is therefore essential to focus on directives to establish a total safety culture through behavior based safety (BBS). Behavior based safety is an approach that appears to be successful and it is gaining more interest across the industry. The directives built into achieving a total safety culture through behavior based safety will bring about significant and certain changes to the industry in the future. Nowhere will this be as apparent as the role of establishing and maintaining an injury free culture by making behavior safe. To be able to use BBS as a guideline for establishing a total safety culture, realistic and clear goals must be set with clear communication between management and employees. This will lead to employee safety ownership through employee investment and participation. There is growing evidence that a safety culture can exist in any industry. BBS is based on people, safety, values and risks and can be implemented successfully in any organization. The existence of a safety culture is evident when employees believe that safety is a value in life and the employer enhances employee safety ownership.

2. BEHAVIOR BASED SAFETY

Behavior based safety is an approach used in the ongoing process of safety management and in order to achieve this, human behavior needs to be understood. Learning occurs over time as a function of positive and negative reinforcement of specific behaviors. This reinforcement can encourage safe behaviors.

Safety within the work environment is too vast a field to cover by management and safety officers alone, therefore it is necessary to emphasize BBS. Employees need to take ownership of their own safety and unsafe behaviors.

BBS is a data-driven, decision-making process. (Kaila, 2008:1). By empowering employees to observe and correct unsafe behaviors, they become the force in measuring safety and making a difference in an organization's safety culture. Actively caring about unsafe and safe behaviors will lead to improved safety behavior since BBS creates changing individual and social characteristics for safety.

Graph 1: Accident pyramid

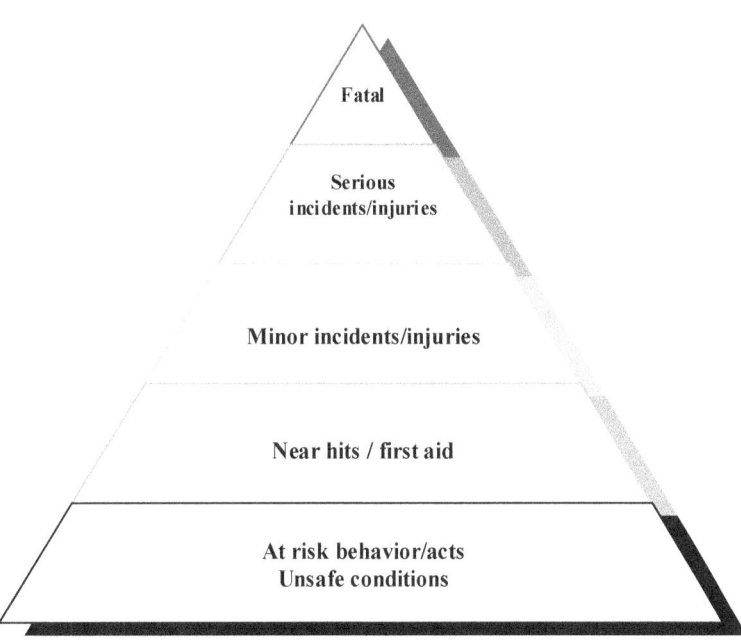

Based on international research on the "Accident pyramid", unsafe behaviors and conditions at the bottom of the pyramid mean that a large number of at-risk behaviors/acts contribute to accidents and incidents. Reducing unsafe behavior would help to eliminate serious accidents/incidents.

Through BBS a positive safety culture will be formed with an increase in production and job satisfaction. BBS therefore aims to implement behavioral observation and feedback as an effective means of reducing incidents and injuries in the work environment. As BBS is a tool, correct motivation is needed to ensure a positive influence in an organization's safety culture. Employees need to be motivated and presented with strategies to enable a total safety culture and to maintain an injury free culture.

2.1 The Process of Behavior Based Safety

The Health and Safety Authority describes BBS as a bottom-up approach with top-down support from safety management. It promotes interventions that are people-orientated and that focus on the observations of employees performing routine tasks, setting goals and giving feedback on observations.

It is therefore clear that the BBS approach is based on motivating, assisting, reinforcing and sustaining safe behaviors among employees. It has to be ongoing, implementing a new way of thinking and working more safely, and must consequently promote continually positive results. BBS aims at increasing safety in the industry by positively influencing the behavior of all stakeholders involved. This is only possible by establishing a safety culture by means of observation, coaching and communication.

Graph 2: Behavior based safety process

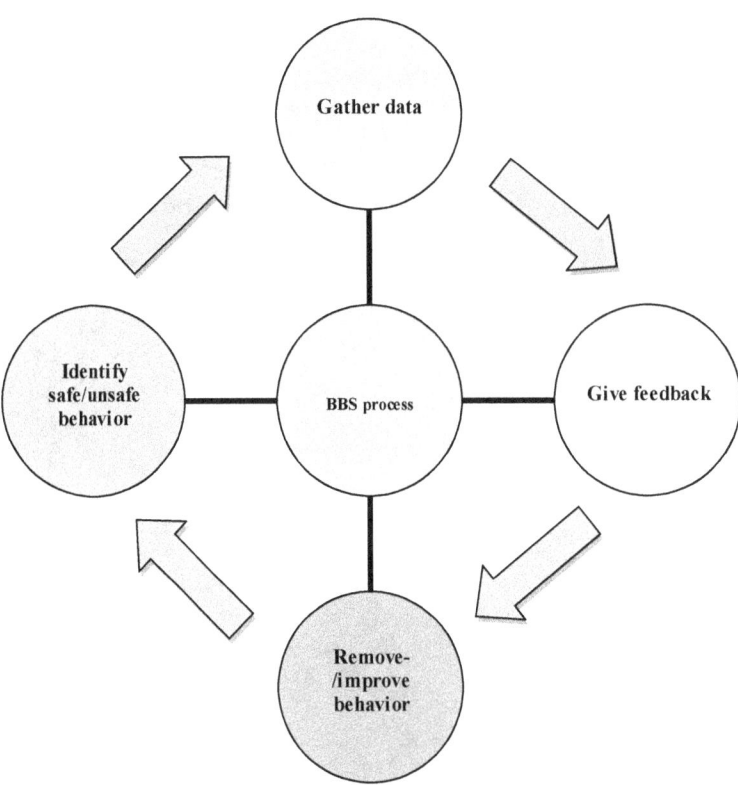

The BBS process is based on people, safety, values and risks and can be implemented successfully by using the ABC model.

2.2 The ABC Model

Behavior based safety aims at increasing safety in industry by positively influencing the behavior of all stakeholders involved. This is only possible by establishing a total safety culture, which implies repeated attention to three aspects: personal, the environment and behavior.

In a total safety culture it is needed to focus on behavior and to understand the ABC model. This understanding of Antecedents, Behavior and Consequence will reduce incidents, improve safety and productive quality of work, prevent injuries, increase profit and save lives.

The letters in the ABC Model refers to the following:

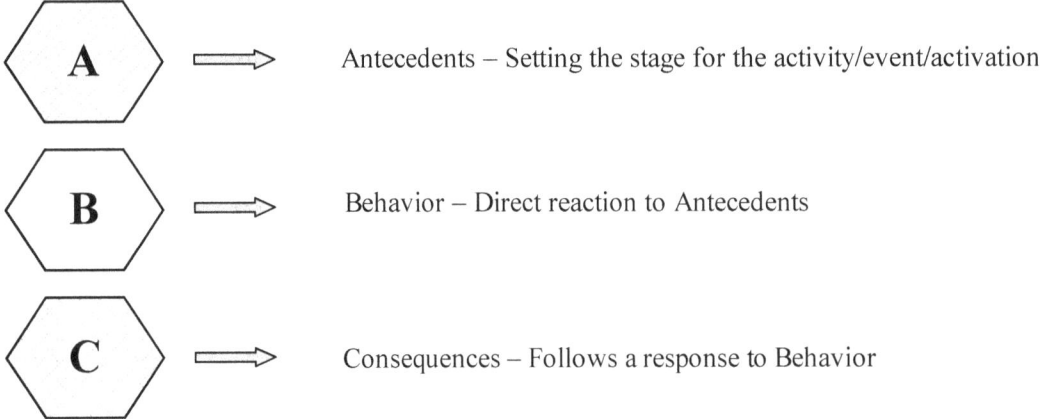

Behavior is activated by an event or activity and the result causes or influences behavior, leading to a consequence. The ABC Model can help motivate employees to develop positive and productive behavior. The most effective way to analyze behavior is to do so while employees function in the work environment.

People cannot be changed, but the way they behave can be influenced or changed. With the ABC Model behavior can be shaped, especially through positive reinforcement or the removal of a negative aspect. By using the ABC Model as a direct observation tool, an observer can record descriptive information in a systematic and organized way. A hypothesis can be formed to remove unsafe behavior or improve safe behavior.

A disadvantage of the ABC Model may be that data may not show an immediate pattern related to antecedent and consequence. A functional analysis with strategies can help overcome this disadvantage. Some consequences observed may be more powerful than others and the most effective consequences could be certain, and sizeable where they are meaningful or significant.

Figure 1: Example of the use of the ABC Model

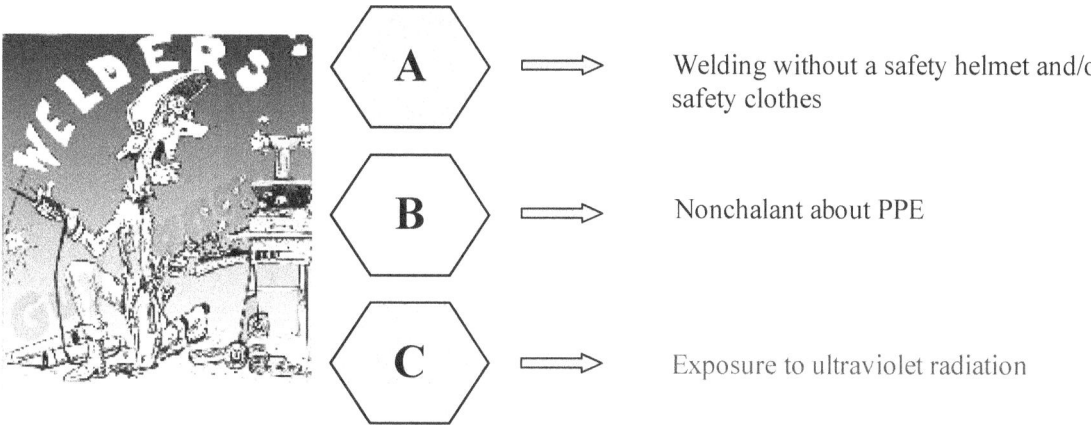

Behavior based safety uses a continuous improvement process, i.e. the DO IT process. This was developed by Geller (2001:13), and it is a process for sustainable behavioral improvement. This will capitalize on an employee safety culture where there is a shift from bad to good behavior.

2.3 The DO IT Process

Figure 2 provides a visual illustration of the DO IT Process.

Figure 2: The DO IT process

The four step improvement process

Define – Determine critical safety behaviors
Observe – Collect data on safety behaviors

Intervene – Change at-risk behaviors
Test – Determine the effect of the intervention(s)

Probabilistic Safety Assessment and Management PSAM 12, June 2014, Honolulu, Hawaii

The Transportation Research Board (2007:7-8) describes the DO IT Process as follows:

Define safety behaviors by employees and target them by means of a behavioral checklist; the target areas should be included. **O**bserve behavior entails collecting data, which is recorded by using the behavioral checklist. These observations will continue until a baseline of the target behaviors is achieved. **I**ntervene to change the at-risk behaviors by influencing the target behaviors. This can be done by studying the baseline rate of target behavior and interventions can be developed to increase and/or decrease target behaviors. **T**est to determine the effect and success of the intervention and whether goals have been achieved. This scientific method of observations, intervention, development and implementation allows continuous improvement. The intervention may be refined at any time to reach the set goals.

3. TOTAL SAFETY CULTURE / INJURY FREE CULTURE

To achieve a total safety culture, employees must take personal accountability and responsibility for their own safety and that of other's. A total safety culture requires an injury free work environment. The ultimate goal of this culture is to create an injury free culture within and through BBS.

An injury free culture requires the reconsideration of safety activities and engagement of employees. Although this culture does not mean that "no injuries" occur, it does imply the design of a work environment where injuries are not acceptable and where everything possible is done to prevent them. An injury free culture is certainly not about statistics and numbers or the prevention of injuries; rather the objective is to change behavior to establish a total safety culture with shared values, justice, fairness and agreements.

The key challenges in establishing an injury free culture are observation, communication and being open minded.

Figure 3: Key challenges of an injury free culture

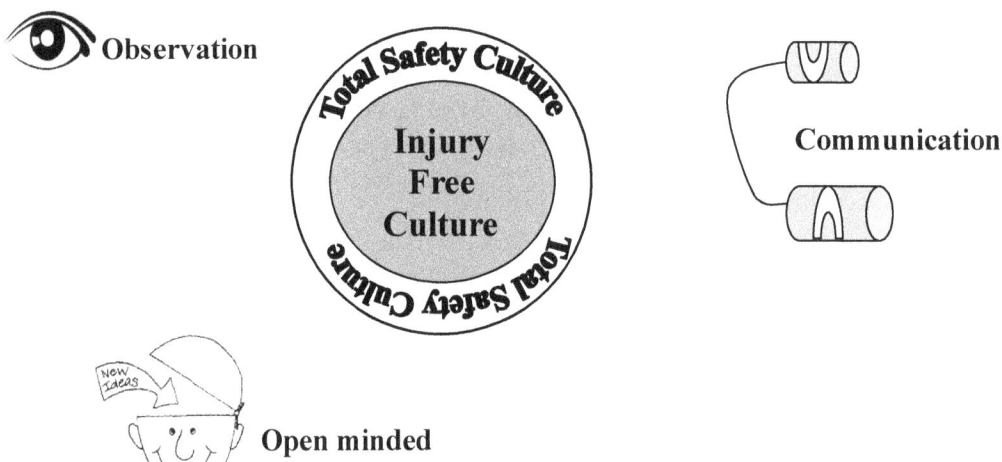

The most effective way to record a pattern or behavior is by means of observation. In BBS, the best way of inducting employees is to allow them to observe specific tasks and behaviors. Employees are observed in their natural surroundings from a holistic perspective.

Communication plays a vital role in how employees will engage in establishing and injury free culture. It comprises three basic elements of the spoken language:

- Verbal – The words used to convey a message

- Vocal – The tone of the voice in which the message is conveyed
- Visual – The body language of the person conveying the message

There are four qualities that contribute to effective communication, which can be seen in Figure 4.

Figure 4: Qualities of effective communication

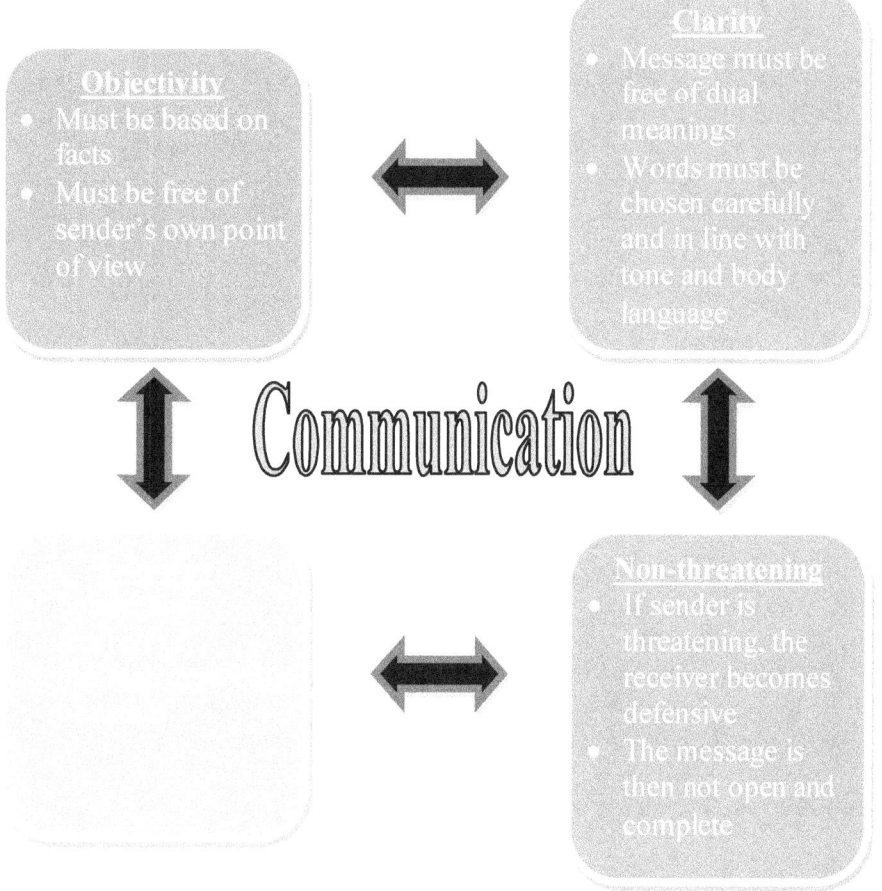

Most people are not easily influenced to accept change. Through BBS, employees are coached to learn and manage new ideas and should be open-minded to visualize future goal setting in achieving an injury free culture.

Any industry's greatest asset is its employees. Creating a workplace with an overall injury free culture that embraces safety and empowers employees can only be done by establishing and maintaining a total safety culture.

3.1 Culture Defined

Understanding the term "culture" will help in the implementation of a total safety culture within an organization. Dr. Edgar Schein (2004:13) formally defines culture as a pattern of shared basic assumptions that the group learned as it evolved its problems of external adaptation and internal integration. Over time this pattern of shared assumptions has worked well enough to be considered valid and therefore, to be thought of by new members as the correct way to perceive, think, and feel in relation to these problems. Culture accordingly includes perception, values, beliefs, and assumptions and can be defined as how individuals see reality and how it affects human behavior.

Culture is an ongoing process and consequently change is difficult, and gaining acceptance of a new idea or change seems to be a challenging task. Culture plays a decisive role in productivity, quality, and safety performance, and once employees are aware of BBS, better choices can be made and changes more readily achieved.

3.2 Accomplish Culture Change

There are basically three strategic elements in achieving a culture change:

- Driver(s) for change – BBS with bottom-up action and management (top-down) support and motivation
- Management support – BBS contributes positively through fewer incidents, higher productivity and quality and cost effective expenses
- Structure for managing change – A dedicated structure is needed and should include the six phases of culture change:
  - Define and communicate the need for change
  - Predict a desired future
  - Assess the current and future culture
  - Do strategic planning
  - Implement changes
  - Evaluate changes

3.3 Existence of an Injury Free Culture

An injury free culture can only exist when every employee shares safety as a value in life, is concerned about the safety of themselves and their colleagues and knows how to assist in creating safe working conditions. Furthermore the employee will demonstrate concern and knowledge in caring about safety behavior on a consistent basis.

This requires rethinking the approach to safety management. Groover (2014) suggests a five-step approach when an organization moves towards an injury free culture.

3.4 Five-step Approach to an Injury Free Culture

- Establish alignment and ownership
  - Focus on what "injury" means and how to establish continuous and sustainable improvement within an organization
  - The aim is thus to have longer injury free periods
  - Management should take ownership of systems, conditions, climate, and culture that influence safety outcomes

- Challenge helplessness
  - This starts with management being aware of the currents among employees
  - When an incident is labeled as an accident, people feel helpless to deal with it and a mind shift needs to be made
  - Consciously express messages that prevent employees from feeling helpless when it comes to safety issues

- Focus on exposures
  - An injury free culture strives to see and understand exposures that could lead to potential injuries within the work environment
  - Employees use the identified patterns, respond to potential exposure and understand the relationship between non-safety behavior and safety performance

- Increase metrics set
  - Standard cover metrics indicators will still carry value, but an organization striving to establish an injury free culture will add to these standard measures in order to gain better safety outcomes

- Engage employees
  - The best safety programs are only successful and effective if supported by employees
  - Organizations should give employees the opportunity to engage meaningfully, help in measuring and managing exposures and identify solutions to safety problems

These steps will help build an injury free culture along with the ability to reach higher goals and meet greater challenges.

3.5 Maintaining an Injury Free Culture

When maintaining an injury free culture, employees need to understand where the risks are and feel the need to change these risks. Employees also need to be involved in avoiding identified risk and improve and reinforce safe work practices of themselves and other employees.

After behavior change coaching and the establishment of an injury free culture, the focus shifts to maintaining this. Management can reward efforts and success with an incentive / reward system. This system also needs to be continuously refined in order to keep employees motivated. Geller (2001:243) states that employees would stay motivated and aligned to the mission if they were frequently reminded of the good they have done and if they were given the opportunity to improve on their interventions.

Hagan (et al) (2009:51) suggests that an injury free culture should be a group effort, supporting the avoidance of lost-time injuries, however, this may cause manipulation to an injury free system by underreporting injuries. Therefore an incentive / reward system must be closely monitored and evaluated, not abused or become an act of deceitful goal concealment of injuries.

The US Department of Energy (2009:3-10) suggests how employees can be shaped to follow an injury free culture:

- Facilitating communication
- Promoting teamwork
- Coaching and reinforcing expectations
- Eliminating organizational weaknesses
- Valuing prevention of error

4. CONCLUSION

Research has shown behavior based safety to be cost-effective. Conventional and easy administrative management is needed for monitoring target behaviors by employees. It furthermore has the major advantage of involving commitment from both the employee and employer. Both a bottom-up approach is needed from the employees and top-down support from management in achieving safety goals. BBS comprises the four steps of the improvement process, which are: define, observe, intervene and test. This will empower an employee safety culture to shift from bad to good behavior.

So often the blame is placed on employees for a poor safety culture, forgetting that poor or weak management can also be the reason for it. Positive, strong and proactive management is required, because leadership will influence the direction of an organization towards a successful injury free culture. Through BBS coaching, a positive safety culture will be formed, which in turn will lead to an increase in production and job satisfaction, and hence an injury free culture.

An injury free culture requires integrating behavior-based and person-based approaches to understand and influence BBS. Behavior based safety is often criticized as an unreliable science that creates fear among employees in reporting injuries and/or near misses. Other critics of BBS implies that it contradicts risk management efforts as many injuries then go unreported and this will influence financial costs. However, it is believed that BBS can be a practical way of correcting a hazardous condition and that it can change human behavior, in continuously striving to maintain an injury free culture.

5. REFERENCES

[1]   E.S. Geller, *Keys to Behavior-Based Safety from Safety Performance Solutions*. Government Institutes. (2001), Maryland: USA

[2]   E.S. Geller, *Working Safe: How to help People Actively Care for Health and Safety*. (2nd Ed.), CRC Press, Lewis Publishers. (2001), Washington D.C.: USA

[3]   P.E. Hagan, J.F. Montgomery and J.T. O'Reilly, *Administration and Programs – Accident Prevention Manual for Business & Industry*, (13th Ed.), National Safety Council. (2009), USA

[4]   H.L. Kaila, *Behavior Based Safety in Organizations – A Practical* Guide, I.K International Publishing House. (2008), New Delhi: India

[5]   T.L. Mathis and S.M. Galloway, *Steps to Safety Culture Excellence*. John Wiley & Sons, (2013), New Jersey: USA

[6]   E.H. Schein, *Organizational Culture and Leadership*. (3rd Ed.), John Wiley & Sons. (2004), San Francisco: USA

[7]   Transportation Research Board of the National Academies, *Commercial Truck and Bus Safety. Synthesis 11 – Impact of Behavior-Based Safety Techniques on Commercial Motor Vehicle Drivers*. TRB Publications. (2007), Washington D.C.: USA

[8]   US Department of Energy, *DOE Standard – Human Performance Improvement Handbook. Vol. 1: Concepts & Principles*. (2009), Washington D.C.: USA

[9]   D. Groover, *Five Steps for an Injury-Free Culture. Article by Behavioral Science Technology*, (2014). Retrieved http://www.bstsolutions.com/en/resources/knowledge-resource/five-steps-for-an-injury-free-culture-

[10]   A.R. French and E.S. Geller, *Creating a Culture where Employees Own Safety*. (n.d.) Retrieved http://www.safetyperformance.com/CreatingaCultureWhereEmployeesOwnSafety.pdf

[11]   Health and Safety Authority, "*Behavior Based Safety Guide*". Retrieved http://www.hsa.ie/eng/Publications_and_Forms/Publications/Safety_and_Health_Management/behaviour_based_safety_guide.pdf

[12]   T-Shirt Designs by Guy Fanguy. Retrieved http://www.guyfanguy.com/images/linked_images/tshirt_welders.jpg

Organising Human and Organisational Reliability

Pierre Le Bot[a][*] **and Hélène Pesme**[a]

[a] EDF Lab, Clamart, France

Abstract: Human Reliability (HRA) and the HRO (High Reliability Organising) approach are two major trends theorising the design, monitoring and improvement of safety in high-risk industries such as the generation of nuclear power. Human Reliability is increasingly requested in current design projects for new reactors or for the renovation of existing reactors in order to incorporate human factors and technical constraints for safety. Based on our observations on simulators and accident analyses, using the MERMOS method we illustrated how human failures in the operation of reactors assessed by Human Reliability for the Probabilistic Safety Assessments (PSA) need to be analysed at an organisational level. An absence of robustness (execution errors), a lack of anticipation (design flaw) or a failure in organisational resilience (lack of reconfiguration based on a new context) generate situations in which safety is threatened. Failure is sure to arise where an organisation is not sufficiently adapted in these situations (lack of recovery). We modelled this logic with the Model of Resilience in Situation that justifies MERMOS. In this paper, we will show how the MRS can be linked to the HRO mindset and how the resulting Human Reliability approach can contribute to High Reliability Organising at the human and organisational level.

Keywords: HRA, Resilience, HRO, Human Reliability.

1. INTRODUCTION

Using the Model of Resilience in Situation (MRS), we justified the collective aspect of our Human Reliability approach by incorporating the organisational aspect. This approach underpins our second-generation MERMOS Probabilistic Assessment of Human Reliability method.

MERMOS was built gradually from our empirical observations on full-scale simulators where we assessed the handling of incident and accident scenarios by EDF power station operators, as well as actual events such as the Three Mile Island nuclear power station accident in 1979. We built MERMOS based on the interpretation of our observations using cognitive psychology, ergonomics and systemic approach theories supported by psychology researchers, engineering researchers and ergonomics researchers. Using MERMOS, we succeeded in creating a failure model that is in keeping with our empirical findings and places human failure leading to a serious nuclear accident at the level of collective operation rather than individual error, and by deviation not from procedures but from functionally required operation.

Secondly, we consolidated this modelling with the MRS [1] based on the sociological theory of Social Regulation, which enabled us to dynamically describe the management of a high-risk situation for an ultra-reliable industrial process by alternating between stable organisational periods of monitoring the rules in place and reconfiguration periods in a situation to change rules in an effort to adapt to developments in the situation (Figure 1).

By working with the HRP (Halden Reactor Project), we were able to examine in detail the MRS processes in sub-functions to refine the functional description of an EOS (Emergency Operation System) [2].

The general principle of the MRS is to describe this management of high-risk situations based on an ongoing process described by functions that are ensured by the control system and the organisation via its management, before and after the occurrence of these situations.

---

[*] Contact e-mail: pierre.le-bot@edf.fr

The aim of this paper is to illustrate how the MRS encounters the HRO (High Reliability Organising) principles as described in the review by Karl Weick et al. [3], as well as the principles of Resilience Engineering explained by Erik Hollnagel et al. [4]. We would also like to suggest some developments for the MRS resulting from its use and to better comply with the HRO principles. In particular, we propose a better distinction between the organisation's different processes (anticipation, adaptation, safe operation, etc.) providing the qualities required for its overall high reliability characteristic (robustness, autonomy, vigilance, etc.), which allows people to trust it.

Figure 1: The dynamics of emergency operation in the Model of Resilience in Situation [1]

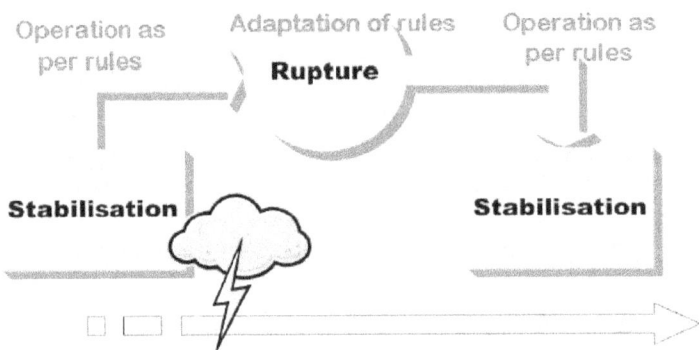

When examining these three approaches, differences in points of view seem to emerge.

- For Resilience Engineering, resilience makes it possible to move beyond the reliability approach to safety: "In contrast, resilience engineering tries to take a major step forward, not by adding one more concept to the existing vocabulary, but by proposing a completely new vocabulary, and therefore also a completely new way of thinking about safety." [4]
- Human Reliability, as we saw with the MRS, considers resilience in a situation as an ability of the organisation that is essential for making the human and organisational contribution reliable enough to ensure the system is safe.
- The HRO approach views resilience as a component of reliability: "We then move to the heart of the analysis and argue that organizing for high reliability in the more effective HROs, is characterized by a preoccupation with failure, reluctance to simplify interpretations, sensitivity to operations, commitment to resilience, and underspecified structuring." [3].

In fact, these points of view seem compatible to us by defining the approach in terms of the organisation's performance and as an ongoing looped process, which actually fits in well with the latest HRO (High Reliability Organising) development described in [3] – increasing the reliability of the organisation's performance. Thus, to sum up, the organisation's reliability is the trust we can put in it because it is safe, and it is safe because it is resilient.

We therefore need to understand Human and Organisational Reliability as described in [5]:
- A quality expected from the organisation: "In the industrial world, human reliability is the behavioural quality companies (and by extension the public and legal authorities regulating industrial operations) expect from the people with whom they entrust the running of a facility." Organisational reliability is not a useful quality first of all for the organisation itself, but for those who wish to trust in the operation of a high-risk system without managing it themselves.
- A requirement for robustness: "The technical approach allows us to manage a facility by anticipating operational situations. The robustness of the performance is understood as the

lack of operator error in following instructions on implementing and managing the process."
Robustness is the collective ability of the operating team to manage a high-risk situation in accordance with the rules built technically by anticipating these situations. Traditionally, the initial Human Reliability approaches have been focused on this concept of robustness, i.e. the foreseeability of performance based on the consistency of behaviours and cognitive processes in collective interactions by ensuring there are no deviations in an effort to guarantee performance.

- A requirement for adaptability: "In contrast, the managerial approach (often called "security management") relies more on humans by delegating the management of situations in keeping with their skills." Robustness is a requirement for compliance with instructions without deviation as much as trust relies on the ability of individuals and groups to take initiative in a situation.

- A capacity for resilience: "The ability to combine the two [i.e. robustness and adaptation] characterises their human and organisational resilience, which comes into play in real-time in actual situations. Most of the time, the facility is run by closely following procedures. If an unexpected situation arises, the team in charge shifts into adaptation mode." Resilience is the ability to interrupt robust operation to make way for a phase of adaptation to developments in the situation in order to start again with a phase of robust operation. Resilience thus makes it possible to resolve the paradox of combining the incompatible approaches of robustness and adaptation.

2. HOW TO ORGANISE RELIABILITY

2.1. A dynamic, active ongoing looped process

As Weick states, reliability is the result of the stability of organisational and cognitive processes (of which we will try to offer a list and a description later in this paper):
"Thus, to understand how organizations organize for high reliability, we need to specify what is done repeatedly – in our case this is cognitive processes – and what varies – in our case this is routinized activity manifest in performance." [3]

We can distinguish between two types of processes:
- The processes directly linked to operation, which make operation of the system reliable (either in anticipation or in a situation), generally constitute what we call High Reliability Organising.
- The processes that feed those mentioned previously, taking account of these operation situations, whether real or simulated, generally constitute Organisational Learning.

Organisational Learning draws upon the results of High Reliability Organising, which it in turn feeds. We thus have a looped, dynamic and ongoing overall process where each individual process is both upstream and downstream from the others. Therefore, depending on the points of view, we can see one of the organisation's qualities offered by these processes, such as robustness or resilience, expertise or knowledge, as dependent on another, or conversely as determining it.

2.2. High Reliability Organising

Our proposition primarily concerns the modelling of High Reliability Organising around two main processes – **anticipation** and **adaptation**†. We also look at the **alert** process. In developing the MRS [1], we have now taken a less in-depth look at Organisational Learning and will therefore briefly offer a few pointers in Section 2.3.

---

† Please note that in the previous descriptions of the MRS, we used the quality of "adaptability" – we prefer to clarify the model by explaining the process of "adaptation", which offers the organisation's qualities of resilience and autonomy.

Anticipation (cf. Figure 2)

The anticipation process organises operational robustness and contributes to resilience in a situation. In general, the means for this organisation will be the following:

- From a Human Reliability point of view, functional organisation as modelled in the MRS for the operating teams in the event of an incident or accident (cf. Table 1)
- From a human factor point of view, through the organisation of the group, the roles and the skills of the operators
- From a technical point of view, through the interfaces, procedures and communication methods

Table 1: MRS Functions [2]

MRS Functions
Information selection and exchange   **Execution**   **Control**   **Verification**   **Reconfiguration**

Organising robustness is the most common means of technically organising reliability. For example, the distribution of the roles of the members of the team in the control room will give the supervisor the task of controlling the application of the procedure by the operators responsible for acting on the system. This control function is a barrier helping to ensure that the rules determining the operation required in a given situation are effectively and correctly applied. This barrier contributes in particular to protecting against operator errors. However, the role of the supervisor is not necessarily the only resource for the control system enabling it to recover from these deviations from the expected operation. A "forgiving" procedure will be able to recover from an operator error by asking them to check the state of a system after having carried out an operation on it. An oversight or an incorrect action can thus be recovered from by noticing that the state of the system does not match the expected state. For example, after starting a pump, the operator will be asked to check the flow rate of the line. This control function, distributed within the control system, thus helps ensure that the operation expected in the engineering offices is correctly implemented.

Engineering, based on general technical knowledge, knowledge of the organisation's specific operation gained via the feedback process, new knowings developed through research (from outside or within the company) and vigilance resulting from the alert process needed to take action in time, is constantly revising the operation rules and the organisation. The goal is to at least maintain, if not improve, the level of prevention against the occurrence of high-risk situations via the correction of technical and organisational deviations observed, and to call upon operators to implement the planned prevention measures in a situation.

Design engineering must also anticipate that a situation may not be anticipated and, as far as possible, put organisational resources in place enabling operators to stop the operation in progress and determine in time which operation should be implemented from that point onwards. For example, when managing an incident, the safety engineer continuously monitors the reactor variables and as soon as they notice any deterioration, they will ask the operators to stop the operation in progress and change procedures or take emergency action immediately. This verification function contributes to the resilience of an organisation which anticipates that it cannot anticipate everything and that an operation worked out in advance, even if initially relevant, may become inadequate for controlling an accident.

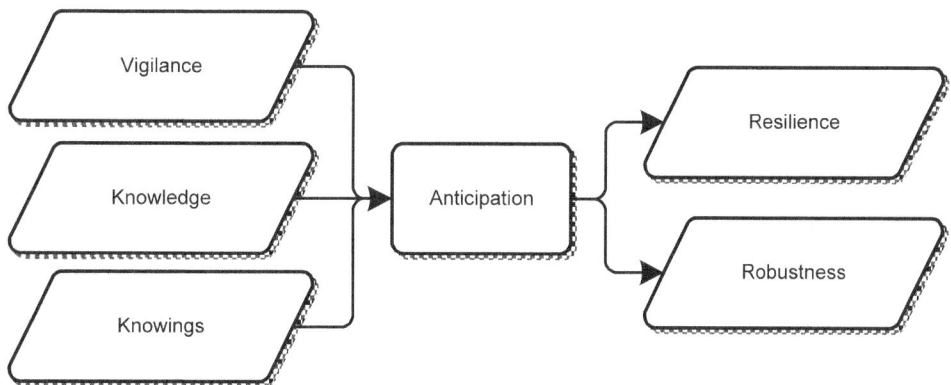

Adaptation (cf. Figure 3)

Adaptation complements the organisational anticipation of situations. Adaptation is a process implemented in a situation to reconfigure the operative system as soon as the operation in effect is diagnosed as unsuitable for the situation. The organisation thus switches from robust operation following shared collective rules [1] to adaptive operation selecting, developing and validating new rules that are adapted to the situation that has unfolded.

Adaptation based on vigilance, expertise and the individual and collective know-how of the operators in a situation functions temporally on the ongoing verification of the suitability of the operators' behaviour for the situation in progress on the one hand, and then when necessary on collaboration, cooperation and negotiation in a situation in order to result in new rules to be applied by collective expert decision through delegation in a situation.

For example, when the Three Mile Island accident occurred in 1979 (cf. [6]), the adaptation process did not work. The operators continued operation for a relatively frequent transient (a known transient increasing the pressuriser level following an emergency shutdown), postponing treatment of the inconsistency between a low reactor coolant pressure and the state of the reactor that they thought had occurred based on what they knew about this transient. They lacked the necessary expertise not only to assess the importance of this discrepancy, but also to decide on the action to be taken in this pressuriser leak situation (had they been able to diagnose it) in order to compensate for the lack of organisational anticipation of this type of transient (the proper procedure arrived shortly afterwards on-site, and thus too late).

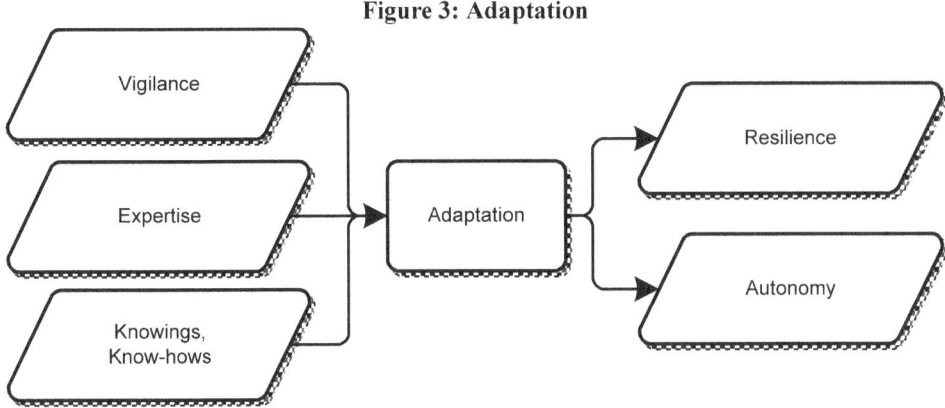

Figure 3: Adaptation

The alert process (cf. Figure 4)

The alert process is supported by the organisation's individual and collective expert expertise. In non-real time, it makes it possible to detect, through the updating of knowledge about itself that the organisation carries out reflexively and continuously, as well as through an industrial watch showing it the behaviour of the other installations, that the way it operates is potentially incorrect in its preparedness for high-risk situations. This diagnosis may concern the following:

- a technical or organisational malfunction diagnosed but not yet corrected (via the anticipation process) even though it is an emergency
- a weak signal, i.e. an event (or a set of several events) that does not call into question operational robustness based on engineering knowledge, but rather through expert intuition with a potentially strong impact, i.e. that significantly calls into question operational robustness without being able to justify it immediately and thus requires investigation

In real time, in a situation, this alert process makes it possible to detect, by checking the suitability of the operation for the situation, that the operation retains its robustness through its suitability for the situation. In this case, it triggers the interruption of the operation in progress and a phase of adaptation to the new situation.

For example, "whistleblowers" are those who trigger the alert process. Generally, there is an organisational inability to process their alerts when they are merely compensation through individual initiative for an organisational malfunction. Indeed, an expert individual may recover, at their own initiative and using their expertise, from a malfunction in the feedback process (in non-real time) or detect a failure in the checking of the situation in progress (in real time, in a situation), and justify it in technical terms. However, whistleblowers are often expected to systematically justify their alerts with the technical rationality of a discrepancy correction request resulting from feedback, even though they may be based on expert intuition. In fact, this type of alert (provided it comes from a legitimate expert) is a weak signal that the organisation needs to process in order to judge whether or not it is a strong signal. Illustrating this with an example is tricky, since retrospective analysis of the alerts issued prior to a serious event is biased by the knowledge we have of the event that occurred. However, examining the organisational process may reveal an organisational failure, if only through the absence of this process of taking alerts into account.

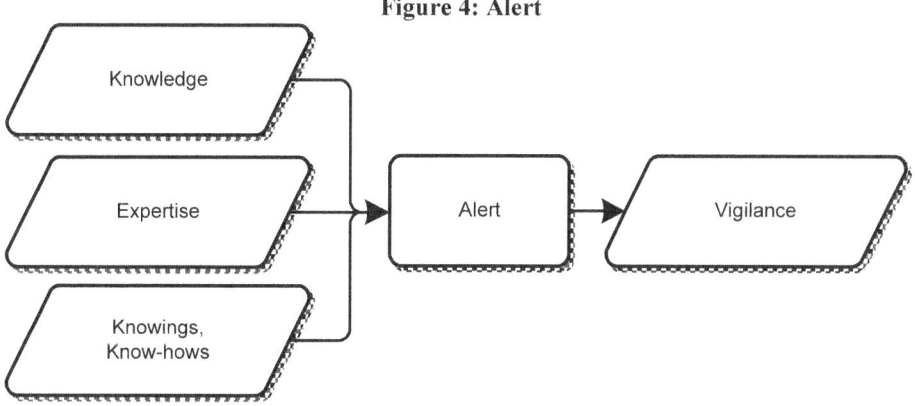

Figure 4: Alert

Safe operation (cf. Figure 5)

In our model, safe operation is thus achieved as far as possible through these organisational qualities of robustness, resilience and autonomy offered by the anticipation, adaptation and alert processes.

This modelling means first of all that a human or organisational failure affecting the safety of operation is potentially possible as soon as one of the processes ceases to function as expected:

- Either the functions it ensures are not in place or do not have the means to reach their objectives – this malfunction will present itself as a failure of one of the expected organisational qualities,
- Or the organisation's upstream qualities enabling these processes to function are lacking (knowledge, knowings and know-how, expertise).

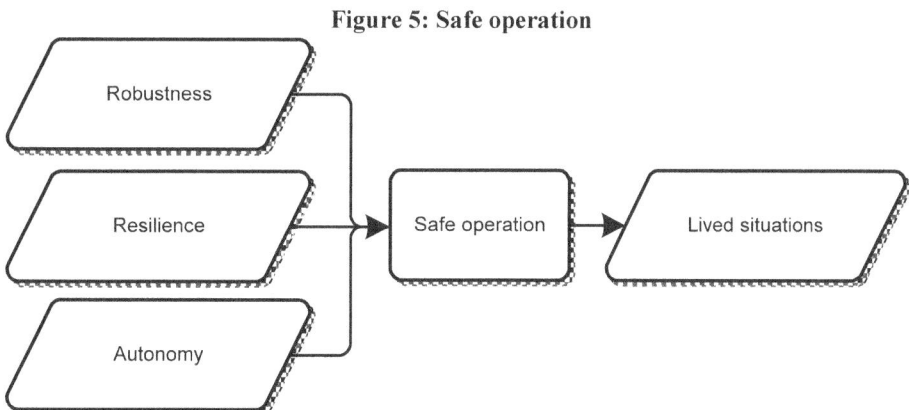

Figure 5: Safe operation

The safe operation of high-risk industries generates lived situations in the sense that it accumulates the near-miss events that it encounters (it memorises their signs, describes them and studies them namely to supply feedback) and recognises where the organisation's robustness, resilience and adaptation have been tested.

2.3. Learning Organisation

We will offer a schematic description of the Learning Organisation drawing on that which is based on deferred regulation ("deferred joint regulation" in the description of the MRS in functional terms, cf. [1]). Here, we present a schematic representation (cf. Figure 6), the aim of which is to illustrate how the organisation process loops starting from lived situations in order to organise its reliability on a continuous basis in accordance with the HRO principles.

Our proposition for modelling draws on four processes – Operational Experience (OPEX); Learning by Experience; Simulation; Scientific and Industrial Watch, and Research.

Figure 6: Learning Organisation

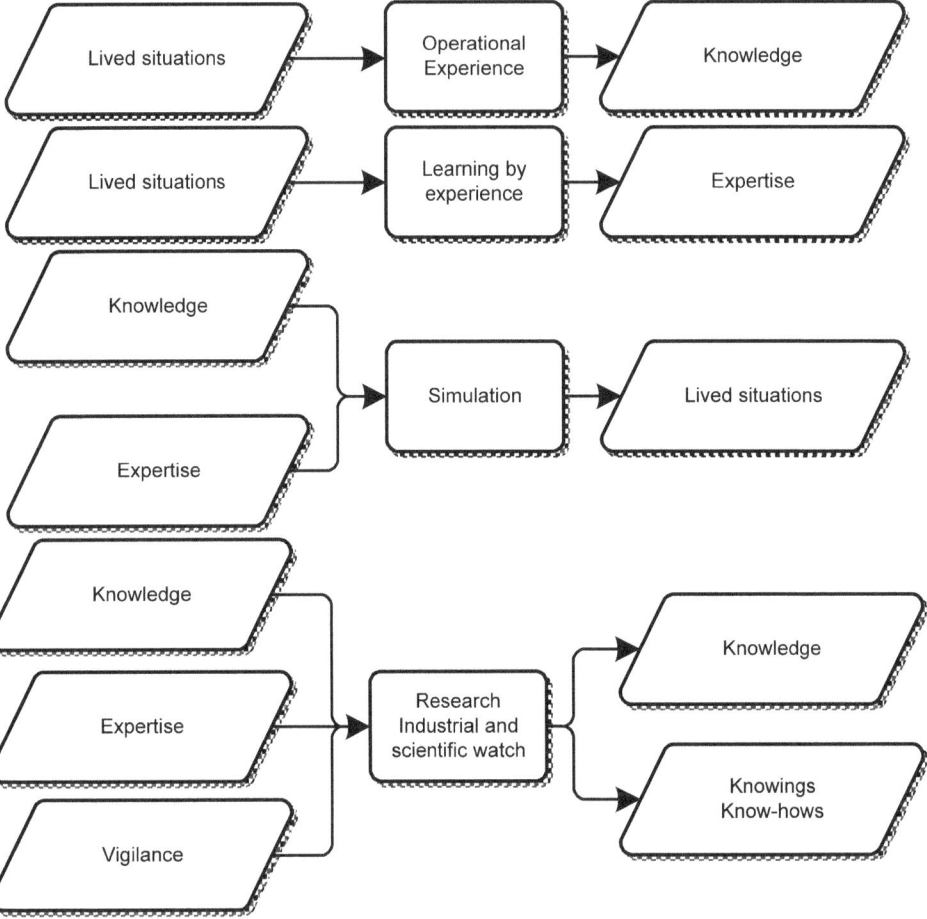

The lived situations are the starting point for the Learning Organisation. Based on these lived situations, the Operational Experience process will offer the organisation knowledge about itself and the Learning by Experience process will generate expertise for the direct (operators) or indirect (managers, designers, researchers) stakeholders. Learning by experience is very different to training related to anticipation, as it does not employ the same principles and is essential namely for decision-making in a situation (all operators must be considered experts), cf. [7].

Simulation makes up for the lack of high-risk situations experienced by the organisation. This necessity has long been highlighted by the HRO approach. Simulation should be considered in a general sense: "Simulation in its most common sense, for accident mode control of nuclear power stations, consists of using control centre simulators with an active operating system (control room simulators for nuclear power stations). PSA (Probabilistic Safety Assessments) can also be deemed to be simulations. These are assessments where system behaviour is evaluated to its limits by modelling. But there are other possibilities: in particular accounts that agents share from memory of an event that they have experienced (war stories, storytelling). Indeed, accounts of past events, through the storytelling effect, transport us to the context of the incident which occurred, like some sort of simulation of this event for the person listening to or reading the account." [1].

The Industrial and Scientific Watch completes this Organisational Learning by examining the knowledge and knowings that can enhance what the organisation knows about itself by learning from others or through innovation.

Ultimately, these processes contribute to High Reliability Organising by offering knowledge, expertise and knowings/know-how.

2.4. Correspondence with Resilience Engineering and the HRO approach

From a Resilience Engineering point of view, as described by E. Hollnagel in [8], anticipation through offering robustness and adaptation contributes to the initial highlighted organisational ability of "Knowing what to *do*": "Knowing what to *do*, that is, how to respond to regular and irregular disruptions and disturbances either by implementing a prepared set of responses or by adjusting normal functioning. This is the ability to address the *actual*."

The process of adaptation through autonomy and alerting through vigilance contribute to the second ability required: "Knowing what to look for, that is, how to *monitor* that which is or can become a threat in the near term. The monitoring must cover both that which happens in the environment and that which happens in the system itself, that is, its own performance. This is the ability to address the *critical*."

Lastly, the Learning Organisation learns lessons from the past and enables anticipation of the future as required in the final two organisational abilities stated by Hollnagel: "Knowing what to *expect*, that is, how to anticipate developments, threats, and opportunities further into the future, such as potential changes, disruptions, pressures, and their consequences. This is the ability to address the *potential*. Knowing what *has happened*, that is, how to learn from experience, in particular how to learn the right lessons from the right experience—successes as well as failures. This is the ability to address the *factual*."

We saw from the modelling that the continuous loop between the High Reliability Organising and Learning Organisation processes corresponded to the dynamic of organising reliability based on the HRO approach.

A fundamental characteristic of our model is to clarify processes distinguished namely by their temporality and their dynamic (in a situation / after or before situations, in real time / in non-real time), and the rationality underpinning them (technical and engineering for anticipation, management and human factors for adaptation, without this list being exhaustive). Paradoxically, this indicates that these processes may contradict or compete with each other, namely anticipation and adaptation, bearing in mind that high-risk organisations know how to manage this contradiction through their resilience. This paradoxical building of reliability was highlighted by Karl Weick: "HROs suggest that the acceptance of paradox continues to create high effectiveness when systems become more tightly coupled and more interactively complex. As we have seen, HROs pursue simultaneous opposites such as rigidity and flexibility, confidence and wariness, compliance and discretion, anticipation and resilience, expertise and ignorance, and balance them rather than try to resolve them." [3]. We can even deduce that ambiguity is contingent on reliability, as is addressed in Stoessel's thesis [9].

This description and explanation of the processes characterising the human and organisational component of reliability offers fundamental knowledge for the design of high-risk socio-technical systems beyond the reliability assessment.

4. CONCLUSION

Our applied research approach in the field of human and organisational reliability adopts a principle alternating between empirical assessment, researching theories that explain our findings, predictive modelling justified by the theories, and back to empirical assessment supported by our models while trying to widen their scope, empirical assessment and new theoretical consolidation, etc.

After having built a human failure model supporting the MERMOS method and transcending reliability approaches focused on human error, we then extended our modelling to the functional characteristics necessary for a safe approach to managing incidents and accidents with the MRS. This was supported by Reynaud's Theory of Social Regulation [11], allowing us to explain the functioning of the groups responsible for risk. In particular, this approach allowed us to maintain the link between the modelling of operator actions and the organisational factors that influence them, and thus to formulate recommendations for designing High Reliability Organisations. This theoretical model has already been applied and has helped design the control system for future nuclear reactors.

In this paper, we have shown how we attempted to extend this modelling of the MRS to the organisational processes of an approach to organising reliability as implemented for the safe operation of a nuclear power station. We rely on the most fundamental aspects of the HRO approach, which describes a process maintaining a constant level of performance in terms of reliability, and of Resilience Engineering, which states how the organisation must be capable of being robust and autonomous based on our interpretation. The next step will consist of dealing with the functions and their description in depth, firstly for High Reliability Organising, as we have begun with the EOS approach [2], which supplies us with a framework for describing a power station control system. Then, we will study the functional aspects of the Learning Organisation with a view to extending our recommendations to this field of the organisations.

References

[1]   P. Le Bot and H. Pesme. "*The Model of Resilience in Situation (MRS) as an Idealistic Organization of At-risks Systems to be Ultrasafe*", PSAM 10, Seattle, Washington, USA, (2010).

[2]   S. Massaiu, P. Ø. Braarud and P. Le Bot. "*Including Organizational and Teamwork Factors in HRA: the EOS Approach*", EHPG 2013, Storefjell Resort Hotel, Gol, Norway.

[3]   K. E. Weick, K. M. Sutcliffe and D. Obstfeld. "*Organizing for high reliability: Processes of collective mindfulness*", Crisis Management, Vol. 3, 81–123, (2008).

[4]   E. Hollnagel, D. D. Woods and N. Leveson, eds. "*Resilience engineering: Concepts and precepts*", Ashgate Publishing Ltd, 2007.

[5]   P. Le Bot. "*The Dictionary - Human (and organizational) reliability*", Laboreal, Vol. IX, (December 2013).

[6]   P. Le Bot. "*Human reliability data, human error and accident models—illustration through the Three Mile Island accident analysis*", Reliability Engineering & System Safety, Vol. 83, 153–167, (2004).

[7]   G. Klein. "*Sources of power*", MIT press, 1999, Cambridge, Mass., London.

[8]   E. Hollnagel. "*Resilience engineering in practice*", Ashgate, 2011, Farnham, Surrey, England, Burlington, VT.

[9]   C. Stoessel. "*Décisions risquées et organisations à risques: autonomie au travail et reconnaissance sociale dans la conduite d'une industrie de process*", Conservatoire national des arts et metiers-CNAM, 2010.

[10] P. Le Bot *et al.* "*Mermos: an EDF project for updating probabilistic human reliability assessment*", REVUE GENERALE NUCLEAIRE-INTERNATIONAL EDITION-, 32–39, (1998).

[11]   J.-D. Reynaud. "*Les Règles du jeu*", A. Colin, 1997, Paris.

On the relation between culture, safety culture and safety management

Teemu Reiman[a*], Carl Rollenhagen[b] and Kaupo Viitanen[a]

[a] VTT Technical Research Centre of Finland, Espoo, Finland
[b] Royal Institute of Technology, Stockholm, Sweden

Abstract: Safety can be considered an emergent phenomenon, making a systems view imperative if the aim is to evaluate or develop the safety of an entire sociotechnical system. This paper deals with one important component of the systems view – the relation between culture and management. Specifically, we will inspect how the concepts of culture and safety culture can be used in conjunction with the concept of safety management in facilitating a more dynamic systems view on safety. The paper proposes a model of eight cultural archetypes and illustrates how these relate to both safety culture and safety management in organizations.

Keywords: Safety culture, Safety management, Organizational culture, Organizational factors.

1. INTRODUCTION

The concept of safety culture has become established in safety management applications in most major safety-critical domains. We have previously argued that in general, safety culture research and practice has often missed the opportunity to integrate with systemic approaches to safety [20]. The interested reader is referred to [20] for a comprehensive critique of safety culture as representing a systems concept. Safety can be considered an emergent phenomenon, making a systems view imperative if the aim is to evaluate or develop the safety of an entire sociotechnical system [17]. This paper deals with one important aspect of the systems view – the relation between culture and management. Specifically, we will inspect how the concepts of culture and safety culture can be used in conjunction with the concept of safety management in facilitating a more dynamic systems view on safety. The paper proposes a model of eight cultural archetypes and illustrates how these relate to both the safety culture and safety management in organizations.

We will base the following paper on two lines of empirical research carried out in parallel by the authors. The first line of research has focused on safety management in the nuclear power industry. In a recent study we conducted and analyzed thirty interviews with managers and safety experts in the nuclear industry and uncovered a number of dilemmas in safety management that need to be resolved by making trade-offs [18, 19]. The original aim of that study was to inspect safety culture in the Nordic nuclear industry based on the idea of taking a closer look at various tensions among values, goals, etc.. This data will be further analysed in this paper. The second line of research has been carried out in the health care domain, where the authors have developed a methodology for evaluating patient safety in hospitals. Based on these projects the authors have developed a preliminary framework of adaptive safety management [22] which has also been tested in an ongoing research project [16]. In this paper, we utilize the framework to illustrate how safety management, safety culture and organizational culture relate to and influence each other.

2. FRAMEWORKS ON CULTURE AND SAFETY

2.1. Safety culture

The concept of safety culture was born in the aftermath of the Chernobyl accident in 1986, when it became clear that nuclear safety should incorporate more than mere technology. Management systems, leadership and a host of other human related factors such as learning, responsibility, values and

---

* Corresponding author. Email: teemu.reiman@vtt.fi

attitudes were taken into consideration (with varying operationalization's) in safety analyses and development initiatives. The concept of safety culture has today become established into safety management applications in safety-critical domains, such as aviation, nuclear power production, petrochemical sector (including offshore oil production), railways, peacetime military operations, maritime, and mining operations. Overviews of the use of the safety culture concept in empirical research have been provided by [2, 5, 6, 26].

The idea that safety culture somehow represents a systemic and holistic view on safety is seldom explicitly spoken out, but nevertheless seem to linger behind many safety culture discourses. A major challenge is however that such a holistic view on safety culture does not leave anything outside culture – everything becomes included and thereby analytical power is lost. Further, it can be argued that conceptualizing technology in cultural terms refers to the social construction of an objective physical reality, not to the physical reality itself [20]. Thus, in order to understand how e.g. safety management and safety culture relate to each other we need to specify what we mean by culture and how culture relates to the overall sociotechnical system which we are trying to manage [cf. 11]. For example, the value of safety often competes with other values and to understand the significance and specific meaning given to safety values one has to understand the whole structure of values in an organization. If safety culture is treated as a constituent part of a sociotechnical system, the overall system will have emergent properties that cannot be deduced from the study of safety culture alone [20].

We argue that in the debate about contents of safety culture one aspect has been largely neglected; namely the role of the safety management system and associated practices. It is true that one often finds that management attention and support is important for safety culture development, but this is usually portrayed in sweeping terms (eg. as items and factors in safety climate assessments) rather than detailed analysis. In order to develop a more detailed framework for analysis we need to have a model of organizational cultures and their relation to safety.

2.2. Competing Values Framework and Organizational Culture

One way of depicting organizational culture is that it refers to values, norms and assumptions concerning an organization's core task and the correct way of carrying it out, measuring success and interacting with each other while doing the work [23, cf. 11]. Development of distinct organizational culture has been considered a source of competitive advantage and even the key ingredient to success [3, 23].

Quinn and Rohrbaugh [15] have suggested a 'competing values approach' to organizational analysis by application of expert judgments and multidimensional scaling. Their study suggests that 'organizational researchers share an implicit theoretical framework, and, consequently, that the criteria of organizational effectiveness can be sorted according to three axes or value dimensions' (p.369). The first dimension is related to organizational focus '…from an internal, micro emphasis on the wellbeing and development of people in the organization to an external, macro emphasis on the wellbeing and development of the organization itself' (p. 369). The second dimension is related to organizational structure, with, at the one end, emphasis on stability and, at the other end, emphasis on flexibility. The third dimension is related to means-end relationships—e.g. planning and goal setting vs outcomes (productivity). Cameron and Quinn [3] developed these findings into Competing Values Framework (CVF) for assessing and profiling the dominant cultures of organizations.

In CVF, the two core dimensions form four quadrants, each representing a distinct cluster of criteria representing what is seen as good, right and appropriate, i.e., the fundamental values that exist in the organization. Safety can be one value – even if it is not explicitly dealt with in the CVF – but its significance and meaning only manifests in connection with the other values existing in the organization. We will return to this later.

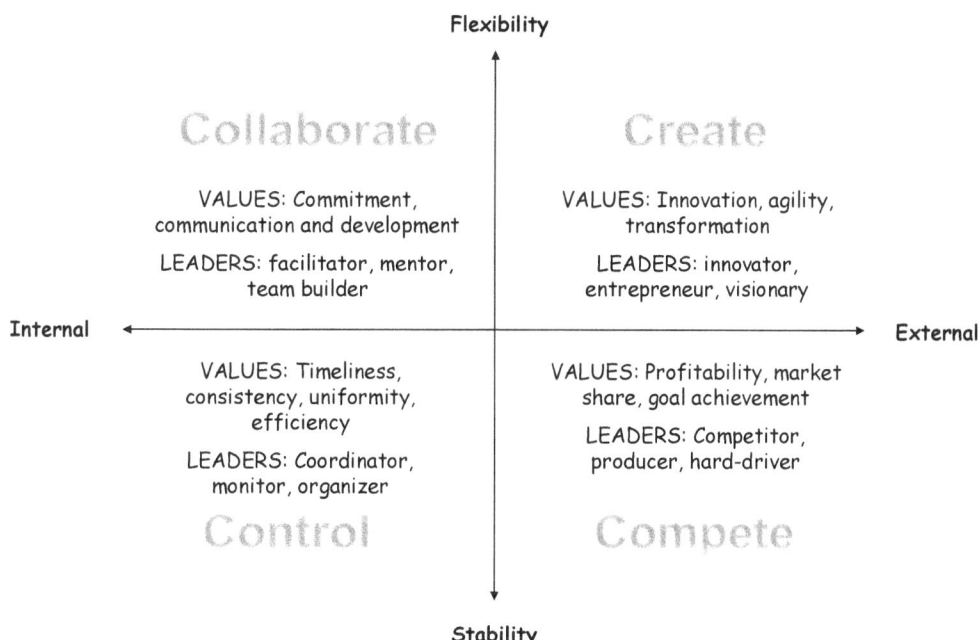

Figure 1: Cameron and Quinn's Competing Values Framework postulates four cultural archetypes with opposing assumptions about success and leadership [3, 14, 15]

The upper left quadrant (Collaborate) identifies value creation and performance criteria that emphasize an internal, organic focus. It is typified as a friendly place to work where people share a lot of themselves, like an extended family with best friends at work. Success is defined in terms of internal climate and concern for people. The organization places a premium on teamwork, participation, and consensus. The lower right quadrant (Compete) identifies value creation and performance criteria that emphasize external, control focus. The glue that holds the organization together is an emphasis on winning. Success is defined in terms of market share and market penetration. The upper right quadrant (Create) identifies value creation and performance criteria that emphasize external, organic focus. The glue that holds the organization together is commitment to experimentation and innovation. The emphasis is on being at the leading edge of new knowledge, products, and/or services. Success means producing unique and original products and services. The lower left quadrant (Control) emphasizes internal, control value creation and performance criteria. The long-term concerns of the organization are stability, predictability, and efficiency. Formal rules and policies hold the organization together. What is notable about these four quadrants is that they represent opposite or competing assumptions. Each continuum highlights value creation and key performance criteria that are opposite from the value creation and performance criteria on the other end of the continuum-- i.e., flexibility versus stability, internal focus versus external focus. The dimensions, therefore, produce quadrants that are also contradictory or competing on the diagonal. [3]

These competing elements in each quadrant give rise to one of the most important features of the CVF, the presence and necessity of paradox. CVF emphasizes that successful managers need to work simultaneously with several contradictory logics and shift their dominant value sets when circumstances so require [3, 14].

2.3. Reconceptualization of the Competing Values Framework

In our previous studies [18, 19], we reanalyzed the thirty interviews mentioned in the introduction from the point of view of tensions and competing values. The specific goal of the particular additional analysis was to look at how tensions, competing values and tradeoffs manifest in the management of nuclear power plants. A second goal was to inspect how existing frameworks, such as CVF, can be used to model the tensions. The analysis identified twelve trade-offs. These were mapped into a framework that combined the Competing Values Framework with the universal value model of Schwartz [24, 25] with some variations made to the combined model. Figure 2 illustrates the main result [see 19].

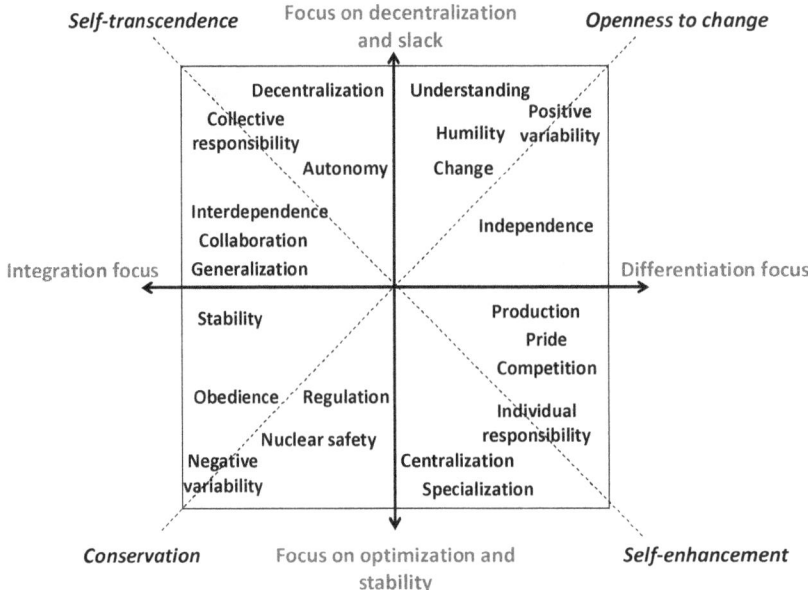

Figure 2. An illustration of the competing values underlying the twelve trade-offs [19]

The framework illustrated in Figure 2 was in later studies [22] expanded into a model of adaptive safety management. This model will be briefly presented next.

2.4. Principles of Adaptive Safety Management

Based on our previous work (see above) we have defined a safety management framework [22]. Figure 3 illustrates the proposed eight principles of managing safety. The underlying idea in the framework is to perceive safety-critical organizations as being complex adaptive systems with inherent features such as emergence, self-organizing and non-linearity [9, 10, 13]. Another underlying idea is that the principles are competing, or even partly in conflict [3, 27], and the managers and other safety professionals have to find the proper way to balance these in daily work.

As illustrated in Figure 3, safety managers need to promote safety as a shared guiding principle according to which situational decisions are made in the organization. This means that safety needs to be a shared value in the organization. In order to guarantee organizational cohesiveness and enough order for the system to both act in a structured manner and yet be flexible when needed, leaders have to facilitate interaction, build connections and build an environment which supports interaction. Novelty and diversity is needed to change and develop the organization. Novelty will lead to self-organized order, potentially contributing to the system's survival. However, in addition to disorder and variance safety-critical systems need other means of encouraging self-organizing. Since a complex adaptive organization cannot be fully controlled in the traditional top-down manner, a capability for self-organizing depending on the situational demands is needed. In complexity science, self-organizing is both a hallmark and the key adaptive mechanism of complex adaptive systems but also something that depends on the other characteristics of the system such as competence and situation awareness.

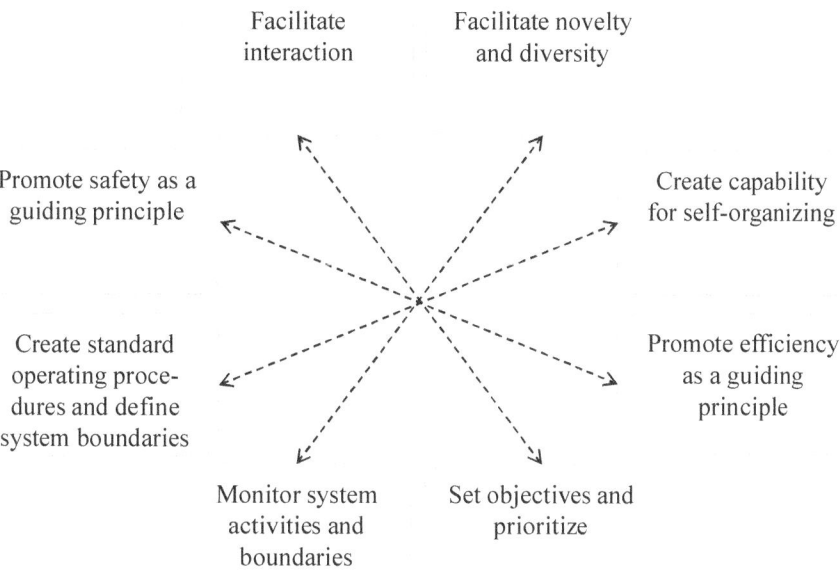

Facilitate
interaction

Facilitate novelty
and diversity

Promote safety as a
guiding principle

Create capability
for self-organizing

Create standard
operating proce-
dures and define
system boundaries

Promote efficiency
as a guiding
principle

Monitor system
activities and
boundaries

Set objectives and
prioritize

Figure 3. Principles of managing safety in complex adaptive organizations, Based on [22]

In addition to the above mentioned tasks, managers need to optimize the efficiency of organizational activities and promote efficiency as a shared goal. This requirement often manifests as a conflicting demand between efficiency and safety [8] but it is also a question of different time-frames [1], i.e., short versus long term goals. Even though complex adaptive organizations cannot be managed in the traditional meaning of the term, leaders in safety critical organizations still need to set objectives and prioritize. This present another consequence of complexity: the need to simplify and prioritize top-down some issues over others while at the same time facilitate interaction and focus on emergent themes coming from interacting groups. Complex adaptive organizations need explicit monitoring of system activities and their boundaries since they are constantly changing and since the change can also endanger safety if it happens unsupervised. Complex adaptive organizations need explicit boundaries since there are no natural all-inclusive boundaries between the various overlapping human systems. In safety-critical domains there is a need for analysis of risk and development of different types of rules and procedures to minimize risk and define the so called safe operating zone [7].

3. SAFETY MANAGEMENT IN DIFFERENT CULTURES

3.1. Challenges Faced in the Nordic Nuclear Industry

The cultural challenges of managing safety were evident also in our interview data. For example, in our interviews a representative of the nuclear industry contemplated decision making in his own organization:

> 'What I always try to say when it comes to safety culture is that, when we have these project managers who would like to go forward and then we have specialists who make demands … we need to reach mutual understanding. That is the highest level of safety culture, we have it and the regulator has it, that you need consensus … we have this [management group] that includes almost all functions of the organization; operations, engineering, quality, safety … it is very extensive group and it needs to reach consensus on what actions to take. You cannot have solitary decision making in the nuclear domain. It makes us a bit slow. Some complain we have too many meetings, but those are for getting people committed, and finding a

> solution that satisfies everyone ... sometimes you need to discuss about issues and
> have a [face-to-face] meeting; it is quite difficult to put everything in writing'

The interview citation illustrated how the organization in question had created a certain type of culture, which is safety oriented and consensus seeking. This culture has been created by emphasizing safety as a guiding principle and by building an environment supporting the interaction of 'almost all functions of the organization'. This kind of culture exhibits traits of a 'people culture', a 'sustainable culture', and also a bit of a 'uniform culture' (see Figure 4). The interviewee also recognized the tension between their approach and that of a more performance focused culture; 'it makes us a bit slow'.

Many of our interviewees pointed out that they need to balance between conflicting principles in their safety management activities. For example, a nuclear industry representative contemplated the significance of work motivation:

> 'If a person is not motivated, he turns indifferent, and that does not go well together
> with maintaining nuclear safety. This is also one of the small dilemmas of nuclear
> power. We need diverse people, but ... there has to be control and there is
> supervision. Some people may be demotivated from the amount of control. Of
> course some people are motivated by control, but we also need those people who
> reflect a bit, who want to think a bit wider. So where does the border for the control
> go when it starts to demotivate. ... But of course the processes must be able to
> handle the issue that somebody is not so motivated. You cannot motivate everyone
> all the time, we are humans and humans have civilian life worries and other things
> that surely reflect to work from time to time. ... But as I said [earlier], we do not
> trust in the one individual, rather it is the system, and the redundancies and
> diversities built into the system, that takes care of nuclear safety, irrespective of
> what the one individual does.'

The citation illustrates how the manager was considering the pros and cons of a culture that is based on uniformity and standardization, and acknowledged the need to counteract the negative effects of uniformity by diversity and autonomy – and the need to again counteract potential side-effects of individual initiative and relying on people.

The challenge of organizational culture change came up many times in the interviews. An interviewee from the power industry was asked about whether practices are adequately reflected upon in their organization and he gave an answer that implied a both yes and no:

> 'When people have been at work for twenty to thirty years, they don't change
> anymore. It is really difficult to get anything to change. Yes they [the practices] are
> reflected upon but to make a change happen is really difficult. Of course we have
> done a lot here and tried to change things and even succeeded in changing things,
> but the change happens through change of personnel. So the answer is yes, they are
> reflected.'

Finally, one interviewee describes how to draw the line between thoroughness and efficiency in decision making [cf. 8]:

> 'And then when everything is taken into account and done and so on, where is the
> line when you have reviewed enough in order to make a decision. This is continuous
> discussion that takes place in an expert organization such as ours; what is the
> adequate level of reviewing so that one dares to make a decision. One should not
> make too hasty decisions, but it is also safety that one does make decisions and
> goes forward.'

These empirical examples illustrate, together with section 2.3, how the management of safety is inherently contradictory activity that requires balancing between several competing demands and values. They also illustrate how the particular facet of organizational culture and organizational values called "safety culture" is situated in a space of different value orientations. It seems reasonable, then, that attempts to diagnose safety culture and manage safety should be sensitive to how individuals and groups in an organization deal with conflicting values. In order to facilitate this, we will next propose a framework of organizational safety culture profiles.

3.2. Organizational Culture Profiles and Their Relation to Safety

In line with the examples in the previous section, we can give labels to eight distinct organizational culture types, or archetypes. We acknowledge that few organizations will fall exactly into any one of the eight categories, but rather exhibit some characteristics of all eight archetypes. Nevertheless, the culture types can be used as a heuristic when thinking about safety management and its relation to culture. The underlying idea is that each cultural type has both pros and cons in terms of safety management. Figure 4 illustrates how the archetypes and show how they are related to the eight principles of adaptive safety management. For example, 'people culture' is related to and congruent with the underlying assumptions of the management principle 'facilitate interaction'.

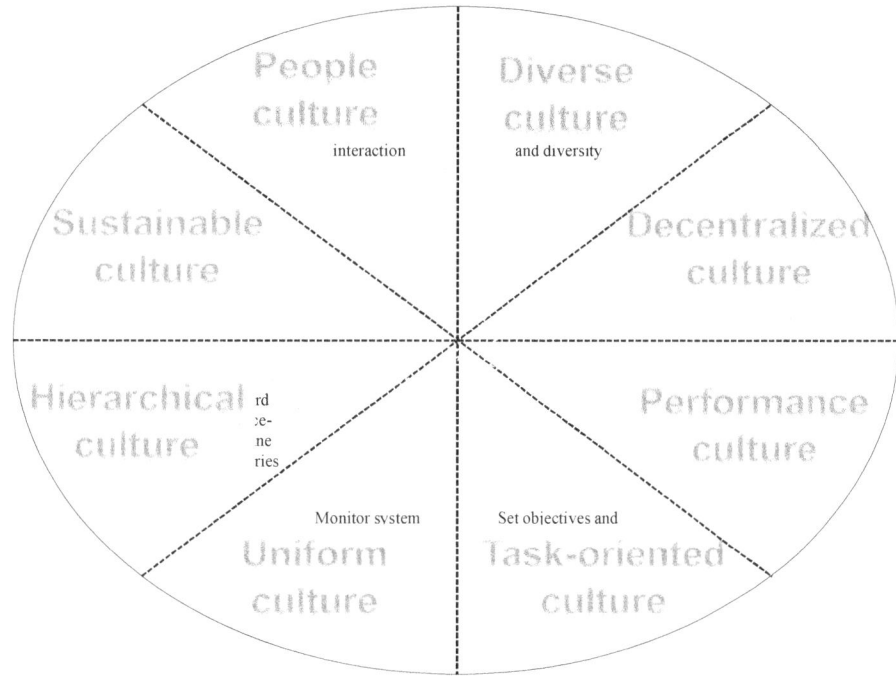

Figure 4. Certain cultural profiles are in line with certain safety management principles [based on 3, 19, 22].

It is important to note that these cultural types are seldom organization wide. Often there are subcultures in organizations and different units and departments can exhibit very different cultural characteristics. This in turn sets further challenges for safety management; what is valued and perceived as important and normal in one unit may not be so in the other department. Safety managers needs to adapt not only based on the external requirements and company culture, but also based on local cultures at different departments, work sites, plants etc.

Table 1 provides a brief description of each culture archetype together with a hypothesized safety manager role that would fit in the culture. Detailed description of the eight proposed cultural archetypes is beyond the scope of the current paper, but they offer a tool for abstracting the dominant culture pattern in each organization under study. Similar to the Competing Values Framework, we do not expect any organization to exhibit characteristics of one archetype only, but rather from all in a certain degree. On the other hand, these competing archetypes can cause tensions that require solving in the organization.

Table 1: Cultural archetypes and the perceived value of safety

Culture type	Brief description	Perceived value of safety
Sustainable	An organization that values long term goals, trust and interacting with personnel and the outside world. Collaboration partners (contractors, regulators, universities etc) are considered important stakeholders.	Safety is an important long-term requirement for a sustainable organization. Safety culture is viewed as a shared vision of safe future, a shared long-term value (an end-state).
People	An organization that values people and personal relations. Trust, transparency, equality, information sharing and collaboration are emphasized. Government institutions, trade unions, and adult educational centers are considered important stakeholders.	Safety is something that is important to personnel wellbeing. Safety is the organization's intellectual capital. Safety culture is viewed as a shared value of the importance of safety.
Diverse	An organization that nourishes novelty and diversity. Multiple and different opinions are tolerated and even embraced. New ideas and solutions are experimented. Universities, consultants and research institutes are considered important stakeholders.	Safety equals positive variability. Safety can be a hindrance to innovation, but innovation and diversity can also contribute to safety. Safety culture is viewed as a mindset supporting requisite variety and safety imagination.
Decentralized	An organization that is constantly evolving and adapting to situational circumstances. Few standards and written rules dictate behavior, and autonomy, initiative and decentralized decision making are valued.	Safety equals adaptability. Hazards that threaten safety create the boundaries that should not be crossed. Safety culture is viewed as a reminder of the safety boundaries and provider of a few simple rules that guide action.
Performance	An organization that values production, speed, keeping of schedules and efficiency. Shareholders are considered an important stakeholder group.	Safety is a prerequisite to operations but also a hindrance to efficiency. Safety is sometimes seen as "necessary evil", something that costs money without providing anything in return. Safety culture acts as a counterforce to production pressures.
Task-oriented	An organization that values productivity and effectiveness. The organization is goal-focused. Customers are considered the most important stakeholder.	Safety is important insomuch as it connects to the task the organization is carrying out, or if the customers require it. Unless customers strongly require safety efforts the commitment to safety can remain superficial.
Uniform	An organization that values consistency, stability and uniformity. The way of doing things is as important as the end result. Technical institutions, standardization agencies and auditing companies are among the key stakeholders.	Safety equals reliability, meaning that things are done in a similar manner and they produce a similar outcome. Safety culture acts as a shared response repertoire. Safety culture provides a uniform response.
Hierarchical	An organization that values hierarchy, rules, standardization and centralized decision making. People are expected to follow the rules and carry out their work within predefined roles and responsibilities.	Safety equals robustness, meaning that the organization is able to anticipate and respond to contingencies without challenging the status quo. Safety culture acts as a barrier against contingencies. Safety culture is viewed as something that should not change.

Table 2 illustrates how the different culture types relate to different safety management roles and what kind of challenges safety manager can encounter in a given culture. Both Table 1 and 2 are based on our empirical research [see above, and 19, 16, 22] as well as findings on organizational culture and safety culture by [1, 3, 23]. However, they are still simplifications and cannot be taken as validated statements but rather as heuristics and hypothesized cultural assumptions, values and norms. Whether a given statement is valid in a specific organization is always an empirical question.

Table 2: Culture affects what kind of safety management is easily accepted and what is resisted

Culture type	Culturally accepted safety manager role	Potential safety challenges
Sustainable	Mentor who reminds people of the importance of safety and acts as an example of a safety-conscious employee. Long-term safety investments are considered acceptable.	Acute tasks are easily neglected in 'sustainable culture'. The lesser importance placed on effectiveness and efficiency can also in some cases have negative safety effects, e.g. if it affects the financial situations of the company.
People	Facilitator who provides support and participates to discussions but lets the personnel decide. People-centered and "soft" management style is expected.	It is not always easy to get 'tough' decisions through in a people culture. A demanding management style may be considered threatening and safety manager can find it difficult to require specific actions.
Diverse	Innovator who brings forth new ideas and who accepts diverse views (and diversity in general) from the personnel. Personal management styles are tolerated but uniformity and constraints are discouraged.	Constant changes and lack of uniform practices can cause anxiety for ordered safety managers. Diversity of people, opinions, practices and ways of working can be a source of risk that is difficult to manage without shared guidelines.
Decentralized	A broker (or a visionary) who provides views on hazards, safety and ways of working, and develops the organization actively, and together with different personnel groups. Autonomy and initiative is expected from both managers and personnel.	It can be difficult to standardize activities across the entire organization since there are multiple autonomous units in the culture. Autonomy can also promote risk taking. Safety manager has a lot of work in keeping up with developments in different units.
Performance	Producer who delivers safety results efficiently and on schedule. Actions that happen fast and are not expensive (or cost-benefit ratio is highly positive) are considered most acceptable.	Safety work is not easy to translate into quarterly performance targets. This easily leads to selection of short term acute tasks that can be completed quickly to the detriment of more strategic long term safety development.
Task-oriented	Hard-driver who listens to how customers view safety and requires that the organization delivers as high (or low) safety as the customer wants. Management style puts task first and people second.	If the customer is not interested in safety issues, it is very hard for the safety manager to get safety improvements made. Also, the lack of people focus can cause stress and decrease wellbeing among the personnel, which affect safety negatively.
Uniform	Monitor (or a stipulator) who checks that everything is in order and points out potential sources of negative variance in performance. The most familiar management style is impersonal and distant.	It is difficult to get ideas accepted if they cannot be applied to the entire organization at the same time. New ideas are easily resisted since variance and novelty is associated with negative events in 'uniform' culture.
Hierarchical	Coordinator (or a specifier) who handles resources and provides clear rules on how to act in different situations. Autonomy or initiative is not expected from managers or personnel.	Personnel expect clear guidance from the safety manager. Ambiguous instructions or giving freedom of choice to the personnel can be felt as anxiety provoking by them. Novel situations easily paralyze the personnel in 'hierarchical' cultures.

Tables 1 and 2 can be used in summarizing the results concerning an organization's dominant culture and comparing it with the organization's, and its managers', salient safety management principles (see also Figures 3 and 4). Differences in profiles may imply challenges in terms of striving to change the company culture or adjusting safety management to better fit the company culture. Often there is a need to conduct a mixture of both approaches.

We argue that it is possible to create a general frame of reference for selection of effective leadership practices by considering the principles and the culture types depicted in this paper. The specifics of selecting the most appropriate safety management strategy lies outside the scope of the current paper, but some factors of importance can be postulated: The current level of safety will most likely influence what type of actions should be taken and how they should be carried out [1]. Also, 'safety culture' maturity, here defined as how safety values are perceived in the organizational hierarchy of values (cf. Table 1) affects both the possibilities for action as well as what should be done. The core task of the organization sets both possibilities and constraints for safety. The inherent hazards differ between various industries, and those set specific requirements for safety management actions. Finally, the culture of the organization is an important factor to acknowledge in safety management, as illustrated in this paper.

It is important to remember that safety management should never focus only on culture but also on the structural aspects of the organization (division of labor, technology, instructions, etc) as well as work practices and personnel issues in general (competence, understanding of hazards etc). Safety emerges from all the elements of the sociotechnical system, not only from safety culture.

Sometimes leadership guides, for example, in the nuclear industry differentiate safety related leadership from production related leadership. There are probably many pragmatic reasons for doing so, and as we have also argued, safety and production can be in conflict. However, we propose that the conflict is not so much a matter of different leadership than different situations. Correspondingly, different situations require different type of leadership and the true quality of leadership is in recognising the type of leadership required at any given moment.

Pidgeon and O'Leary [12] call for 'safety imagination' to overcome the rigidity in beliefs about risks. They write: 'Avoiding disaster … involves an element of thinking both within administratively defined frames of reference (to deal with well-defined hazards that fall within an organization's prior worldview) and simultaneously stepping outside of those frames (to at least consider the possibility of emergent or ill-defined hazards that have not been identified in advance – or which perhaps fall outside of an organization's strict administrative or legal remit)' [12, p. 22]. Adaptive safety management should seek to benefit from the cultural characteristics of the organization yet transcend them and seek to dip into other belief systems to overcome excess rigidity – to maintain cultural adaptive capacity.

It can thus be argued that there is a further meta-dimension in addition to the eight principles – dimension of dynamic (or adaptive) versus static leadership. A dynamic leadership is able to shift from one principle to another based on the situational circumstances at hand. A static leadership is stuck in one role and is unable to adapt even when circumstances change. An adaptive leader is able to balance between different management principles [3, 4], and thus may overcome some of the blind spots created by the organization's dominant culture.

The safety management principles and the culture profiles depicted in this paper can help us in understanding the dual role of managers as both creators of culture [cf. 23] and agents of culture. Safety managers simultaneously lead and influence the system and act in the system. This means that safety managers need to balance not only between different safety management principles but also between actions targeted towards creating preconditions for others to act in a certain manner, and actions that manifest this type of wanted behavior – sometimes going against the dominant values of the culture in question. For example, a manager can on the one hand define system boundaries and create rules and standard operating procedures for the personnel to follow, and on the other hand himself obey the rules and boundaries of the organization. On the other hand, a safety manager who perceives their organization as too proceduralized and dependent on written instructions and guidelines may decide to introduce some variance in the system by provoking discussion on hazards, potential blind spots as well as the role of rules in general. Changing cultural values is slow, and usually requires change in other elements of the sociotechnical system as well. The safety manager needs to work with all the elements, including but not limited to working with safety culture.

4. CONCLUSION

We can conclude the relation between safety culture, organizational culture and safety management by the following definitions:
- Safety culture refers to shared safety values and assumptions about safety
- Organizational culture refers to shared values, norms and assumptions concerning issues such as leadership, effectiveness etc. Some of these values and norms deal with safety and form the above mentioned safety culture, a subsystem of the organizational culture.
- Safety management needs to take put the safety values into the context of other values and shape the culture as well as practices, structures and technology to better facilitate overall safety. This requires trade-offs as well as adaptation to situational circumstances
Thus, culture defines how safety management is carried out, yet safety management should aim at influencing the culture by contextualizing safety as a value, reflecting the potential cultural blind spots and simultaneously building on the strengths of the culture.

The models presented in this paper may help in identifying the dynamics and specific values of the culture in question, and in defining how to proceed with increasing the importance of safety in the culture and other organizational preconditions for safety.

Acknowledgements

The work reported in this paper has been conducted in several different projects. The authors would like to acknowledge the funding organizations of these projects, including VTT, the Finnish Work Environment Fund (TSR), Nordic Nuclear Safety Research platform (NKS), and the Finnish National Nuclear Safety Research Programme.

References

[1]   R. Amalberti, *"Navigating safety. Necessary compromises and trade-offs – Theory and practice,"* Springer, 2013.
[2]   S. Antonsen, "Safety Culture: Theory, Method and Improvement", Ashgate, 2009, Farnham.
[3]   K. S. Cameron and R. E. Quinn, *"Diagnosing and Changing Organizational Culture. Based on the Competing Values Framework, Third Edition"*, Jossey-Bass, 2011, San Francisco.
[4]   G. H. Eoyang and R.J. Holladay, *"Adaptive action. Leveraging uncertainty in your organization"*, Stanford University Press, 2013, Stanford.
[5]   F. W. Guldenmund, *"The nature of safety culture: a review of theory and research"*, Safety Science, 34, pp. 215–257, (2000).
[6]   F. Guldenmund, *"The use of questionnaires in safety culture research—an evaluation"*, Safety Science, 45, pp. 723–743, (2007).
[7]   A. Hale and D. Borys, Working to rule, or working safely. In Bieder, C. & Bourrier, M. (Eds.), *"Trapping Safety into Rules. How Desirable or Avoidable is Proceduralization?"* Ashgate, 2013, Farnham.
[8]   E. Hollnagel, *"The ETTO principle: Efficiency-thoroughness trade-off"*, Ashgate, 2009, Farnhamn.
[9]   R. R. McDaniel and D. J. Driebe, *"Complexity science and health care management"*, Advances in Health Care Management, 2, pp. 11-36, (2001).
[10] E. McMillan, *"Complexity, management and the dynamics of change"*, Routledge, 2008, London.

[11]   D. J. Meyers, J. M. Nyce, S. W. A. Dekker, *"Setting culture apart: Distinguishing culture from behavior and social structure in safety and injury research"*, Accident Analysis and Prevention, in press.

[12]   N. Pidgeon and M. O'Leary, *"Man-made disasters: why technology and organizations (sometimes) fail"*, Safety Science, 34, pp. 15-30, (2000).

[13]   P. E. Plsek and T. Greenhalgh, *"The challenge of complexity in health care"*, BMJ, 323, pp. 625-628, (2001).

[14]   R.E. Quinn, *"Beyond rational management"*, Jossey-Bass, 1988, San Francisco.

[15]   R.E. Quinn and J. Rohrbaugh, *"A spatial model of effectiveness criteria: towards a competing values approach to organisational analysis"*, Management Science, 29, pp. 363-377, (1983).

[16]   T. Reiman and E. Pietikäinen, *"The role of safety professionals in organizations – developing and testing a framework of competing safety management principles"*, Probabilistic Safety Assessment and Management PSAM 12, June 2014, Honolulu, Hawaii.

[17]   T. Reiman and C. Rollenhagen, *"Human and organizational biases affecting the management of safety"*, Reliability Engineering & System Safety, 96, pp. 1263-1274, (2011).

[18]   T. Reiman and C. Rollenhagen, *"Competing values, tensions and tradeoffs in management of nuclear power plants"*, Work, 41, pp. 722-729, (2012).

[19]   T. Reiman and C. Rollenhagen, *"Reconceptualization of the competing values framework tailored for management of nuclear power plants"*, 11th International Probabilistic Safety Assessment & Management Conference, 25-29 June 2012, Helsinki, Finland.

[20]   T. Reiman and C. Rollenhagen, *"Does the concept of safety culture help or hinder systems thinking in safety?"*, Accident Analysis and Prevention, in press.

[21]   T. Reiman, E. Pietikäinen, P. Oedewald and N. Gotcheva, *"System modeling with the DISC framework: evidence from safety-critical domains"*, Work, 41, pp. 3018-3025, (2012).

[22]   T. Reiman, C. Rollenhagen, E. Pietikäinen and J. Heikkilä, *"Principles of adaptive management in complex safety critical organizations"*, submitted manuscript.

[23]   E. Schein, "Organizational Culture and Leadership". Jossey-Bass, 1985, San Francisco.

[24]   S. H. Schwartz, *"Universals in the Content and Structure of Values: Theory and Empirical Tests in 20 Countries"*, In M. Zanna (ed.), Advances in Experimental Social Psychology (Vol. 25). Academic Press, New York, 1-65, 1992.

[25]   S. H. Schwartz, *"Are there universal aspects in the content and structure of values?"*, Journal of Social Issues, 50, pp. 19-45, (1994).

[26]   J. N. Sorensen, *"Safety culture: a survey of the state-of-the-art"*, Reliability Engineering and System Safety 76, pp. 189–204, (2002).

[27]   D. D. Woods and M Branlat, *"How human adaptive systems balance fundamental trade-offs: Implications for polycentric governance architectures"*, in Proceedings of the Fourth Resilience Engineering Symposium, 2011, Sophia Antipolis, France.

Toward Monitoring Organizational Safety Indicators by Integrating Probabilistic Risk Assessment, Socio-Technical Systems Theory, and Big Data Analytics

Justin Pence[a,1,*], Zahra Mohaghegh[a], Cheri Ostroff[b], Ernie Kee[c], Fatma Yilmaz[d], Rick Grantom[e], and David Johnson[f]

[a] Department of Nuclear, Plasma, and Radiological Engineering, University of Illinois at Urbana-Champaign, Urbana, USA
[b] University of South Australia, Adelaide, Australia
[c] YK.risk, LLC, Bay City, USA
[d] South Texas Project Nuclear Operating Company, Bay City, USA
[e] C.R. Grantom PE & Assoc. LLC, West Colombia, USA
[f] ABS Consulting, Irvine, USA

Abstract: Many catastrophic accidents have organizational factors as key contributors; however, current generations of Probabilistic Risk Assessment (PRA) do not include a comprehensive representation of the underlying organizational failure mechanisms. This paper reports on the current status of new research with the idealistic goal of monitoring organizational safety indicators. Because of the evolving nature of computational power and information-sharing technologies, 'Internet of Things' has been adopted as a metaphor to describe the authors' vision for combining multiple levels of organizational analysis into a real-time application for monitoring the changing landscape of risk. The short-term objectives are: (1) identifying the organizational root causes of failure and their paths of influence on technical system performance, utilizing theoretical models of social phenomena, (2) quantifying the models using advanced measurement and predictive methodologies, big data analytics, and uncertainty analysis, and (3) proposing preventive approaches. Socio-Technical risk analysis deals with wide-ranging, incomplete, and unstructured data. Therefore, this research focuses on developing hybrid predictive technologies for PRA that are not only grounded on Socio-Technical Systems theory, but also serve to expand the classical approach of data extraction and execution for risk analysis by incorporating techniques such as text mining, network data analytics, and data curation.

Keywords: Organizational Safety Indicators, Socio-Technical Risk Analysis, Big Data, Probabilistic Risk Assessment, Real-Time Safety Monitoring

1. INTRODUCTION

Organizational failure mechanisms are widely recognized as key contributors to some of the world's worst accidents (e.g., Piper Alpha, Columbia, Three Mile Island, Chernobyl, Bhopal, Challenger, Deepwater Horizon, and Fukushima). Recently, the International Atomic Energy Agency (IAEA) reported that in Nuclear Power Plants (NPPs), 80% of significant events are caused by human error, and of those human error events, 70% are due to organizational weaknesses (e.g., lack of line-management ownership, weak self assessments, insufficient training), and 30% are due to individual mistakes [1]. This means that only 20% of all significant events are caused by equipment failure [2, 3], and 56% of all significant events at NPPs are caused by organizational weaknesses [1]. Therefore, the ability to detect organizational weaknesses is critical for preventing catastrophic technological accidents, and maintaining public health and safety.

1.1. Organizational Failure Mechanisms in Probabilistic Risk Assessment

Safety methods continue to be developed based on the understanding that "the performance of a complex socio-technical system is dependent on the interaction of technical, human, social, organizational, managerial and environmental factors" [4]. There is considerable evidence from

---

1. Currently employed by Argonne National Laboratory

Probabilistic Safety Assessment and Management PSAM 12, June 2014, Honolulu, Hawaii
*jpence2@illinois.edu

organizational psychology and management science that organizational factors (e.g., safety culture and climate, leadership style, leader priorities, reward practices) are strongly related to safety, injuries, and accidents [5-8], however these organizational factors have not been fully incorporated into complex technological risk models. One potential reason for this is that, through training and socialization in engineering and technical disciplines, researchers may have developed cognitive-frameworks that constrain them from viewing organizations as an external or implicit source of influence [9], thereby reducing the likelihood that organizational factors will be incorporated in the analysis of system failure scenarios. The primary purpose of this paper is to begin to develop a process for better incorporating the *socio* aspect in social-technical system models of risk through the development of a multi-level model. Moving beyond current Probabilistic Risk Assessment (PRA) and dynamic frameworks in human reliability analysis, a networked monitoring system of organizational safety indictors is proposed to theoretically link together a range of technical, human, and organizational factors.

Human Reliability Analysis [10, 11], which is the study of the nature, causes, and probability of human actions in the design and operation of systems and processes, concentrates on individual error and typically ignores the effects of organizational factors in a formal and comprehensive way. As Reason [12] states, "while cognitive psychology can tell us something about an individual's potential for error, it has very little to say about how these individual tendencies interact within complex groupings of people working in high-risk systems". Although PRA [13] is an effective tool for calculating risk due to the interactions of equipment failure and human error, (1) it is not possible to explicitly assess the risk due to the specific organizational weaknesses, (2) it is not feasible to locate the organizational root causes of failures in order to take effective corrective action, and (3) there is the possibility of underestimating the risk associated with human error [14]. There are two major challenges that hinder the full integration of organizational factors into PRA. First is the requirement of a common multidisciplinary language among social scientists and engineers for understanding deep organizational factors. Second, is that the regulatory agency and industrial organization are deeply embedded in the socio-technical system, making it difficult to internally assess organizational factors in an unbiased or comprehensive way.

Over the past 20 years, a number of models have begun to integrate organizational factors. Mohaghegh [15], conducted a thorough review of literature related to the incorporation of organizational factors into risk models and categorized them in two generations with respect to both *theories* and *quantitative techniques*. The first generation included Reason's Swiss Cheese Model [12, 16], a well-known example of organizational accident theories describing the process of organizational effects on human error, and, consequently, on the rate of accidents. There are also a number of first-generation models that attempt to quantify the impact of organizational factors on system risk, including; MACHINE [17], WPAM [18, 19], SAM [20], Omega Factor Model [21], ASRM [22], and Causal Modeling of Air Safety [23]. The nature of first-generation theories and techniques can be characterized in terms of "deviations from normative performance" [24].

The second generation of models began to incorporate the *actual behavior* of individuals and the social system structures of organizations. Second-generation "organizational models" of risk frameworks are still evolving. These models are beginning to represent the underlying organizational mechanisms of accidents, with a focus on the systemic and collective nature of organizational behavior, as opposed to focusing on individual actors alone. On the theoretical side, Rasmussen [24] cites the self-organizing nature of High Reliability Organizations [25] and Learning Organizations [26, 27] as concepts useful in analyzing the managerial and organizational influences on risk. The Normal Accident Theory [28], which views accidents caused by interactive complexity and close coupling, can be considered to be a second-generation perspective on organizational safety. More recently, Wahlström and Rollenhagen [29] utilized the control metaphor for assessing safety management systems in a combined Man, Technology, Organization, and Information (MTOI) framework. From a quantitative stance, second-generation techniques primarily tackle the dynamic aspects of organizational influence. For example, Biondi [30] uses the qualitative model developed by Bella [31] to describe the changes in the reliability of a system due to organizational dynamics. Other researchers, e.g., Cooke [32] and Leveson [33], used the System Dynamics (SD) approach [34] to describe the dynamics of organizations, but these models do not include detailed, PRA-style models of the technical system. Yu et al. [35] and Gajdosz et al. [36] also used the SD approach to assess the effects of organizational factors on risk models; however, the interconnections between PRA and SD have not been clarified. Some of the newer aviation safety models such as the Traffic Organization and Perturbation AnalyZer (TOPAZ) use Monte Carlo simulation [37], while Bosse, T., et al. [38] have included Agent Based Modeling (ABM) for modeling organizational performance. In the most recent iteration of the National Aerospace Laboratory of the Netherlands' 'Safety Methods Database,' there are approximately 137 safety methods associated with the nuclear domain, and 124 documented theoretical models that include Organizational Factors [39].

Based on the analysis of the existing research in this field, Mohaghegh [15] concluded that in the absence of a comprehensive theory, or at least a set of principles and modeling guidelines backed by theory, it would be difficult to assess the validity and quality of the existing modeling techniques. Integrating concepts from multiple disciplines, she introduced a set of thirteen principles [40] for the field of Socio-Technical Risk Analysis. A new framework, called Socio-Technical Risk Analysis (SoTeRiA) [15, 41], was then developed based on these thirteen principals, to more comprehensively incorporate the organizational factors into risk models. Section 1.2 in this paper explains the SoTeRiA theory and its areas of advancement.

1.2. Combining Socio-Technical Systems Theory with Big Data Analytics

SoTeRiA has foundations in Social-Technical Systems (STS) theory and PRA, and further can be used with big data analytics. STS theory emerges from an 'eclectic empiricism' [42] that addresses the interactions of people and technology in the workplace [43] as well as the differences between complex living systems and complex mechanical systems with respect to their failure mechanisms [44]. STS theory has been operationalized in the SoTeRiA framework [15] (See Figure 1), which explicitly recognizes causal relationships among organizational constructs at multiple levels of analysis [45], and also in an integrated modeling technique [46, 47], which combines probabilistic tools (Bayesian Belief Network (BBN)), deterministic and dynamic simulation techniques (SD) with classical PRA methods (Event Tree (ET) and Fault Tree (FT)).

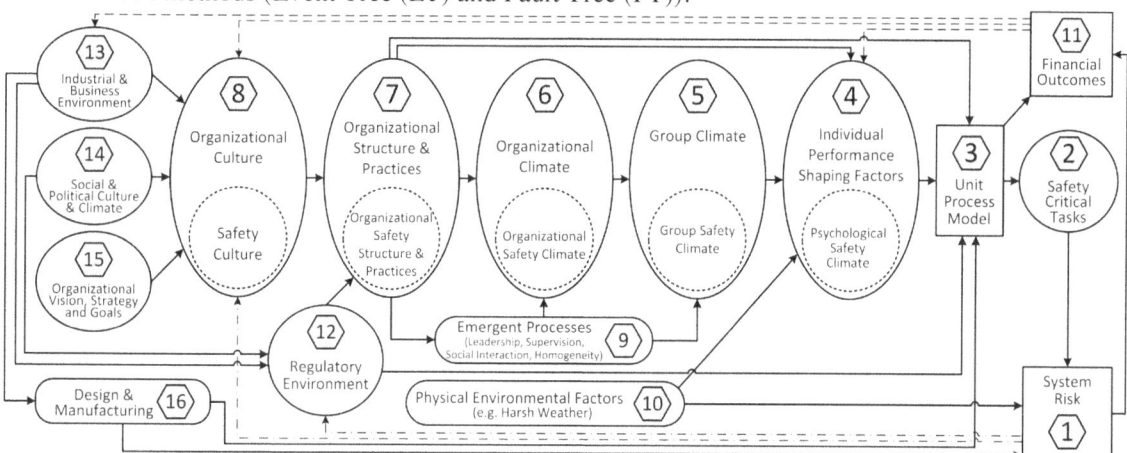

Figure 1. The Socio-Technical Risk Analysis (SoTeRiA) Theoretical Framework

While SoTeRiA represents a significant improvement in integrating organizational factors in technical system risk models, additional work is needed to enhance its theoretical details, quantified methodological features, and practicality in order to develop a more accurate decision-making tool for high-risk organizations. Thus, the focus of this paper is on the advancement of the theory, quantification and measurement of the SoTeRiA framework.

Recently, Bar-Yam [48] asserted that (a) big data is critical for addressing complex systems, (b) theory is essential for understanding complex systems, and (c) that theory makes data more useful. Organizations produce, process and store a large volume of 'information assets' used for regular business activities and compliance purposes. After these incomplete, unstructured, information assets are used, they are stored (often without indexing), and become part of the underutilized [49] 'long tail' of organizational *dark data*[†]. The big data movement is quickly uncovering areas of analytics (i.e., predictive analytics, forensic data analytics, human resource analytics, team analytics, people analytics [50]) and methodologies (e.g., data mining, text mining) capable of interpreting dark data with a focus on organization-level constructs (e.g., reporting, performance indicators, business intelligence, and budgeting). Also, the growing application of 'network thinking' (e.g., Knowledge Management Networks [51], Policy Network Analysis [52]) is advancing research on complexity modeling.

Compared to the equipment and human errors modeled in PRA, the sub-level interactions of organizational failure mechanisms are not well understood and, currently, safety oriented organizational-level (meso-level) datasets beyond safety climate surveys do not exist. The SoTeRiA framework has many constructs, and aside from safety climate surveys, there are not many reliable measurements that can be included into organizational models. Furthermore, such models need to explicitly consider the dynamic interactions and interdependencies that create the collective social systems. This paper proposes a hybrid combination of (1) big data (analytics) and (2) theory (Socio-Technical Risk Analysis, i.e., an integration of STS theory and PRA) to efficiently extract organizational metrics and executes them in a network of organizational safety indicators. This hybrid integration would help (a) avoid the potential misleading results of solely data-informed approaches (big data methods, "top-down" approaches) and, at the same time, (b) overcome the challenges of "bottom-up" methods in quantifying socio-technical risk theories. Section 2.2 of this paper explains this hybrid measurement approach in more detail.

1.3. Toward Monitoring Organizational Safety Indicators

This paper reports on the current status of new research with goal of developing a networked monitoring system for organizational safety indicators. Due to the evolving nature of computational power and information-sharing technologies, the concept of Internet of Things (IoT) [53] has been adopted as a metaphor to describe the authors' vision for combining multiple levels of organizational analysis into a real-time application for monitoring the changing landscape of risk. The IoT commonly refers to networked physical objects (e.g., embedded sensors), and the IoT metaphor provides a way to describe interconnected 'objects' communicating (e.g., machine-to-machine) in real-time. Therefore, in the context of this paper, the IoT is utilized as a metaphor for the purpose of monitoring organizational safety indicators to; (1) explicitly define the varying safety/risk contributing factors among multiple levels of analysis, (2) bridge and quantity their dynamic interactions in a causal model capable of considering the underlying social and physical phenomena, and (3) conduct real-time analysis of multivariate data streams.

The continuously evolving IoT architecture has been described in many different ways; for example, Unit IoT (man-like nervous system model), Ubiquitous IoT (Social Organization Framework model) [54], and Social IoT (i.e., integration of social networking concepts into IoT) [55]. The 'objects' being linked in the IoT primarily include 'smart objects' (e.g., embedded sensors, GPS, and cellular communications to provide equipment operations data), and 'social objects' [55] (i.e., social behavior of smart objects). These models are limited for assessing safety of social phenomena, because they are not based in theory and do not explicitly include broader organizational constructs.

This research proposes to bring objects of 'organization-level constructs' into the IoT metaphor. A networked monitoring system of organizational safety indicators would integrate a wide range of organization-level safety metrics into a 'theory-data dialectical network'. Such a system would be capable of monitoring and detecting leading and lagging organizational safety indicators by bringing together real-time results of multiple levels of analysis for monitoring the changing landscape of safety and risk. Taking the concept one step further, and drawing a specific example for NPPs, a full scope

---

[†] Gartner Inc. Gartner IT Glossary - Dark Data Available from: http://www.gartner.com/it-glossary/dark-data.

'internet of safety indicators' would span the entire socio-technical system, from the micro-level (components; embedded sensors; existing control room indicators), to the meso-level (the proposed organization-level constructs), and to the macro-level (PRA of complex socio-technical systems). Although the research is still at the level of conceptualization of this long-term goal, the authors assert that the path toward monitoring organizational safety indicators begins with the enhancement of a *big theory* [56] capable of handling big data (dark data) and the complexity of socio-technical systems risk theory.

2. RESEARCH STRATEGY

Section 2, Research Strategy, has been broken down into 8 Tasks as listed below. The first Task relates to the development of a strong theory-based understructure (See Section 2.1). The research will require several levels of methodological discovery (Tasks 2-7; See Sections 2.1-2.7) before being applied to NPPs (Task 8; Section 2.8) and, being used as an input for the further implementation of the concept of monitoring organizational safety indicators.

> *Task 1. Develop Factors, Sub-Factors, and Causal Relationships in SoTeRiA*
> *Task 2. Develop Measurement Techniques for Factors, Sub-Factors and their Causal*
> *   Relationships in SoTeRiA*
> *Task 3. Develop a Predictive Socio-Technical Causal Modeling Technique*
> *Task 4. Uncertainty and Sensitivity Analysis*
> *Task 5. Verification and Validation of the Causal Model*
> *Task 6. Integrate SoTeRiA with PRA*
> *Task 7. Estimate Risk Importance Measures*
> *Task 8. Implement with the Nuclear Power Plant Specific Data*

2.1. Develop Factors, Sub-Factors, and Causal Relationships in SoTeRiA (Task 1)

SoTeRiA provides theory-based, multi-level causal paths for influence of organizational factors on the elements of the PRA models. While it provides a roadmap for organizational risk analysis; it requires an increased level of granularity before it can be operationalized and is capable of giving industries a way to effectively manage their organizational risk. Each of the elements of SoTeRiA (i.e., Nodes 1-16 in Figure 1) need to be populated by relevant constructs, their associated factors and sub-factors, taking into consideration certain levels of validity and adequacy. In this research, the goal is to expand the level of detail of the causal elements of the SoTeRiA framework by mapping (1) theoretical (e.g., academic literature), (2) regulatory (e.g., regulations, standards and recommendations) and (3) industrial (e.g., knowledge management, management systems, training materials, logs, and expert opinion) perspectives. In other words, using these three perspectives, related constructs will be identified and placed within the sixteen elements of SoTeRiA. In addition, each construct will be broken down into factors and sub-factors. Sub-factors will be further decomposed according to their attributes, features, and dimensions. After developing the factors and sub-factors, each construct will undergo another detailed level of theoretical and practical validation to determine the adequacy of its placement within the key elements of SoTeRiA.

Another aspect of the theoretical advancement of SoTeRiA relates to developing details in the causal relationships of the model. For instance, the evaluation of the relationship between safety/risk performance (Node 1 in Figure 1) and the financial outcome (Node 1 in Figure 1) of an organization is the topic of an ongoing research [57]. As another example, organizational structure and practices (Node 7 in Figure 1) are strongly interrelated, but in the initial development of SoTeRiA, the relation between safety structure (e.g., centralization, formalization) and safety practices (e.g., human resource practices, procedure-related activities) was left unanalyzed. Research needs to uncover the detailed relations between organizational safety structure and safety practices.

An ongoing literature review is being conducted for the three perspectives (i.e., theoretical, regulatory, and industry) to develop a taxonomy of organizational factors supporting the SoTeRiA theoretical framework. High-level classification of elements includes; external environment (i.e., regulatory, governmental, international agency, industry and business), internal environment (i.e., processes, procedures, training manuals, performance records), financial outcomes, and risk models (i.e., PRA). The result of *Task 1* will theoretically map the key elements of SoTeRiA (i.e., Nodes 1-16 in Figure 1) to sub-nodes (factors, sub-factors, attributes), and their causal relationships. For example, organizational culture (as a high-level element of SoTeRiA) has factors of artifacts, espoused values and basic assumptions. Sub-factors of organizational culture include different culture orientations (i.e., hierarchy, clan, adhocracy, market, safety [58]), and the associated attributes are consensus on cultural values across employees; strength of the culture; degree of socialization of new employees to culture; and the highest priority among values and culture.

2.2. Develop Measurement Techniques for Factors, Sub-Factors and their Causal Relationships in SoTeRiA (Task 2)

Organizational factors have been widely recognized as major contributors to risk, yet their sub-level interactions are not well understood, and have been excluded from accepted risk quantification methodologies. *Task 1* (Section 2.1) of this research addresses this need by theoretically developing the factors and sub-factors and their relationships. *Task 2* of this research focuses on developing a new 'theory-data dialectical network' of complex organizational systems.

There are two distinct epistemological approaches for developing models, (1) the *bottom-up* (Baconian) approach and (2) the *top-down* (data-informed) approach. The bottom-up approach is based on developing theoretical models and quantifying them using observations and experiments. The bottom-up approach is the most common practice in engineering and science because there are available quantified equations and theories that have been developed and validated (empirically and/or analytically) over the years. The top down approach is through data-informed discovery, where large data sets are used to observe patterns of causality without having an underlying theory.

In social science, the key challenge of the bottom-up approach is that, the quantification and validation of the theoretical models require reduction of factors at the beginning of analysis in order to handle the scope of factors relating to complex systems. Since there are not enough quantified underlying theories for organizations (unlike engineering and science), in order to quantify the models (e.g., using correlations), the contextual factors need to be reduced at the beginning of the analysis. This reduction of factors would decrease the completeness of the depiction of contexts. The main criticism of top-down approaches is the potential for being misled by data due to lack of the underlying theories [59], and the misuse of "associations" instead of "causations" [60].

This paper proposes a *hybrid* measurement approach that integrates the bottom-up and top-down approaches. This hybrid approach will remove the need for the initial reduction of factors (common in the bottom-up approach) by providing a method to scientifically reduce the number of factors, while providing a theory-based network to direct the analysis. In the hybrid approach, big data can be used in a structured way using the underlying theory. The hybrid approach will create initial networks of factors extracted directly from documents (e.g., root cause analysis reports, near-miss reports, corrective action program reports), and place them into a causal model (SoTeRiA). Once the critical factors and paths in the model are determined, the analysis can be re-focused, using scientific reduction to change the scope and to measure and quantify the most important factors and causal paths. The hybrid approach takes advantage of both data-informed approaches and an underlying theory by introducing a SoTeRiA network-generating algorithm. The algorithm provides a methodology for the extraction and analysis of socio-technical data, placing and 'folding' data into the SoTeRiA network structure, and operationalizing the network to provide input for the predictive causal model. The algorithm has the following steps:

1. Extract data and develop an initial network:
   a. Using a combination of text mining with network data, extract factors from documents (e.g., NPP incident reports) and develop a data network. A candidate

software code is ConText [61]. Popova and Sharpanskykh [62, 63] also demonstrate a framework for semantic organizational modeling, which uses Temporal Trace Language (TTL) (first order predicate logic) to formalize *Nodes* (attributes; *n*) and their relationships (*Edges*; *q*). This candidate technique could be used to represent the relationships "between the descriptions of different elements" [64].

   b. By comparing the underlying theory (SoTeRiA) and the data network, calibrate and adjust the network.

   c. For unavailable or missing data relationships in the network, generic causal relationships will be applied from theory.

   d. If data are unavailable (e.g., for new organizational design), use theory to generate experimental (simulated) scenarios and create observational documents for experiments (i.e., documented interpretations of the experiments). Then, extract network data from the new documented observations utilizing the tool from step *1.a*.

Comparatively, the difference between the SoTeRiA network (developed from Step 1) and traditional network analysis or, the more recent Knowledge Network Analysis [51, 65], is that the SoTeRiA network enables a theory-data dialectical network for risk analysis, meaning that the network developed from data analytics will be adjusted and directed by the SoTeRiA theory.

2. Scientific reduction of contextual network factors:

   a. Determine the value of causal Edges (e.g., frequency of occurrence) in the network data (i.e., $q_1, q_2 \ldots q_k$), where "q_k" represents frequency of Edge k in the network.

   b. Determine the Node values (e.g., frequency of occurrence) in the network data (i.e., $f_1, f_2 \ldots f_n$), where "f_n" represents the frequency of Node n in the network.

   c. Use ($q_1, q_2 \ldots q_k$) to quantify the causal links in a predictive modeling tool (e.g., BBN; Jensen, et al. [66]) and also, utilize ($f_1, f_2 \ldots f_n$), to quantify the aggregate probability of causal factors (e.g., Node 1, Node 2… Node *n*) in the BBN.

   d. Sensitivity analysis (See Section 2.4) and Importance Measure analysis (See Section 2.7) will be used to rank the importance of nodes and causal paths in the predictive model. This will provide an opportunity to reduce the scope (number of factors and causal paths) and re-focus on the important factors.

   e. Once the important high-level factors have been identified, the associated sub-factors (identified in Section 2.1) will be used to create a network that leads to the higher-level important factors. In other words, for each important high-level factor identified, a network should be developed, repeating Steps 1 & 2 to develop a network of sub-factors, and to identify the most important sub-factors associated to high-level important factors of the SoTeRiA network.

   f. The network of important sub-factors will be integrated with the network of high-level important factors to create one main three-dimensional network. The Importance Ranking analysis will be implemented in the main network to run another round of scientific reduction and to pass the important sub-factors to Step 3 for measurement.

3. Measurement of Important Factors:

Once the important sub-factors have been identified (from Step 2), all classical measurement approaches (subjective, objective), i.e., common measurement approaches in bottom-up approaches, should be utilized. It is a common practice in social science to develop surveys to measure the limited list of contextual factors. Kongsvik [67] addresses that the variation of scales, questionnaires and dimension in safety climate surveys (due to differences in developmental histories between industries), is "seen as a sign of the immaturity of this field of research". Also, there is much ambiguity and inconsistency in the literature on assessing safety culture and climate among the safety community. To avoid the automatic use of surveys, Steps 1 and 2 will be used to find the most important factors and sub-factors before moving on to Step 3, where measurement techniques will be applied only to the most important sub-factors remaining after scientific reduction. In other words, the purpose of this SoTeRiA hybrid algorithm is to rely less on surveys by directly embedding sub-factors into the main network and, dedicating the focus of measurement (e.g., if necessary, to develop surveys) on the identified important sub-factors.

Technical bases for the development of measurement approaches of important sub-factors will be built based on the advancement of a multidimensional measurement perspective [68]. According to this perspective, measurement methods should be selected based on the type and level of model constructs (i.e., individual, global, configural, and shared), the required level of accuracy, and the availability of data. The sub-factors can be measured using subjective (perception-based), objective (compliance-based) and hybrid (i.e., combination of *subjective* and *objective*) methods. Mohaghegh and Mosleh [46], also demonstrated that the Bayesian approach is an effective technique to operationalize multidimensional measurements. In this research, data analytics will introduce a new suite of textual and numerical data into the *hybrid* measurement methods.

4. Developing a large-scale SoTeRiA network:

Once the important factors and sub-factors have been quantified, the entire SoTeRiA network should be populated with a combination of detailed information for important nodes (from Step 3), and generic information (i.e., solely based on the theoretical model) for the rest of the nodes. The algorithm will proceed, with the large-scale SoTeRiA network, to Tasks 3 to 7.

Figure 2 visualizes a three-dimensional sub-network and surface-level network based on the SoTeRiA network-building algorithm. For technical systems, PRA has been used as a structured and formal method for developing high-level risk scenarios (using ETs). PRA implements reductionism, or a *breaking down* of the high-level events of ETs into basic events (component-level) using FTs to the point where; (1) reliability data can be more available, and (2) more understanding of common root causes of failures can be possible. Much like the classical PRA philosophy, the high-level causal network in SoTeRiA (i.e., the surface causal relationships among the grey nodes in Figure 2) can be visualized as the ETs and the associated network of each factor (i.e., the cones in Figure 2) can be considered as the FTs (under the elements of ETs) connecting the sub-factors to the high-level factors.

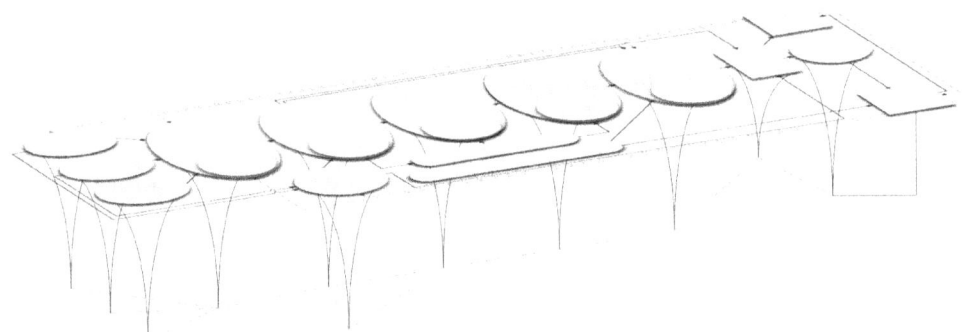

Figure 2. Visual Example of the Expanded Three-Dimensional SoTeRiA Network

2.3. Develop A Predictive Socio-Technical Causal Modeling Technique (Task 3)

The resulting network developed from *Task 2* will represent data of different varieties, therefore, the final choice of predictive modeling technique must relate to the type of analytics used, and the nature and the scope of the organization-level phenomena. Because of the multidisciplinary nature of the SoTeRiA theoretical framework, a single modeling technique is not adequate for its quantification, and so the integration of probabilistic and deterministic approaches is required. The SoTeRiA modeling approach applies a combination of BBN (a probabilistic method), and SD (a deterministic method), to quantify the SoTeRiA theoretical framework [15, 47]. BBN can establish explicit probabilistic relations among elements of the model, where objective data are lacking and use of expert opinion is necessary. This, of course, is very important for the quantification of socio-technical models, when dealing with the soft nature of human and causal organizational failure mechanisms. BBN can be mathematically linked to classical PRA techniques, and is also capable of incorporating the positive features of regression-based techniques and process-modeling methods. However, BBN alone is inadequate for representing dynamic aspects such as feedback loops and delays. Combining SD with BBN enables BBN with dynamic features.

2.4. Uncertainty and Sensitivity Analysis (Task 4)

This task relates to the development of efficient uncertainty and sensitivity analysis methodologies for propagation of uncertainty in the organizational causal model. Uncertainty and sensitivity analysis will be conducted for all assumptions, input variables and models, including the integrated predictive causal modeling techniques, as well as data analytics techniques. Uncertainty propagation can be accomplished by assuming that the input parameters are random variables with distributions derived from historical data, experimental data, expert elicitation, or a combination of these sources. These values would be propagated through the organizational causal model to yield an output, an estimator of a key performance measure, such as the probability of a subsystem failure, which is then passed to the PRA model. A rich simulation environment will be developed that will allow uncertainty and sensitivity investigations of arbitrary model inputs and measurements.

2.5. Verification and Validation of the Causal Model (Task 5)

This research has multiple layers of validation:

(A) Theoretical validation: the development of constructs, factors, and sub-factors (*Task 1*) will be based on the theoretical grounds of the SoTeRiA framework, and in accordance with the analysis of the related theories and literature.

(B) Calibration of theories with up-to-date regulatory results: regulatory perspectives will be incorporated using samples from agencies that consider safety management and safety culture in their operations and oversight (e.g., NRC Safety Culture Common Language [69]). Perspectives will be derived from domestic and international rules, regulations, policies, guidelines, and associated documents. Potential sources for construct mapping include guidelines for safety management systems, safety culture policy statements, human factor engineering and human reliability guidance documents, etc.

(C) Qualitative & quantitative validation with industry observations: industrial perspectives will be incorporated using indicators from safety management systems, training programs and procedures, incident reporting systems, administrative documents, and expert elicitation. Industrial observations will be used to provide a practical and realistic context to the entire SoTeRiA ontology and theory-data dialectical network, while regulatory perspectives will be used to identify gaps in theoretical and industrial perspectives.

2.6. Integrate SoTeRiA with PRA (Task 6)

This task is dedicated to developing a methodology for the quantitative interface of the organizational causal model (SoTeRiA) and the PRA to create an *Integrated PRA*. At this stage, we need to find the associated basic events in PRA that are going to receive updated probabilities from the organizational causal model, and determine if any basic events need to be added to the PRA. In this integrative modeling approach, the classical PRA of the plant would be used, but the organizational performance phenomena would be modeled in a separate module (developed and quantified using the SoTeRiA framework and algorithm) and the module would then be linked to the classical PRA. The linkage will convert the outputs of the organizational causal modeling module into distributions, and connect them to the basic events of the FT (See Figure 3). This linkage will help analyze the changes in the key elements of PRA due to the changes in the quality of key factors in the organization. The SoTeRiA hybrid modeling technique uses a proven methodology [70] to 'fold' phenomenological uncertainties associated with organizational failure mechanisms into plant performance metrics in order to support risk-informed decision-making.

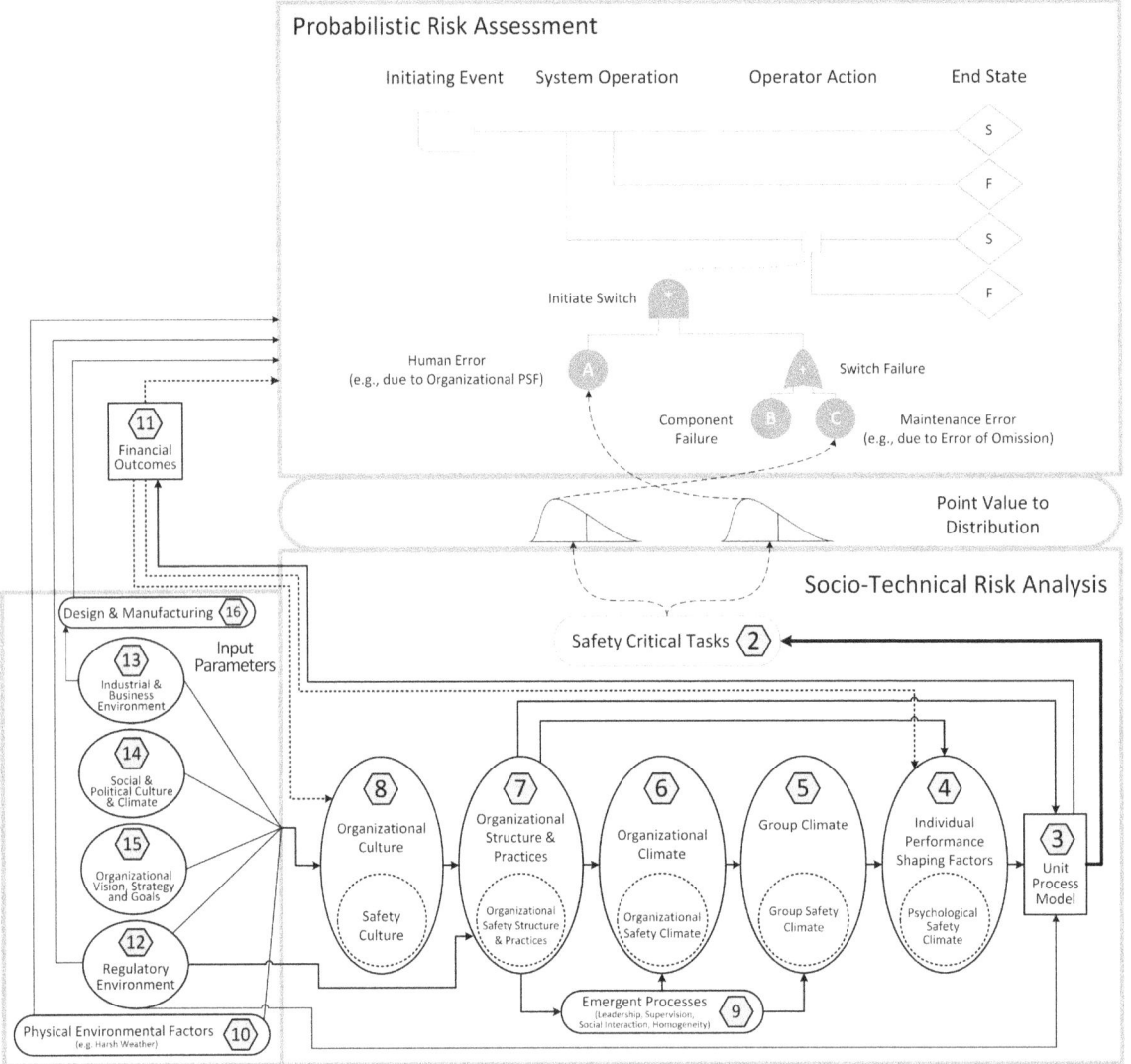

Figure 3. Illustration of the Integrated PRA Framework

2.7. Estimate Risk Importance Measures (Task 7)

This task focuses on the comparative studies of Risk Importance Measure (RIM) methodologies and the selection of an RIM approach for the SoTeRiA frameworks. The ultimate goal when using a RIM method is to find the most critical organizational failure mechanisms with respect to NPP risk (i.e., Core Damage Frequency) and to rank them based on the degree of their contribution to the risk. Doing so would allow an organization to dedicate more resources and time to the accuracy and validity of the values (and their associated uncertainties) for the most critical factors. Several types of RIMs have been developed and used for classical PRA. Sakurahara et al. [71] conducted a literature review on the methods of importance measures including both local measures (e.g., Fussel-Vesely (FV) and derivative-based importance measures) and global analysis (e.g., variance-based and moment-independent approaches). Based on the characteristics of the *integrated* PRA framework (e.g., uncertainty quantification and propagation, dynamic interactions, and dependencies among parameters and variables in the simulation module), the authors selected a Global Importance Measure (GIM) as the most suitable RIM method for *integrated* PRA frameworks. In this research, we will analyze the application of GIM for the SoTeRiA-PRA model in order to rank the critical organizational failure mechanisms. This ranking will also help us find the most critical factors and sub-factors of SoTeRiA (with respect to risk) and, then run a more detailed measurement for them (i.e., Step 3 in section 2.2). The difference between the initial ranking in Step 2.d and the ranking of this section is that, in the initial ranking, the output performance is the output node of the network, but in this section the output is the total risk of the system and the factors are ranked based on their contributions to system risk. This possible feedback of the analysis will certainly increase the reliability of the quantifications and measurements.

2.8. Implement with the Nuclear Power Plant Specific Data (Task 8)

This section relates to the implementation of the developed theoretical model (*Task 1*) and the methodological approaches in *Tasks 2 to 7* for NPP risk analysis. It will lead to the building of a causal model for plant-specific organizational failure mechanisms of a NPP, and the linkage of the model as a separate module to the existing PRA. The causal model will be able to highlight the root organizational weaknesses of a 'Significant Event' and their paths of influence on the technical system risk scenario. We will quantify the organizational failure mechanism module and will quantitatively link it to PRA. This linkage will provide an estimate of the total risk associated with organizational failure mechanisms. It will also help to develop a precursor analysis and a risk importance measure analysis (*Task 7*) to find the most critical organizational failure mechanisms requiring high priority for correction at a given NPP and, will provide indicators of important factors for other high-risk industries. The integrated model (integration of the organizational module and PRA) will facilitate the analysis and quantification of the effects of change in total risk, due to the change in the root organizational weaknesses, and will help managers make decisions regarding optimum corrective actions. It is important to identify the controllable characteristics of an organization in order to prevent incidents or accidents.

3. CONCLUSION

Illusive organizational factors remain a significant hazard of complex socio-technical systems and the current generations of PRA do not include a comprehensive representation of the underlying organizational failure mechanisms. The SoTeRiA theoretical framework [15, 41] provides a roadmap for the analysis of organizational risk-contributing factors by integration of STS theory and PRA. The focus of the research, summarized in this paper, is to enhance SoTeRiA's theoretical details, quantified methodological features, and practicality in order to develop a more accurate decision-making tool for high-risk organizations.

The *Theoretical* contributions of this research include the advancement of the detailed factors, sub-factors, and their casual relationships in the SoTeRiA framework, based on theoretical, industrial and regulatory perspectives. This improvement would help identify in-depth organizational root causes of failure and their paths of influence on technical system performance, utilizing theoretical models of social phenomena.

The *Methodological* contributions of this research relate to quantifying the SoTeRiA framework using advanced measurement and predictive methodologies, big data analytics, and uncertainty analysis. Socio-technical risk analysis deals with wide-ranging, incomplete, and unstructured data. Therefore, this research is dedicated to developing new predictive technologies for PRA that are not only grounded on socio-technical system theories, but also serve to expand the classical approach of data management for risk analysis by incorporating techniques such as text mining, data mining, data analytics, network analysis, as well as data curation and storage. In other words, this study proposes a hybrid combination of (1) big data (analytics) and (2) theory (SoTeRiA; an integration of STS theory and PRA) in order to efficiently extract organizational metrics and execute them in a network of organizational safety indicators. This hybrid integration would help (a) avoid the potential misleading results of solely data-informed approaches (big data methods, top-down approaches) and, at the same time, (b) overcome the challenges of bottom-up approaches in quantifying socio-technical risk theories.

The *Practical* contributions of this research include the identification of organizational factors influencing technical system risk of NPPs, the aggregation and analytics of organizational data from NPPs in a causal modeling framework, and the development of a dynamic model for risk-informed organizational decision-making. By applying SoTeRiA to the PRA of a NPP, this research will quantify the effects of organizational factors on system risk and, analyze the criticality of organizational decision parameters. The research will provide a technology for industry that can quantitatively rank the importance of organizational risk contributing factors, and report on areas of organizational strengths and weaknesses based on their influence on risk, and propose effective preventive approaches.

The *Future* contributions of this research include the long-term goal of monitoring organizational safety indicators, capable of harmonizing the automated SoTeRiA network-generating algorithm and hybrid modeling technique to provide a real-time analysis that monitors the changing landscape of risk. By operationalizing a system capable of monitoring organizational safety indicators, real-time information can help set management decision-making thresholds to limit harmful organizational factors, detect root causes of organizational weaknesses, advance knowledge management for high-risk industries, and contribute to the development of organizational requirements for the safe operation of NPPs.

References

[1]   IAEA, *Managing Human Performance to Improve Nuclear Facility Operation*, in *Nuclear Energy Series*. 2014, IAEA.

[2]   IAEA, *Managing Human Resources in the Field of Nuclear Energy*, in *Nuclear Energy Series*. 2009, IAEA.

[3]   DOE, *Human performance improvement handbook volume 1: concepts and principles*, in *Standard AREA HFAC* 2009, US Department of Energy.

[4]   Gordon, R.P., *The contribution of human factors to accidents in the offshore oil industry.* Reliability Engineering & System Safety, 1998. 61(1): p. 95-108.

[5]   Hofmann, D.A. and Morgeson, F.P., *Safety-related behavior as a social exchange: The role of perceived organizational support and leader–member exchange.* Journal of applied psychology, 1999. 84(2): p. 286.

[6]   Nahrgang, J.D., Morgeson, F.P., and Hofmann, D.A., *Safety at work: a meta-analytic investigation of the link between job demands, job resources, burnout, engagement, and safety outcomes.* Journal of Applied Psychology, 2011. 96(1): p. 71.

[7]   Beus, J.M., Payne, S.C., Bergman, M.E., and Arthur Jr, W., *Safety climate and injuries: an examination of theoretical and empirical relationships.* Journal of Applied Psychology, 2010. 95(4): p. 713.

[8]   Zohar, D. and Luria, G., *A multilevel model of safety climate: cross-level relationships between organization and group-level climates.* Journal of Applied Psychology, 2005. 90(4): p. 616.

[9]   Le Coze, J.-c., *Outlines of a sensitising model for industrial safety assessment.* Safety Science, 2013. 51(1): p. 187-201.

[10]   Swain, A.D. and Guttmann, H.E., *Handbook of human-reliability analysis with emphasis on nuclear power plant applications. Final report.* 1983, Sandia National Labs., Albuquerque, NM (USA).

[11]   Mosleh, A. and Chang, Y., *Model-based human reliability analysis: prospects and requirements.* Reliability Engineering & System Safety, 2004. 83(2): p. 241-253.

[12]   Reason, J., *Human error.* 1990: Cambridge university press.

[13]   NRC, *Reactor Safety Study: An Assessment of Accident Risks in US Commercial Nuclear Power Plants, WASH-1400 (NUREG-75/014).* October. Available from National Technical Information Service, Springfield, VA, 1975. 22161.

[14]   Mohaghegh, Z. and Pence, J. *PANEL: INFLUENCES OF ORGANIZATIONAL FACTORS AND SAFETY CULTURE ON RISK OF TECHNICAL SYSTEMS*. in *American Nuclear Society Embedded Topical Meeting on Risk Management for Complex Socio-Technical Systems*. 2013. Washington D.C.: Transactions of the American Nuclear Society.

[15]   Mohaghegh, Z., *On the theoretical foundations and principles of organizational safety risk analysis*. 2007: ProQuest.

[16]   Reason, J., *Managing the risks of organizational accidents*. Vol. 6. 1997: Ashgate Aldershot.

[17]   Embrey, D.E., *Incorporating management and organisational factors into probabilistic safety assessment*. Reliability Engineering & System Safety, 1992. 38(1): p. 199-208.

[18]   Davoudian, K., Wu, J.-S., and Apostolakis, G., *Incorporating organizational factors into risk assessment through the analysis of work processes*. Reliability Engineering & System Safety, 1994. 45(1): p. 85-105.

[19]   Davoudian, K., Wu, J.-S., and Apostolakis, G., *The work process analysis model (WPAM)*. Reliability Engineering & System Safety, 1994. 45(1): p. 107-125.

[20]   Paté-Cornell, E.M. and Murphy, D.M., *Human and management factors in probabilistic risk analysis: the SAM approach and observations from recent applications*. Reliability Engineering & System Safety, 1996. 53(2): p. 115-126.

[21]   Mosleh, A. and Golfeiz, E.B., *An approach for Assessing the Impact of Organizational Factors on Risk*, in *Technical Research Report*. 1999, Center for Technology Risk Studies, University of Maryland at College Park.

[22]   Luxhøj, J.T. *Building a safety risk management system: a proof of concept prototype*. in *FAA/NASA Risk Analysis Workshop, Arlington, VA, USA*. 2004.

[23]   Roelen, A., Wever, R., Hale, A., Goossens, L., Cooke, R., Lopuhaä, R., Simons, M., and Valk, P., *Causal modeling for integrated safety at airports*. Proceedings of ESREL 2003, Safety and Reliability, 2003. 2: p. 1321-1327.

[24]   Rasmussen, J., *Risk management in a dynamic society: a modelling problem*. Safety science, 1997. 27(2): p. 183-213.

[25]   Rochlin, G.I., La Porte, T.R., and Roberts, K.H., *The self-designing high-reliability organization: Aircraft carrier flight operations at sea*. Naval War College Review, 1987. 40(4): p. 76-90.

[26]   Weick, K. and Sutcliffe, K.M., *Managing the Unexpected: Assuring High Performance in an age of complexity*. 2001, Jossey Bass Publishers: San Francisco, CA.

[27]   Senge, P.M., *The fifth discipline: The art and practice of the learning organization*. New York, 1990.

[28]   Perrow, C., *Normal accidents: Living with high risk systems*. 1984, New York: Basic Books.

[29]   Wahlström, B. and Rollenhagen, C., *Safety management–a multi-level control problem*. Safety Science, 2013.

[30]   Biondi, E.L., *Organizational factors in the reliability assessment of offshore systems*. 1998.

[31]   Bella, D.A., *Organized complexity in human affairs: The tobacco industry*. Journal of Business Ethics, 1997. 16(10): p. 977-999.

[32]   Cooke, D.L., *The dynamics and control of operational risk*. 2004.

[33]   Leveson, N., *A new accident model for engineering safer systems*. Safety Science, 2004. 42(4): p. 237-270.

[34]   Sterman, J., *Business dynamics*. 2000: Irwin-McGraw-Hill.

[35]   Yu, J., Ahn, N., and Jae, M., *A quantitative assessment of organizational factors affecting safety using system dynamics model*. JOURNAL-KOREAN NUCLEAR SOCIETY, 2004. 36(1): p. 64-72.

[36]   Gajdosz, M., Bedford, T., and Howick, S., *Understanding and modeling organizational factors within probabilistic risk analyses*, in *Transactions of the American Nuclear Society and Embedded Topical Meeting: Risk Management for Complex Socio-Technical Systems*. 2013, American Nuclear Society, Inc.: Washington, D.C. p. 2127.

[37]   Blom, H., Bakker, G., Blanker, P., Daams, J., Everdij, M., and Klompstra, M., *Accident risk assessment for advanced air traffic management*. Progress in Astronautics and Aeronautics, 2001. 193: p. 463-480.

[38] Bosse, T., Blom, H.A., Stroeve, S.H., and Sharpanskykh, A. *An Integrated Multi-agent Model for Modelling Hazards within Air Traffic Management*. in *Web Intelligence (WI) and Intelligent Agent Technologies (IAT), 2013 IEEE/WIC/ACM International Joint Conferences on*. 2013. IEEE.

[39] Everdij Mariken H.C. (NLR), H.A.P.B.N., Michael Allocco (FAA), David Bush (NATS), Mete Çeliktin (Eurocontrol), Barry Kirwan (Eurocontrol), Patrick Mana (Eurocontrol), Jochen Mickel (Goethe University), Keith Slater (NATS), Brian Smith (NASA), Oliver Sträter (Eurocontrol), Edwin Van der Sluis (NLR), *Safety Methods Database*, T.N.A.L.o.t. Netherlands, Editor. 2013: Web.

[40] Mohaghegh, Z. and Mosleh, A., *Incorporating organizational factors into probabilistic risk assessment of complex socio-technical systems: Principles and theoretical foundations*. Safety Science, 2009. 47(8): p. 1139-1158.

[41] Mohaghegh, Z., *Socio-Technical Risk Analysis*. 2009, VDM Verlag. ISBN.

[42] Kelly, J.E., *A reappraisal of sociotechnical systems theory*. Human Relations, 1978. 31(12): p. 1069-1099.

[43] Emery, F. and Trist, E., *Sociotechnical systems. IN Management sciences models and techniques. Vol. 2*. 1960, London: Pergamon Press.

[44] Von Neumann, J., *The general and logical theory of automata*. Cerebral mechanisms in behavior, 1951: p. 1-41.

[45] Ostroff, C., Kinicki, A.J., and Muhammad, R.S., *Organizational culture and climate*, in *Handbook of psychology*, I.B. Weiner, N.W. Schmitt, and S. Highhouse, Editors. 2013, John Wiley & Sons: Hoboken, NJ. p. 643-676.

[46] Mohaghegh, Z., Kazemi, R., and Mosleh, A., *Incorporating organizational factors into Probabilistic Risk Assessment (PRA) of complex socio-technical systems: A hybrid technique formalization*. Reliability Engineering & System Safety, 2009. 94: p. 1000-1018.

[47] Mohaghegh, Z. *Combining System Dynamics and Bayesian Belief Networks for Socio-Technical Risk Analysis*. in *Intelligence and Security Informatics (ISI), 2010 IEEE International Conference on*. 2010. IEEE.

[48] Bar-Yam, Y., *The Limits of Phenomenology: From Behaviorism to Drug Testing and Engineering Design*. arXiv preprint arXiv:1308.3094, 2013.

[49] Heidorn, P.B., *Shedding light on the dark data in the long tail of science*. Library Trends, 2008. 57(2): p. 280-299.

[50] Waber, B., *People Analytics: How Social Sensing Technology Will Transform Business and what it Tells Us about the Future of Work*. 2013: FT Press.

[51] van Reijsen, J., Helms, R., Batenburg, R., and Foorthuis, R., *The impact of knowledge management and social capital on dynamic capability in organizations*. Knowledge Management Research & Practice, 2014.

[52] Evans, M., *Understanding dialectics in policy network analysis*. Political Studies, 2001. 49(3): p. 542-550.

[53] Ashton, K., *That 'internet of things' thing*. RFiD Journal, 2009. 22: p. 97-114.

[54] Ning, H. and Wang, Z., *Future Internet of things architecture: like mankind neural system or social organization framework?* Communications Letters, IEEE, 2011. 15(4): p. 461-463.

[55] Atzori, L., Iera, A., and Morabito, G., *From" smart objects" to" social objects": The next evolutionary step of the internet of things*. Communications Magazine, IEEE, 2014. 52(1): p. 97-105.

[56] West, G., *Big data needs a big theory to go with it*. Scientific American, May, 2013. 15.

[57] Abolhelm, M., Pence, J., Mohaghegh, Z., Kee, E., Yilmaz, F., and Johnson, D., *Toward Demonstrating the Monetary Value of Probabilistic Risk Assessment for Nuclear Power Plants*, in *Proceedings of 12th International Topical Meeting on Probabilistic Safety Assessment and Analysis (PSAM12)*. 2014.

[58] Cameron, K.S. and Quinn, R.E., *Diagnosing and changing organizational culture: Based on the competing values framework*. 1999, Addison-Wesley (Reading, Mass.).

[59] Lazer, D., Kennedy, R., King, G., and Vespignani, A., *The Parable of Google Flu: Traps in Big Data Analysis*. Science, 2014. 343(6176): p. 1203-1205.

[60]   Roelen, A.L.C., *Causal risk models of air transport: comparison of user needs and model capabilities*. 2008: IOS Press.

[61]   Diesner, J. *ConText: Software for the Integrated Analysis of Text Data and Network Data*. in *Conference of International Communication Association (ICA)*. 2014. Seattle, WA.

[62]   Popova, V. and Sharpanskykh, A., *Formal analysis of executions of organizational scenarios based on process-oriented specifications*. Applied Intelligence, 2011. 34(2): p. 226-244.

[63]   Popova, V. and Sharpanskykh, A., *Modeling organizational performance indicators*. Information Systems, 2010. 35(4): p. 505-527.

[64]   Jonker, C.M., Popova, V., Sharpanskykh, A., Treur, J., and Yolum, P., *Formal framework to support organizational design*. Knowledge-Based Systems, 2012. 31: p. 89-105.

[65]   Helms, R. and Buijsrogge, K. *Application of knowledge network analysis to identify knowledge sharing bottlenecks at an engineering firm*. in *ECIS*. 2006.

[66]   Jensen, F.V., Lauritzen, S.L., and Olesen, K.G., *Bayesian updating in causal probabilistic networks by local computations*. Computational statistics quarterly, 1990. 4: p. 269-282.

[67]   Kongsvik, T., Almklov, P., and Fenstad, J., *Organisational safety indicators: Some conceptual considerations and a supplementary qualitative approach*. Safety Science, 2010. 48(10): p. 1402-1411.

[68]   Mohaghegh, Z. and Mosleh, A., *Measurement techniques for organizational safety causal models: Characterization and suggestions for enhancements*. Safety Science, 2009. 47(10): p. 1398-1409.

[69]   Keefe, M., Frahm, R., Martin, K., Sigmon, R., Shoop, U., Morrow, S., Sieracki, D., Powell, R., Shaeffer, S., Rutkowski, J., and Ruesch, E., *Safety Culture Common Language*, in *NUREG-2165*. 2014, Nuclear Regulatory Commission.

[70]   Kee, E., Mohaghegh, Z., Kazemi, R., Reihani, S., Letellier, B., and Grantom, R., *Risk-Informed Decision Making: Application in Nuclear Power Plant Design & Operation*. in *American Nuclear Societal Embedded Topical*. 2013.

[71]   Sakurahara, T., Reihani, S., Ertem, M., Mohaghegh, Z., and Kee, E., *Analyzing Importance Measure Methodologies for Integrated Probabilistic Risk Assessment*, in *Proceedings of 12th International Topical Meeting on Probabilistic Safety Assessment and Analysis (PSAM12)*. 2014.

www.ingramcontent.com/pod-product-compliance
Lightning Source LLC
Chambersburg PA
CBHW081143180526
45170CB00006B/1907